PHYSICAL PRINCIPLES OF MAGNETISM

To my grandchildren
JAMES and LOUISE

Physical Principles of Magnetism

F. BRAILSFORD
Professor of Electrical Engineering
in the
University of London

D. VAN NOSTRAND COMPANY LTD
LONDON
TORONTO NEW YORK
PRINCETON, NEW JERSEY

D. VAN NOSTRAND COMPANY LTD.
358 Kensington High Street, London, W.14

D. VAN NOSTRAND COMPANY INC.
120 Alexander Street, Princeton, New Jersey
24 West 40 Street, New York 18

D. VAN NOSTRAND COMPANY (CANADA) LTD.
25 Hollinger Road, Toronto 16

Library of Congress Catalog Card No. 66-21547

Printed in Great Britain by
Butler & Tanner Ltd, Frome and London

PREFACE

The aim of this book is to present a fairly complete account of the subject of ferromagnetism, in particular, in terms as simple and direct as this very intricate topic will allow. It is hoped the book will be found useful by undergraduate and graduate engineers, physicists and metallurgists. The subject of magnetism is such a unified whole that any exposition of ferromagnetism must include a discussion of diamagnetism and paramagnetism, and in the book antiferromagnetism and ferrimagnetism are also dealt with. Moreover to explore these topics a good knowledge of the electronic structure of atoms and of the chief metal and ionic crystal structures is required. An attempt has been made to cover all this ground, at least in logical sequence, while not involving the reader in mathematics beyond the capacity of the average second-year undergraduate. The rationalized metre-kilogram-second system of units has been used throughout. References to the literature and problems for solution are also given with each chapter.

F. BRAILSFORD

CONTENTS

Chapter 4 Diamagnetism and Paramagnetism

Chapter 5 Ferromagnetism and Antiferromagnetism

Chapter 6 Magnetic Anisotropy

CHAPTER 10 TECHNOLOGICAL ASPECTS OF MAGNETISM

Chapter 1
INTRODUCTORY

1.1 Some Magnetic Parameters

If a long solenoid has n turns per unit length and the current in the turns is I, the magnetic field strength or magnetic intensity inside the solenoid is H where $H = nI$. In the rationalized metre-kilogram-second (MKS) system of units, used throughout this book, I is in amperes, the unit of length is the metre and H will be in ampere-turns per metre, or more briefly, in amperes per metre (A/m). Magnetic fields may be produced also in the neighbourhood of magnetized bodies but all magnetism is ultimately attributable to electric currents, that is to the movement of electric charges.

A magnetic intensity H in free space produces a magnetic induction or magnetic flux density in space B where $B = \mu_0 H$. The unit of magnetic flux is the weber (Wb) and the flux density B is therefore expressed in webers per square metre (Wb/m^2). The factor μ_0 is called the permeability of free space, or sometimes the primary magnetic constant, and $\mu_0 = 4\pi \times 10^{-7}$ henrys per metre (H/m).

If the space in which H is acting is now filled with any material substance, whether gas, liquid or solid, there will be, as well as the free-space component of flux density $\mu_0 H$, a further component due to the material itself called the intensity of magnetization M, also expressed in webers per square metre. Hence in this case

$$B = \mu_0 H + M$$

The material is now said to have an absolute permeability μ in henrys per metre where $\mu = B/H$. The permeability μ obviously contains the contribution due to free space as well as that from the material itself. We thus have $B/H = \mu_0 + M/H$. The quantity M/H is called the absolute susceptibility of the material. It is denoted by κ and is also in henrys per metre. Hence

$$\mu = \mu_0 + \kappa$$

The ratio of the absolute permeability of the material to the absolute permeability of free space is called the relative permeability, denoted by μ_R. Thus $\mu/\mu_0 = 1 + \kappa/\mu_0$ or

$$\mu_R = 1 + \kappa_R$$

$\kappa_R = \kappa/\mu_0$ may be called the relative susceptibility and μ_R and κ_R are simply numerics.

Consider now a magnetic dipole, for example, in the form of a small rod of length l and cross-sectional area a, which, due to internal forces whose origin we can disregard at the moment, is uniformly and permanently magnetized in the direction of its axis. The rod will have magnetic poles, of equal strength m but of opposite polarities, at the two ends, respectively. The pole strength m, in MKS units, is in webers. The uniform intensity of magnetization in the rod is then given by $M = m/a$, that is as pole strength per unit area. Or we may put $M = ml/al$. The magnetic moment of the rod is ml (Wb.m), and al is the volume, so that the intensity of magnetization M may also be defined as the magnetic moment per unit volume. This is a very useful concept for future reference.

We now bring the dipole from an infinite distance into a magnetic field H. The field strength H is supposed small compared with the internal forces in the dipole so that the 'permanent' magnetism of the dipole is not affected. If finally the axis of the dipole is perpendicular to the direction of H, as shown

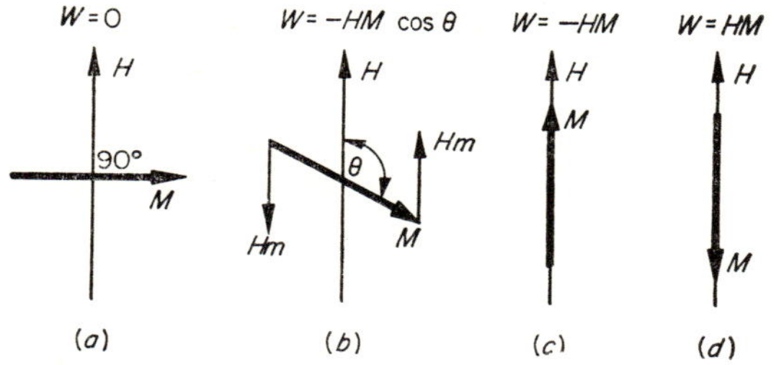

FIG. 1.1. Magnetic potential energy of a magnetic dipole in a field H.

in Fig. 1.1(a), the sum total amount of work done on the equal and opposite poles will be zero. The magnetic potential energy of the magnet in the field H for the condition shown at (a) will therefore be zero. If the magnet is now turned, as shown at (b), so that the angle between H and M is θ, then since there is a force Hm on each pole in the directions shown, the counterclockwise torque acting is $Hml \sin(\pi - \theta) = HM\, al \sin\theta$. The work done on the magnet, which is the magnetic potential energy of the magnet for any angle θ is therefore, per unit volume, $W = \int_{\pi/2}^{\theta} HM \sin\theta \,.d\theta$, or

$$W = -HM\cos\theta$$

The magnetic potential energy of the dipole, per unit volume, in the field H is therefore $-HM$ when the directions of H and M coincide ($\theta = 0$), and $+HM$ when H and M are in opposite directions ($\theta = 180°$).

1.2 Classification of materials

The atoms of any substance have an electronic structure in which electrons may be regarded as circulating in orbits about a central heavy nucleus. The electrons also have a gyroscopic spin. Both the orbital and spin motions of the electron charges constitute currents in closed circuits, and with these permanently circulating currents magnetic moments are associated. This subject is dealt with in some detail in Chapter 2. It is known that the free atoms of some elements have zero magnetic moment because the various orbital and spin components completely cancel each other, while the atoms of other elements have a resultant magnetic moment. The magnetic properties of any material in bulk then depend, not only on the magnetic moments of the free atoms or molecules, but also particularly on temperature, that is on the thermal energy, and in solids on complicated interatomic forces in the crystals.

We may, however, make the following broad classification of material substances.

(a) *Diamagnetic*. Materials made up of atoms or molecules having no magnetic moment might be expected to show no magnetic response to an applied magnetic field, but this, in fact, is not quite the case. An orbital electron, having mass and a high angular velocity about the axis of the orbit, has angular momentum as well as a magnetic moment. On this account it behaves like a gyroscope or spinning top. Each electron also has angular momentum and magnetic moment due to spin about its own axis. The application of a magnetic field produces a torque upon the system. This torque does not, of itself, align the axes with the field direction but, as in the case of spinning tops, causes the orbital and spin axes to precess about the field direction with an angular velocity proportion to the field strength. The direction of the precessional motion of the electron charge is such as to produce a component of magnetic moment in opposition to the direction of the field, and this opposing effect is irrespective of the original and unchanging angle between the applied field and the precessing axes. Every electron of the material in bulk is affected in this way so that the application of a field H produces a reversed or negative component of intensity of magnetization M. The absolute susceptibility κ, and the relative susceptibility κ_R of the material on this account are therefore negative, and the relative permeability μ_R will be less than unity. Such materials are said to be diamagnetic, although the effect is very feeble and κ_R is very small compared with unity. Copper, gold, mercury, water and hydrogen gas (H_2) are examples, having the approximate room temperature values of κ_R given in Table 1.1.

(b) *Paramagnetic*. When the atoms or molecules of a material each have a magnetic moment, and a field is applied, the magnetic potential energy of the system is reduced if the elementary magnets turn towards the direction of the applied field, so producing a component of intensity of magnetization in the field direction. The violent thermal motions of the particles at a given temperature, however, tend to keep the magnetic axes in disorder. The equilibrium condition in a magnetic field depends on the relative values of magnetic potential energy and thermal energy. This is a statistical problem which has been successfully solved and which is discussed in Chapter 4. Such materials,

TABLE 1.1

Material	Relative Susceptibility κ_R
Diamagnetic:	
copper	$-1{\cdot}0 \times 10^{-5}$
gold	$-3{\cdot}6 \times 10^{-5}$
mercury	$-3{\cdot}2 \times 10^{-5}$
water	$-0{\cdot}9 \times 10^{-5}$
hydrogen (n.t.p.)	$-0{\cdot}2 \times 10^{-8}$
Paramagnetic:	
aluminium	$+2{\cdot}2 \times 10^{-5}$
manganese	$+98 \times 10^{-5}$
platinum	$+36 \times 10^{-5}$
tungsten	$+8 \times 10^{-5}$
air (n.t.p.)	$+36 \times 10^{-8}$
	$(\mu_R = 1 + \kappa_R)$ (approximate highest values)
Ferromagnetic:	
iron (dynamo sheets)	5000
(laboratory specimens)	$8{\cdot}8 \times 10^{4}$
(crystals)	$1{\cdot}4 \times 10^{6}$
silicon-iron (transformer sheets)	7×10^{4}
(crystals)	$3{\cdot}8 \times 10^{6}$
nickel-iron (commercial)	9×10^{4}
(laboratory specimens)	10^{6}
Ferrimagnetic:	
Fe_3O_4	100
ferrites	5000

said to be paramagnetic, have a positive susceptibility which is a function of the absolute temperature, and which outweighs the diamagnetic component present in all materials. The observed room-temperatures values of κ_R for a number of paramagnetic materials are given in Table 1.1. The values are again small compared with unity but, in general, large compared with those obtained in the diamagnetic materials, as may be seen from the table.

(c) *Ferromagnetic*. Ferromagnetism occurs at normal working temperatures in the elements iron, cobalt and nickel and in a number of other elements at lower temperatures. It occurs also in many alloys containing iron, cobalt and nickel and in a few made up entirely of non-ferromagnetic elements. In all cases these metals revert to a paramagnetic condition if the temperature is raised above the magnetic change point or Curie temperature.

The most remarkable characteristics of a ferromagnetic metal are that the relative susceptibility is very great compared with unity, so that $\mu_R = \kappa_R$ very nearly, and that magnetic saturation can be achieved in fields lower than those required for the paramagnetics by a factor of the order of 10^4. It will be seen from the table that values of μ_R up to about $3{\cdot}8 \times 10^6$ have been observed.

In ferromagnetic materials there are powerful internal fields of force which pull the neighbouring atomic magnets into line. These spontaneously-produced fields overcome the disordering effect of temperature, provided this is not too high, and are basically responsible for the remarkable magnetic properties.

(d) *Antiferromagnetic.* In antiferromagnetic substances there are atoms or ions with magnetic moments and again, as in ferromagnetism, strong internal fields, but in this case the internal forces cause the magnetic moments of near-neighbour atoms to point in opposite or antiparallel directions. Thus there is an ordered arrangement of the atomic magnets, but with two equal and opposite systems, so that the resultant magnetic moment over a finite volume is zero. The materials are characterized by a relation between susceptibility and temperature which has a sharp discontinuity at a temperature called the Néel point. Below this temperature the material is ordered and antiferromagnetic, while at higher temperatures it is disordered and behaves as a paramagnetic substance. The elements manganese and chromium and many chemical compounds behave in this way.

(e) *Ferrimagnetic.* Ferrimagnetism may be regarded as a special case of antiferromagnetism (or perhaps vice versa). The commonest example of a ferrimagnetic substance is the magnetic iron oxide Fe_3O_4 in which the iron ions have magnetic moments in an ordered arrangement. There are, however, in this case two unequal and antiparallel systems of ionic magnets. A resultant intensity of magnetization thus occurs due to the difference in magnetic moment between them. Many artificial oxides, known as ferrites, can be made up which are ferrimagnetic and which have magnetic properties similar to those of the ferromagnetic metals.

1.3 Typical Ferromagnetic Characteristics

The general behaviour of a ferromagnetic material in a magnetic field may be exemplified by the curves for a polycrystalline sample of soft iron in Fig. 1.2. The way in which the flux density B increases when an increasing field H is applied to the material, originally in a demagnetized condition, is shown in Fig. 1.2(*a*) by the initial magnetization curve *OCD*. This curve is not reversible, for, on decreasing the field, a magnetic hysteresis loop is traced out in the manner shown, and a major loop may have minor or subsidiary loops superimposed upon it, depending on the way in which the applied field is varied. If the initial magnetization curve is taken to relatively high field strengths the form of the curve is as shown at (*b*) where both the flux density B and the intensity of magnetization M are plotted. The following points may be noted. A relatively small field strength is sufficient to raise the flux density to the region E which may be referred to as the knee of the curve. Reference to the scales of H in Figs. 1.2(*a*) and (*b*) shows that to increase B beyond the

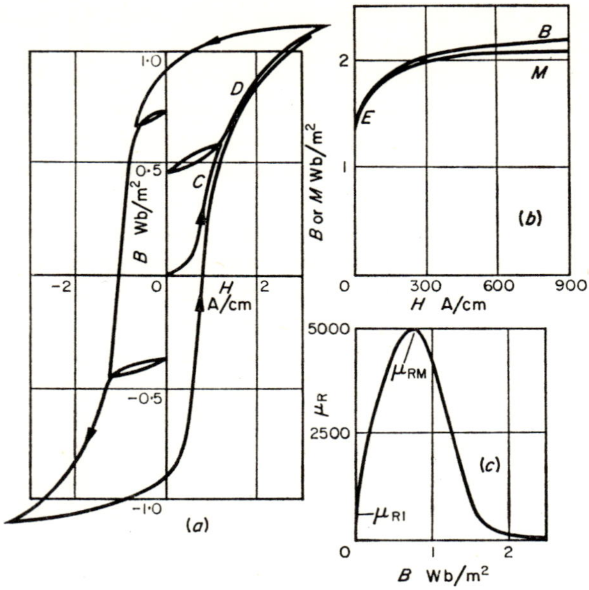

FIG. 1.2. Magnetic characteristics of soft iron.

knee requires a very much greater applied field, and indicates some change in the magnetizing process at this point. Eventually the intensity of magnetization approaches a saturation value M_S where, for a normal sample of iron containing some impurities, M_S is about 2·1 Wb/m². It may be noted that, since $B = \mu_0 H + M$, the flux density B cannot in any event reach a saturation value since the component $\mu_0 H$ increases continuously with increasing H.

The relation between the permeability and the flux density in this case is shown in Fig. 1.2(*c*), where the relative permeability values $\mu_R = B/(\mu_0 H)$ are taken from the initial magnetization curve. The initial relative permeability μ_{RI}, that is the value in very weak fields, obtainable from the slope of the B–H curve at the origin, is about 250, and the maximum value of the relative permeability μ_{RM}, occurring at a point of the B–H curve where a tangent from the origin touches the curve, is about 5000. At very high flux densities μ_R would tend towards unity. The permeability therefore varies over a very wide range due to the non-linear B–H relationship, and of course, in any calculations involving alternating fields and alternating flux densities, further complications are introduced due to the form of the hysteresis loops.

While the curves in Fig. 1.2 have been given for soft iron, similar magnetization and permeability curves and hysteresis loops may be drawn for any ferromagnetic or ferrimagnetic material. However, the scales of B, H and μ may be greatly different for different materials. A comparison of the values of μ_R given in Table 1.1 will illustrate this point.

If in the case of Fig. 1.2(*a*) a slowly alternating field is applied, and repeated a number of times, between limits $\pm H_{max}$, the material settles into a 'cyclic state', the successive loops then coinciding. On the descending side of the loop when H is reduced to zero there is a remanent flux density or remanence in the material. When H reverses and is increased in the negative direction, B will decrease to zero when the reversed field has a value referred to as the coercive force. In the particular case when the tips of the hysteresis loop have been taken well into the saturation region the corresponding values of remanence and coercive force are known as the retentivity B_R and the coercivity H_C, respectively. This is illustrated in Fig. 1.3(*a*). That part of the

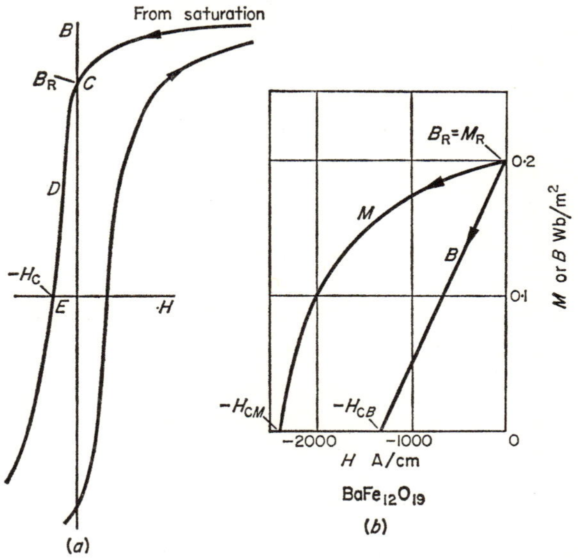

FIG. 1.3. Illustrating retentivity and coercivity.

hysteresis loop CDE between B_R and $-H_C$ is called the demagnetization curve, a characteristic which is important in permanent magnet materials.

In most cases in practice the intensity of magnetization M and the flux density B, when plotted on a hysteresis loop, except of course in the saturation region, are nearly the same. However, there are some materials where B_R is relatively small and H_C, as just defined, is relatively large. M and B may then be greatly different for a given value of H in the hysteresis loops. The value of the coercive force $-H_{CM}$ which reduces M to zero in a saturation loop may then be very much greater numerically than the value $-H_{CB}$ which reduces B to zero. Actual values for a barium ferrite, permanent magnet material ($BaFe_{12}O_{19}$) are shown in Fig. 1.3(*b*). Either value, H_{CM} or H_{CB}, may be quoted by an investigator of a new material but it is the lower numerical value H_{CB} which is important in practical, technological applications. In either case, however, $B_R = M_R$ when $H = 0$.

1.4 Demagnetizing Factors

If a ferromagnetic specimen of finite size is placed in a uniform magnetic field, the magnetic field effectively acting internally on the material to produce a magnetic flux is less than the externally-applied field. An exception to this would be if the specimen formed a closed magnetic circuit, for example, if it were in the form of an anchor ring. A circumferential applied field produced by a toroidal winding on the ring would then be the same as the field effectively acting on the material in the ring. There would be no magnetic flux leaving the specimen in this particular case and so no free magnetic poles induced at any point on the surface of the material, or indeed inside the material either.

The simplest example to consider is that of a flat plate of infinite extent with a magnetic field H' applied normal to the plate as shown in Fig. 1.4(a).

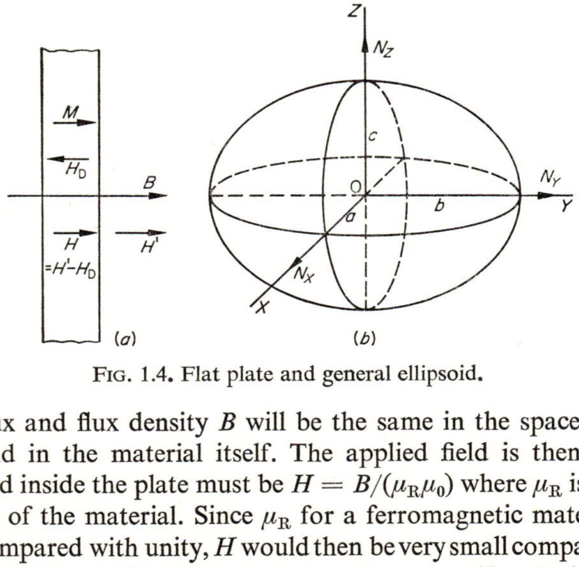

FIG. 1.4. Flat plate and general ellipsoid.

The total flux and flux density B will be the same in the space outside the specimen and in the material itself. The applied field is then $H' = B/\mu_0$ while the field inside the plate must be $H = B/(\mu_R\mu_0)$ where μ_R is the relative permeability of the material. Since μ_R for a ferromagnetic material may be very great compared with unity, H would then be very small compared with H'.

Referring to Fig. 1.4(a), free magnetic poles are uniformly induced over the two surfaces of the infinite plate. Externally to the plate a magnetic field strength is induced by each infinite surface sheet of magnetic pole, but these components of field strength are equal and opposite so that the applied field H' is undisturbed by the polarization of the plate. However, inside the material the poles on both surfaces act in the same direction to produce an internal demagnetizing field H_D. The resultant field acting on the material is then $H = H' - H_D$. The intensity of magnetization M in the material is also the pole strength per unit area on the surfaces, and the demagnetizing field is proportional to this. Therefore $H_D = N_D M$ where N_D is called the demagnetizing factor.

The value of N_D is readily deduced for this particular form of specimen. We have $B = \mu_0 H + M = \mu_0(H' - N_D M) + M$ or $B = \mu_0 H' + M(1 - \mu_0 N_D)$. But $B = \mu_0 H'$. Therefore $\mu_0 N_D = 1$, or $N_D = 1/\mu_0$.

Another simple hypothetical form of specimen is an infinitely long rod of uniform cross-section in a longitudinal field. Since in this case free poles at the end of the rod are at an infinite distance the demagnetizing field in the rod $N_D M$ would be zero so that $N_D = 0$. The same applies, as already stated, to an isotropic specimen of uniform cross-section in a closed magnetic circuit.

The long rod and the flat plate are extreme forms of specimen with values of $\mu_0 N_D$ of zero and unity, respectively. Intermediate shapes of specimen have demagnetizing factors between these limiting values. It may be noted that, in the MKS system of units, $\mu_0 N_D$ is a numeric and that N_D has the dimensions of $(\text{permeability})^{-1}$ or $(\text{henrys})^{-1}$.metre.

If the magnetic specimen is of irregular shape and is placed in a uniform magnetic field then the intensity of magnetization and the demagnetizing field within the specimen will vary from point to point. However, it was shown by MAXWELL [1] that, for an isotropic body having a surface of the second degree in a uniform applied field, the internal values of H, M and B are the same at every point. The only such surface which is finite, and hence which will enclose a volume of finite size, is that of the ellipsoid. Thus a homogeneous, ferromagnetic ellipsoid in a uniform field H' will have uniform internal magnetic conditions, and the same demagnetizing field at all points, although the directions of M or B in the material will not coincide with the direction of H' except in special cases, for example, when H' is parallel with one of the principal axes of the ellipsoid. A sphere is a special case of the general ellipsoid, and an infinite flat plate or an infinitely long rod of circular or elliptical cross-section may also be considered as special cases.

Since uniform conditions are theoretically obtainable in an ellipsoid, specimens of this form are sometimes useful in magnetic investigations. Any such specimen has demagnetizing factors appropriate to the directions of the principal axes and it is useful to know their numerical values. Simple formulae have been derived for the demagnetizing factors of ellipsoids of revolution and for other special cases and some of these formulae are given below [2]. OSBORN [3] has also given, very completely, numerical values for the general case.

We consider the general ellipsoid, as shown in Fig. 1.4(*b*), with the principal axes coinciding with the axes of rectangular coordinates, OX, OY, OZ, and having the ellipsoidal semi-axes a, b, c in the three directions, respectively. If N_X, N_Y and N_Z are the respective demagnetizing factors in the X, Y, Z directions, then a very useful relation, in MKS units, is

$$\mu_0(N_X + N_Y + N_Z) = 1 \tag{1.1}$$

Thus for a sphere $\mu_0 N_X = \mu_0 N_Y = \mu_0 N_Z = 1/3$. For an infinitely long rod of circular cross-section with its axis along OX, $\mu_0 N_X = 0$, $\mu_0 N_Y = \mu_0 N_Z = 1/2$. For an infinite plate in the YZ plane, $\mu_0 N_Y = \mu_0 N_Z = 0$, and $\mu_0 N_X = 1$ as already deduced.

We may note the following useful cases of ellipsoids of revolution:

(a) A prolate or elongated spheroid: $b = c$ and $a > b$.
The demagnetizing factor for a direction along the long axis of this pencil-shaped specimen is given by

$$\mu_0 N_X = (1/e^2 - 1)\{(1/2e)\log(1+e)/(1-e) - 1\} \tag{1.2}$$

where the eccentricity of the ellipse is $e = (1 - c^2/a^2)^{\frac{1}{2}}$. For directions parallel to a diameter of the specimen $\mu_0 N_Y = \mu_0 N_Z$ and using equation (1.1)

$$\mu_0 N_Y = (1 - \mu_0 N_X)/2 \tag{1.3}$$

where $\mu_0 N_X$ is obtained from equation (1.2).

Equation (1.2) is represented by curve A in Fig. 1.5, and equation (1.3) by the curves B in Figs. 1.6 and 1.7.

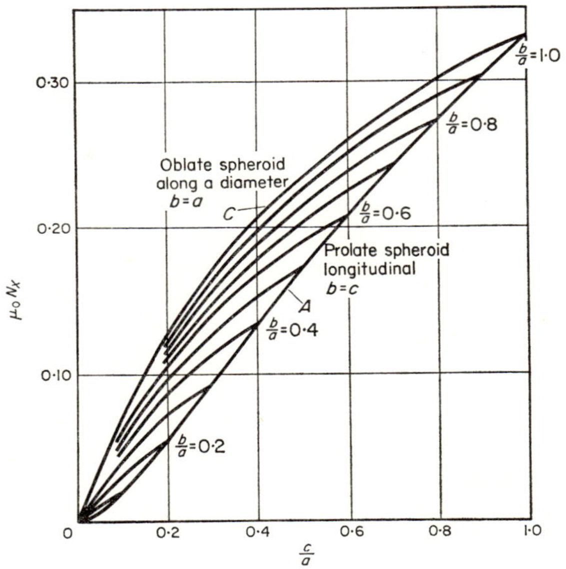

FIG. 1.5. Demagnetizing factors of the general ellipsoid along the OX axis.

(b) An oblate or flattened spheroid: $a = b$ and $b > c$.
In this disc-shaped specimen the demagnetizing factor in the direction of a diameter is given by

$$\mu_0 N_X = [\{(1 - e^2)^{\frac{1}{2}}/e^3\}\sin^{-1} e - (1 - e^2)/e^2]/2 \tag{1.4}$$

where as before $e = (1 - c^2/a^2)^{\frac{1}{2}}$. Using equation (1.1) and noting that $\mu_0 N_X = \mu_0 N_Y$ we have, for direction OZ normal to the plane of the circle

$$\mu_0 N_Z = 1 - 2\mu_0 N_X \tag{1.5}$$

where $\mu_0 N_X$ is obtained from equation (1.4).

Equation (1.4) is given by the curves C in Figs. 1.5 and 1.6 and equation (1.5) by curve D in Fig. 1.7.

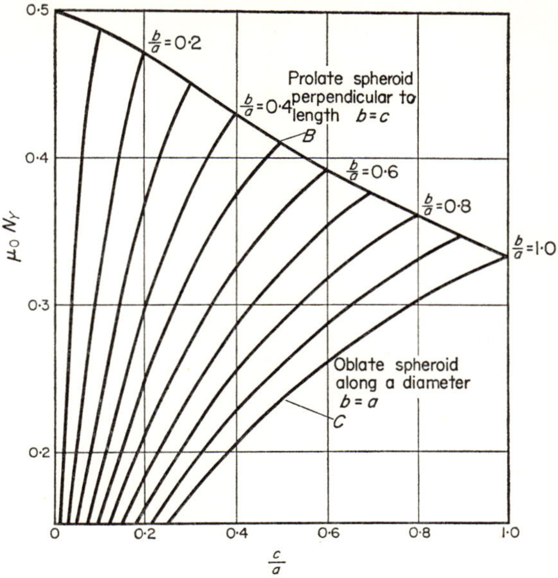

FIG. 1.6. Demagnetizing factors of the general ellipsoid along the OY axis.

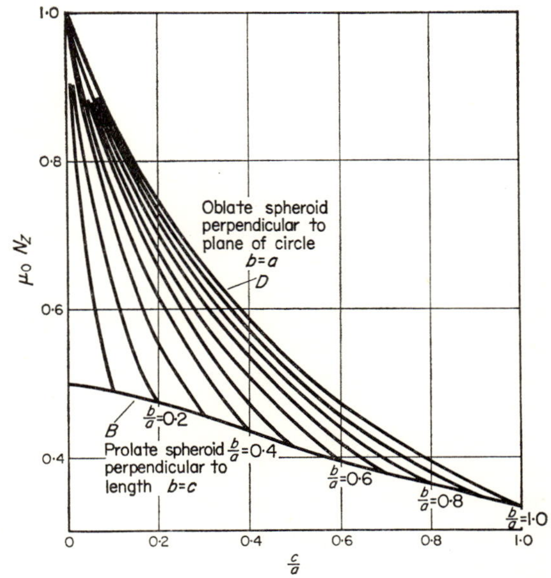

FIG. 1.7. Demagnetizing factors of the general ellipsoid along the OZ axis.

Osborn's results for the general ellipsoid are also given in Figs. 1.5, 1.6 and 1.7.

1.5 Magnetization Curve and Magnetostatic Energy of an Ellipsoid

When measurements are made on small samples, for example of ferromagnetic single crystals, the specimens are frequently made in the form of prolate ellipsoids. Oblate spheroids are also often used. However, consider an ellipsoid, elongated in the OX direction and having a demagnetizing factor N_X in this direction, brought into a magnetic field H' parallel with OX. We then have $H' = H + N_X M$ where H is the effective field acting on the material of

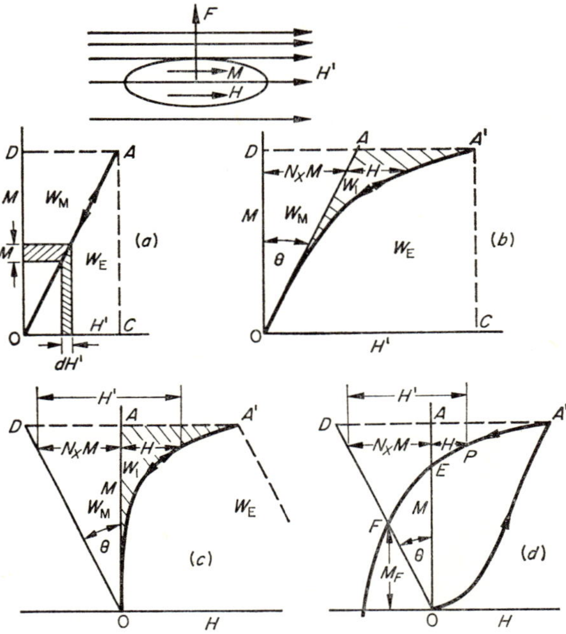

FIG. 1.8. Magnetization and energies in an ellipsoid.

the specimen and M is the intensity of magnetization in it. If the permeability of the specimen under these conditions is high and the factor N_X is not too small, then H will be small compared with $N_X M$ and $H' = N_X M$ approximately, giving a linear relation between H' and M as shown by OA in Fig. 1.8(a). In these circumstances the intensity of magnetization is determined by the shape of the specimen, that is by the value N_X, and hardly at all by the permeability of the material.

If the specimen moves through an increment of field dH' with an increment of the intensity of magnetization dM, the magnetic potential energy of the system per unit volume changes from $-H'M$ to $-(H' + dH')(M + dM)$. The potential energy therefore decreases by an amount $H'\,dM + M\,dH'$, represented by the shaded area in Fig. 1.8(*a*). This energy is (a) partly used up in doing mechanical work against any force restraining the inward movement of the specimen, and (b) partly stored as magnetostatic energy in the system. The increment dM occurring in the presence of the internal demagnetizing field $N_X M$ gives a component of internal stored energy $N_X M\,dM = H'\,dM$, and the mechanical work done is then $M\,dH'$. Thus when the specimen moves from zero field into a field H' represented by point A in the figure, the work done by the specimen against external forces is $W_E = \int_0^{H'} M\,dH'$, equivalent to the area of the triangle OAC, and in addition the magnetostatic energy stored is $W_M = \int_0^M H'\,dM$ represented by the area OAD. Therefore in this particular case, half the total change in magnetic potential energy is converted to external work, and half is stored internally in the specimen. The magnetostatic energy is then $W_M = \int_0^M H'\,dM = \int_0^M N_X M\,dM = N_X M^2/2$.

In Fig. 1.8(*a*) it was assumed that the permeability of the material was so high that H could be neglected in comparison with $N_X M$. However, in any practical case, particularly when M rises above the knee of the magnetization curve of the material, the observed relation between H' and M will not be linear but follow a curve such as OA' in Fig. 1.8(*b*). If the value of N_X is accurately known, from its shape, then the line OA may be drawn at an angle θ to the ordinate axis such that $\tan\theta = N_X$. Then since $H = H' - N_X M$ the true magnetization characteristic of the material, that is the relation between M and H, can be determined. This is not in general an accurate method of measuring the M–H characteristic of a material in the high permeability region since H is found by taking a difference of two much larger quantities. Figure 1.8(*b*) may be replotted as shown at (*c*) by tilting the figure through the angle θ.

As before the mechanical work done per unit volume is $W_E = \int_0^{H'} M\,dH'$, and this is represented by the area $OA'C$ under the curve OA' in Fig. 1.8(*b*). The total stored energy is given by

$$\int_0^M H'\,dM = \int_0^M N_X M\,dM + \int_0^M H\,dM = W_M + W_1$$

and consists of two components, the magnetostatic energy $W_M = N_X M^2/2$ as before, and the energy W_1 represented by the shaded areas in Fig. 1.8 at (*b*) and (*c*). On the hypothetical basis that the magnetization curve OA' is reversible and so involves no energy losses the component W_1 is energy stored in the magnetic material itself. This would be approximately the case for a sample of soft iron, for example, magnetized well above the knee of the magnetization curve, for the losses would then be relatively small compared with W_1 in the figure.

However, more generally, and notably in the case of a permanent-magnet type of material having a high coercivity, the magnetic conditions in an ellipsoidal specimen may be as shown in Fig. 1.8(*d*). On magnetizing the specimen from the initial demagnetized condition at O to the point A' and then decreasing the applied field H', the hysteresis curve $A'EF$ is traced out. At any point P on the return path $H' = H + N_X M$ as shown, so that when H' is reduced to zero the final condition of the specimen is represented by the point F where the hysteresis loop cuts the line OD. The specimen partially demagnetizes itself but retains an intensity of magnetization M_F and thus has permanent magnet properties. It will be seen that for a very long specimen, when $N_X \to O$, M_F approaches the remanent point E in the figure, while shortening the specimen, and so increasing N_X, reduces the value of M_F. It may also be seen that M_F increases as the coercive force of the particular hysteresis loop increases.

1.6 Ring Specimen with small Air Gap

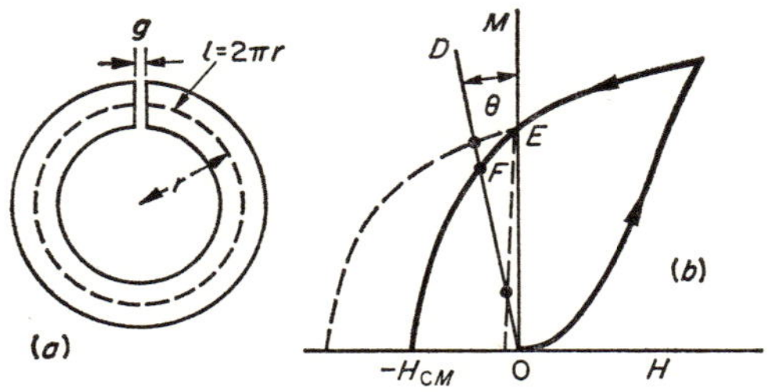

FIG. 1.9. Demagnetizing effect of a small air gap.

Figure 1.9(*a*) represents a ferromagnetic ring specimen of uniform cross-sectional area forming a closed magnetic circuit of mean length l. If the ring is magnetized to saturation, by a suitable winding on the ring, and the magnetizing field is then removed, the material will retain the remanent magnetization $B_R = M_R$ corresponding to the point E in Fig. 1.9(*b*). If now a small air gap of length g is cut in the ring the intensity of magnetization in the material of the ring will fall along the demagnetization curve to some point F as shown. For the purpose of this simple analysis we will suppose that with a small gap g the flux density remains constant at all points in the ring and in the air gap, that is, leakage effects in the neighbourhood of the gap will be neglected.

Given uniform conditions in the material, we may suppose that the appearance of free poles at the air gap produces a demagnetizing field $N_D M$ in the material, where N_D is the effective demagnetizing factor. The relation between the demagnetizing field strength and M is represented by the line OFD in the figure.

Thus corresponding to the point F we have in the material $B_F = \mu_0 H_F + M_F$ and in the gap $B_F = \mu_0 H_G$ where H_G is the field strength in the air gap. The total externally-applied ampere turns, or magnetomotive force, acting round the circuit is zero, but the m.m.f. between the poles of the gap is $H_G g$ and round the remainder of the circuit is $H_F(l - g)$. Hence $H_G g + H_F(l - g) = 0$ and, from the above $\mu_0 H_G = \mu_0 H_F + M_F$. Whence

$$M_F g + \mu_0 H_F l = 0 \tag{1.6}$$

Remembering that H_F is negative we also have $-H_F = N_D M_F$ (1.7)

Therefore from equations (1.6) and (1.7) the effective demagnetizing factor of the ring with a small gap is given by $\mu_0 N_D = g/l$.

It will be clear from the broken lines in Fig. 1.9(*b*) that if the material has a low coercivity the introduction of a small gap would reduce the retained intensity of magnetization to a low value, but if the coercivity is high the intensity of magnetization falls only a little.

1.7 Ballistic Demagnetizing Factors

Magnetic measurements are often made on ferromagnetic specimens in the form of long, open-ended, cylindrical rods or wires. A long solenoid may be used to apply a uniform, axial field H' to the specimen. While it is easy to measure the flux density B at any point along the length of the specimen, determination of the internal field H presents difficulties. This is because of the uncertain value of the demagnetizing field strength produced by a nonuniform distribution of free poles along the length and ends of the specimen. Satisfactory measurements of permeability, particularly if this is high, can then only be made at the centre of rod specimens, and indeed only if there is a high length/diameter ratio. The subject has been very fully investigated by Bozorth and Chapin [4].

We consider a long cylindrical specimen of length l and diameter d in a uniform field H' applied parallel to the axis of the rod, as shown in Fig. 1.10.

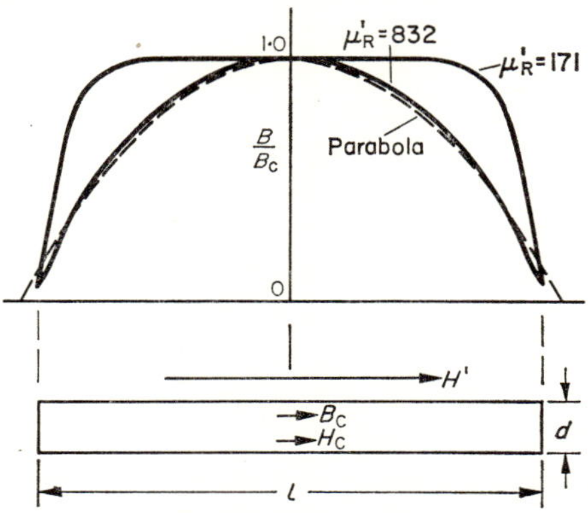

Fig. 1.10. Distribution of flux density along the axis of a cylindrical rod placed in a uniform field H'. $l/d = 52{\cdot}4$.

If B_C is the flux density and H_C the internal field strength at the centre of the rod, then the apparent absolute permeability at this point is $\mu' = \mu_0\mu'_R = B_C/H'$. The true value of permeability is $\mu = \mu_0\mu_R = B_C/H_C$ and μ will be greater than μ' since H_C is less than H', but H_C is not directly measurable. The flux density falls from the maximum value B_C at the centre of the rod to a low value at the ends, and when the permeability is high tends to have a parabolic distribution, as shown by the broken line in Fig. 1.10, but a flatter distribution occurs for lower permeabilities. Experimental results due to Bozorth and Chapin for a nickel-iron rod with $l/d = 52{\cdot}4$ are given in the figure for the cases when $\mu'_R = 832$ and 171, respectively.

Because of the nonuniform conditions in a cylindrical specimen it is not possible to assign a demagnetizing factor to it which would be applicable at any point in the specimen. However, if M_C is the intensity of magnetization at the centre of the rod we may put

$$H' = H_C + N_B M_C \tag{1.8}$$

and N_B has been termed the ballistic demagnetizing factor. We also have

$$B_C = \mu_0 H_C + M_C \tag{1.9}$$

Whence, by eliminating M_C between equations (1.8) and (1.9) and dividing through by B_C, we obtain

$$1/\mu'_R = (1/\mu_R)(1 - \mu_0 N_B) + \mu_0 N_B$$

In any practical case $\mu_0 N_B$ is small compared with unity, and hence approximately

$$1/\mu'_R = 1/\mu_R + \mu_0 N_B \tag{1.10}$$

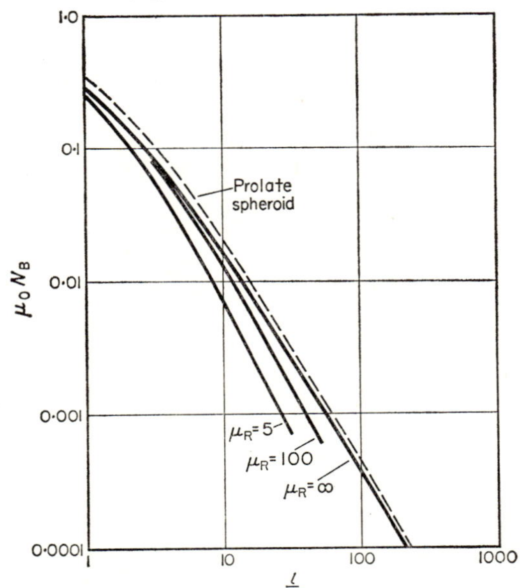

FIG. 1.11. Ballistic demagnetizing factors of cylindrical rods in relation to ratio l/d and μ_R.

If N_B is known then the true permeability μ_R may be arrived at from the measurement of $\mu'_R = B_C/(\mu_0 H')$, although it is apparent from equation (1.10) that the value of N_B for a rod of a particular l/d ratio depends also on the permeability μ_R of the material of the rod.

Figure 1.11 gives values of N_B in relation to l/d for several values of μ_R. The curves may be compared with the curve for a prolate spheroid having ratios of the semiaxes $a/c = l/d$.

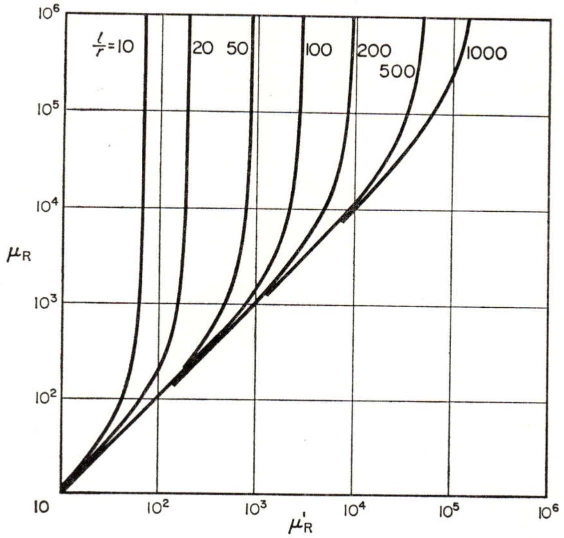

FIG. 1.12. Relation between true permeability μ_R and apparent permeability μ'_R for rods of different l/r ratios.

Figure 1.12 shows how, for a cylindrical specimen of a given l/d ratio, the true permeability μ_R may be read off from the observed apparent value μ'_R. Examination of the curves shows that when μ_R is large it cannot accurately be found from a measurement of μ'_R unless the ratio l/d is large.

1.8 Crystal Planes and Directions

The magnetic materials dealt with in this book are mostly crystalline solids and, in general, the individual crystals in the solids are strongly anisotropic, particularly in their magnetic properties. The directional properties are associated with the crystal structure and it will be necessary frequently to specify particular planes and directions in the crystals. This is normally done by using the Miller Indices.

The atoms or molecules in a crystal are arranged in a regular, three-dimensional, geometric pattern, and to describe the structure and symmetry of crystals Bravais in 1848 introduced the idea of a space lattice [5]. In the metals and in the magnetic oxides we are concerned with atoms rather than molecules. The centres of the atoms lie on lattice points which in turn are situated in planes and on straight lines.

Crystals are classified on the basis of the unit cell. The true unit cell is the smallest parallelepiped in the space lattice such that, when moved a distance equal to one of its sides, the movement will cause any atom in the cell to occupy identical surroundings. The true unit cell would then in fact contain the equivalent of one whole atom only. For convenience in practice, however, the unit cell adopted may be larger and include several such parallelepipeds, as for example in the cases of the hexagonal and cubic crystal structures given below.

Any parallelepiped may be specified by three axes, not necessarily at

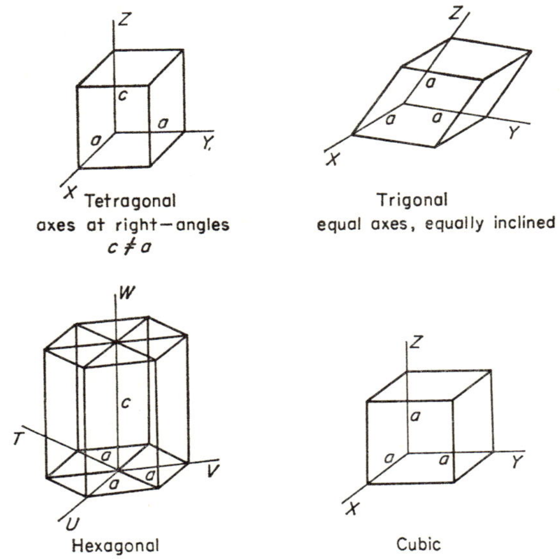

FIG. 1.13. Four crystal systems in metals.

right-angles, and by three axial distances, *a*, *b*, *c*, corresponding to the sides of the unit cell. Bravais showed that there are seven such systems of axes where equal parallelepipeds will fit together to fill all space. The systems are called triclinic, monoclinic, orthorhombic, tetragonal, trigonal, hexagonal and cubic. Metals crystallize in the last four named forms of which the unit cells for the simple systems are shown in Fig. 1.13, but we shall be concerned only with the hexagonal and cubic structures. Figure 1.13 shows for the cubic system the simple lattice but in this system there is also a body-centred and a face-centred lattice, as described in Chapter 3. There are indeed fourteen lattice types for the seven crystal systems.

The method of specifying a plane and a direction in a cubic crystal is illustrated in Fig. 1.14, where the unit cell—a cube of side *a*—has one corner at the origin of rectangular coordinates *OX*, *OY*, *OZ* which coincide with

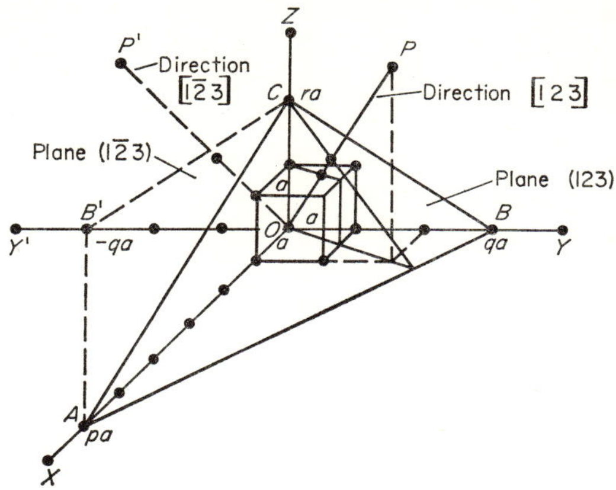

FIG. 1.14. Miller indices in the cubic system.

cube edges. *ABC* is a plane intercepting the axes at distances *pa*, *qa* and *ra*, respectively. The plane is specified by the lowest whole numbers which are proportional to the reciprocals of the intercepts. For the example drawn in Fig. 1.14, $p = 6$, $q = 3$, $r = 2$, so that if h, k, l are the indices of the plane

$$h : k : l = 1/p : 1/q : 1/r = 1/6 : 1/3 : 1/2 = 1 : 2 : 3$$

ABC is then a (123) plane. The designation (123) in round brackets signifies any plane parallel with the plane *ABC*. In the figure the plane *AB'C* is a similar plane except that the intercept *OB'* is $-qa$. Plane *AB'C* and planes parallel with it would then be given as $(1\bar{2}3)$ where the bar over the figure indicates a negative intercept. When the figures are enclosed in braces, or curly brackets, thus $\{123\}$, the symbol then represents all planes in the crystal which are crystallographically similar to the plane *ABC*.

A direction in the crystal may also be given in terms of three numbers but using square brackets thus [*hkl*]. The direction may be found by taking a point *P*, as in Fig. 1.14, whose coordinates with respect to the *XYZ* axes are $x = ha$, $y = ka$, $z = la$, and joining *P* to the origin. In the figure *P* has been placed with the coordinates *a*, 2*a*, 3*a*, and the direction from *O* to *P*, and any line in the crystal parallel with it and in the same direction, is designated [123]. In the cubic system the indices of a plane are the same as the indices of a direction perpendicular to the plane. Thus [123] is at right-angles to plane (123), or *OP'* in the figure represented by $[1\bar{2}3]$ is perpendicular to plane *AB'C* given by $(1\bar{2}3)$. This may be confirmed by studying the geometry of Fig. 1.14. All directions which are crystallographically similar to [*hkl*], whether the integers are negative or not, may be represented generally using caret brackets thus $\langle hkl \rangle$.

Directions and planes in the cubic system of particular interest in magnetism are the cube edges, cube-face diagonals, the body diagonals and the planes perpendicular to them.

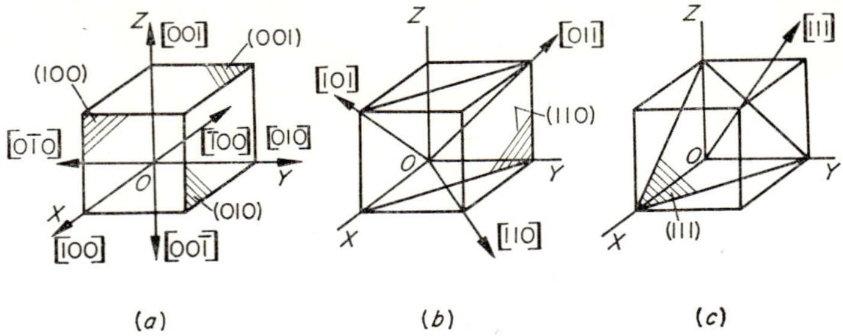

FIG. 1.15. Principal directions in the cubic system. Planes perpendicular to them have the same indices.

Figure 1.15(*a*) shows the six cube-edge or ⟨100⟩ directions. These are [100], [010], [001], [$\bar{1}$00], [0$\bar{1}$0] and [00$\bar{1}$]. Planes perpendicular to these directions will, as already indicated, have the same indices. For example, the intercepts of the (100) plane on the coordinate axes are 1, ∞, ∞, and the reciprocals are thus 1, 0, 0. The axis OX is a [100] direction and is normal to this plane. The planes (100), (010) and (001) are shown partly shaded in the figure. The cube itself, with six sides, may be represented by the symbol {100}.

There are twelve cube-face diagonals, or ⟨110⟩ directions, of which three are shown in Fig. 1.15(*b*). The (110) plane is also shown in the figure. Figure 1.15(*c*) shows one of the eight body diagonals, or ⟨111⟩ directions, together with a (111) plane.

Figure 1.16 gives the Miller–Bravais indices used for the hexagonal structure. Two systems are in use. The system illustrated at (*a*) has a hexagonal unit cell with three equal axes OU, OV, OT, 120° apart in the basal plane, and the vertical or *c*-axis OW. Any plane may then be specified in terms of the reciprocals of the intercepts, measured in units of *a* or *c*, on the four axes.

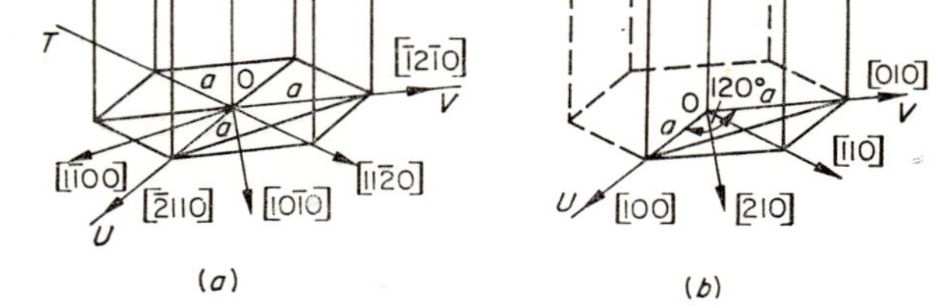

FIG. 1.16. Miller–Bravais hexagonal indices: (*a*) four-indices system, (*b*) three-indices system.

Thus the plane marked $(11\bar{2}0)$ in Fig. 1.16(*a*) has the intercepts, in the order U, V, T, W, of $a, a, -a/2, \infty$. The reciprocals will then give $(11\bar{2}0)$. Similarly the plane (0001) in the figure has the intercepts $\infty, \infty, \infty, c$, the reciprocals of which give the indices (0001). In general, if the indices are *hkil* it is easy to show from the geometry of the hexagon that $h + k + i = 0$, or $i = -(h + k)$. The third figure is therefore superfluous and frequently a dot is substituted for it. Thus, for example, $(11\bar{2}0) = (11{\cdot}0)$ or $(0001) = (00{\cdot}1)$.

The indices for directions are found by moving from the origin by amounts *ha, ka, ia, lc*, parallel to each of the four axes, respectively, the distances moved then being reduced to the smallest whole numbers. Joining the origin to a point *P* thus reached is the corresponding direction in the crystal. Such a point may, in fact, be reached by alternative routes, using different combinations of the coordinates *h, k, i*, but the route chosen *must* be one that makes $h + k + i = 0$. A number of crystal directions are given in Fig. 1.16(*a*), and here again, as just mentioned, the third figure may have a dot substituted for it.

The alternative system, using three indices, is illustrated in Fig. 1.16(*b*). The heavy lines outline the unit cell which has a basal parallelogram of length of side *a*, and a vertical height *c*. There are three axes, *OU* and *OV* 120° apart, and *OW* mutually perpendicular. Planes are specified by the reciprocals of the intercepts on the three axes, and directions from the coordinates of a point obtained with respect to the three axes. Several planes and directions are shown in Fig. 1.16(*b*). Comparing the two systems, given in the figure at (*a*) and (*b*), shows for example that (0001) = (001), (1010) = (100), $[10\bar{1}0] = [210]$, $[11\bar{2}0] = [110]$, and so on.

PROBLEMS

1.1. A soft iron sphere is placed inside a long solenoid uniformly wound with 10 turns/cm and carrying a steady current of 10 A. Calculate the magnetic field strength, intensity of magnetization and flux density in the sphere

(a) if the iron is of very high permeability, and

(b) if the relative permeability of the iron is 10.

1.2. A ferromagnetic ellipsoid has semi-axes *a, b, c* of 5, 3 and 2 cm, respectively. Find by means of Figs. 1.5, 1.6 and 1.7 the demagnetizing factors of the ellipsoid for each axial direction and check the answers by using equation (1.1).

1.3. Show that, in a crystal of cubic structure, a direction given by the Miller Indices $[h\,k\,l]$ is perpendicular to a plane having the same indices.

1.4. A prolate spheroid of high permeability with the longitudinal axis parallel with *OX* has demagnetizing factors $\mu_0 N_Y = \mu_0 N_Z = 0{\cdot}45$. A field $H = 300$ A/cm is applied in a direction having direction cosines *l, m, n* with respect to the coordinate axes and where $l = m = n = 1/3^{\frac{1}{2}}$. Calculate the intensity of magnetization M in the spheroid and the direction cosines l_1, m_1, n_1 of the M vector. (*Hint*: use components of H resolved along the three axes.)

1.5. If the volume of the spheroid in Problem 1.4 is 500 cm^3 calculate

(a) the torque acting on the spheroid,
(b) the magnetic potential energy of the system and
(c) the magnetostatic stored energy.

(*Note*: the angle θ between two straight lines is given by $\cos\theta = l\,l_1 + m\,m_1 + n\,n_1$.)

1.6. If the spheroid in Problem 1.4 now rotates against a restraining force until the longitudinal axis and the field H are in line, thereby doing external mechanical work, calculate

(a) the final magnetic potential energy,
(b) the new stored magnetostatic energy, and
(c) from the answers to (a) and (b) and to Problem 1.5, deduce the mechanical work done.

(The student should satisfy himself that, in this example, half the total change in magnetic potential energy is converted to external work, and half is stored internally in the specimen. See Section 1.5.)

1.7. A sample of iron has a relative permeability of 11·5 when the flux density is 2·3 Wb/m^2. Calculate the strength of the externally-applied field required to produce this flux density in the material if it is in the form of

(a) a sphere, (b) a flat plate normal to the direction of the field.

1.8. If in Problem 1.7(b) the direction of the applied field is at 10° to the normal to the plate, and the flux density is 2·3 Wb/m^2, calculate the magnitude of the applied field and the direction of the flux vector in the plate.

1.9. A prolate spheroid of the same material as in Problem 1.7 has a demagnetizing factor for the a-axis given by $\mu_0 N_X = 0{\cdot}2$. An external field is applied at an angle of 60° to the a-axis and the flux density in the spheroid is 2·3 Wb/m^2. Calculate the magnitude of the applied field and the direction of the flux vector in the spheroid.

REFERENCES

[1] MAXWELL, J. C. *Electricity and Magnetism*, Vol. 2, Art. 437, 67, The Clarendon Press, Oxford (1892).

[2] EWING, J. A. *Magnetic Induction in Iron and Other Metals*, p. 24, Electrician Publishing Co., London (1900).

[3] OSBORN, J. A. 'Demagnetizing factors of the general ellipsoid', *Phys. Rev.* **67,** 351 (1945).

[4] BOZORTH, R. M. and CHAPIN, D. M. 'Demagnetizing factors of rods', *J. Appl. Phys.* **13,** 320 (1942).

[5] BARRETT, C. S. *Structure of Metals* (Chapter 1), McGraw-Hill (1952).

Chapter 2

ELECTRONS AND ATOMS

2.1 **Optical Spectra**

An atom of any element consists of a central heavy nucleus with which is associated a number of extra-nuclear electrons. The number of electrons in a normal, free atom is equal to the atomic number Z of the element. For hydrogen $Z = 1$. There is one proton, with one unit of positive charge, forming the nucleus, and one electron, and the mass is 1·008 units. One atomic mass unit (amu) $= 1{\cdot}6598 \times 10^{-24}$ g. For iron, to take another example, $Z = 26$. The atomic mass is 55·953 amu and there are 26 electrons. In an atom containing many electrons, the electrons are arranged in a regular way about the nucleus. As well as carrying negative charge they are in motion and they interact upon one another, or are *coupled*, in a quite complicated manner. The electrical, magnetic and chemical properties of the elements depend upon outer electrons in the atoms. A very complete picture of atomic structure has been built up mainly from a detailed examination of the light emitted from atoms when the element is in a gaseous state at a high temperature, that is, from a study of optical spectra.

If a beam of white light is passed through a transparent prism the light is dispersed into a spectrum of varying colours. A spectrograph for producing a pure spectrum, that is one in which the colours do not overlap, is shown in Fig. 2.1. White light from a suitable source is focused on a thin slit A, the length of which is perpendicular to the plane of the paper in the diagram. The slit A is in the first focal plane of an achromatic lens B so that a parallel

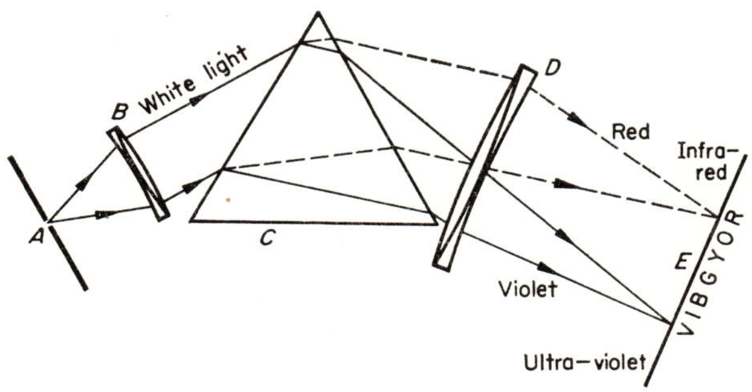

FIG. 2.1. Arrangement of a spectrograph.

beam of white light falls upon the prism C which may be of glass or quartz. The colours are dispersed in passing through the prism, but rays of any one colour emerge as a parallel beam and may be brought to a focus by a second achromatic lens D on a screen or photographic plate E which is in the second focal plane of lens D. Red rays are shown, for example, by the broken lines and violet rays by the full lines. All the colours of the spectrum from red to violet are thus focused on the screen as well as invisible rays, infrared or heat rays on one side of the visible spectrum and ultraviolet light on the other.

Light, according to the wave theory, has the character of an electromagnetic vibration. At any point in space it consists of an electric and a magnetic field mutually perpendicular and at right-angles to the direction of propagation, each field alternating simultaneously in relation to time with a frequency which is characteristic of the colour. The velocity of propagation of light of any frequency is a basic physical constant which has been determined experimentally by several methods, and is known with considerable accuracy. A review of the observed values indicates that the most probable value of the velocity of light in a vacuum is $c = 2{\cdot}99790 \times 10^{-8}$ m.s^{-1}. In dry air at normal temperature and pressure the value is a few parts in 10,000 lower. If λ is the wavelength for a train of waves of monochromatic light the number of complete waves passing a particular point in free space in one second will be c/λ. The number of complete alternations of the electric and magnetic fields associated with the wave in one second at the point, or the frequency ν of the wave, is therefore given by the relation $\nu = c/\lambda$. The frequency of a light wave is often given in terms of a wave number ν' where $\nu' = \nu/c$. The wave number is therefore the number of complete waves in a wave train of unit length, or the reciprocal of the wavelength. The complete electromagnetic spectrum is shown in Fig. 2.2 and it will be seen that visible light forms only a small part of it.

If the source of light is a solid at a high temperature, for example a piece of iron at a white heat, white light is emitted and, referring again to Fig. 2.1, a continuous spectrum is obtained. However, if the source is an arc or spark discharge in a gas or vapour the spectrum is no longer continuous but is made up of electromagnetic radiations of a number of discrete frequencies.

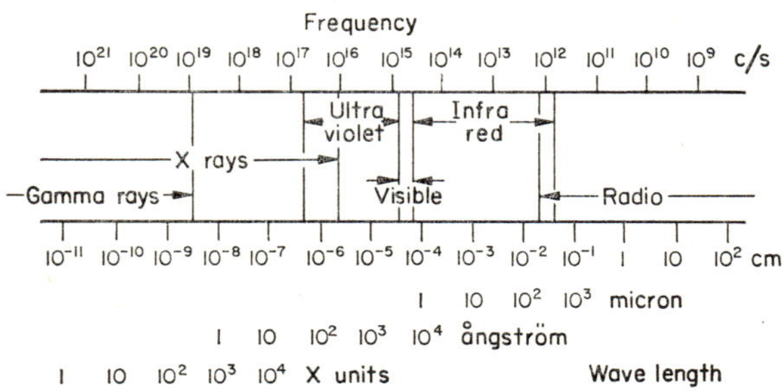

FIG. 2.2. Range of frequencies and wavelengths in the electromagnetic spectrum.

With a narrow slit and a prism the spectrum is then observed as many lines, or images of the slit, of different colours. Such an emission spectrum may be recorded as sharp lines on a photographic plate or film. For the heavier elements there may be many thousands of such lines even in the visible range, the lines being displaced on the plate relatively to one another according to the frequencies of the radiations, and in general forming overlapping series of lines characteristic of the element. Some optical spectra are shown in Fig. 2.3.

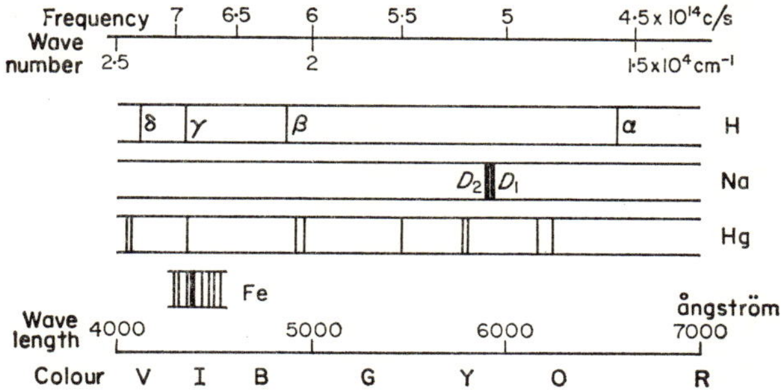

FIG. 2.3. Some spectral lines of hydrogen, sodium and mercury, and a small part of the iron spectrum.

Wavelengths may be measured with very great accuracy by employing the principle of interference. The phenomena of reflection, refraction and diffraction of light can be explained on the supposition that a beam of light is made up of many waves and by applying Huygens' principle that any point on a wavefront is a new point source of waves. Waves of the same wavelength travelling in a particular direction reinforce one another if they are in phase

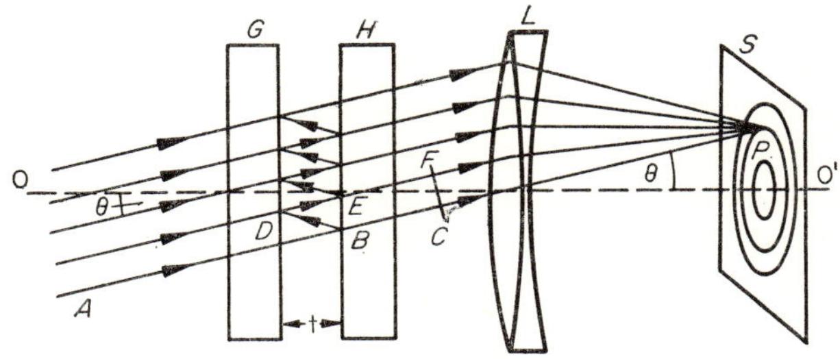

FIG. 2.4. Principle of the Fabry–Pérot interferometer for measuring wavelength.

and cancel out if they are in phase opposition. The principle of the Fabry–Pérot interferometer used to measure wavelengths is illustrated in Fig. 2.4. *G* and *H* are two thick glass or quartz plates with their inner surfaces lightly silvered and parallel. Consider parallel rays of monochromatic light derived from a broad source and inclined at an angle θ to the axis OO' of the apparatus. The light undergoes multiple reflections in the air space between the plates as well as being transmitted. It may be shown, considering for example the rays *BC* and *BDEF*, that the waves are in phase at the wavefront *FC*, and thus reinforce one another, when $2t \cos \theta = n\lambda$, where t is the distance between the plates, λ is the wavelength and n is an integer. The light may then be brought to a focus at P on the screen S by the lens L. If the condition just stated is met, all rays coming conically about the axis OO' at an angle θ from the broad source are brought to a focus in a bright ring passing through P. A series of bright rings will be formed, as θ changes, when n changes by unit amounts. In this way wavelengths may be measured in terms of linear dimensions in the apparatus and these dimensions, in turn, may be compared ultimately with the standard metre kept in France. A comparison between the wavelength of the red line of cadmium and the standard metre has been made many times. A comparatively recent determination, by Sears and Barrell about 1933, gives the wavelength of this line, in a vacuum, as $\lambda = 6440{\cdot}2510 \times 10^{-10}$ m, with a probable accuracy of $\pm 0{\cdot}001 \times 10^{-10}$ m [1].

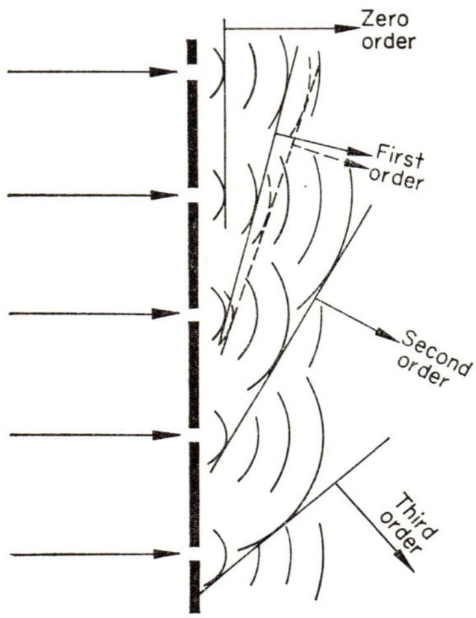

FIG. 2.5. Principle of the transmission grating.

The International Committee of Weights and Measures, however, now defines the metre in terms of a very sharp line obtained from the krypton 86 atom (radiation corresponding to the transition between the $2p_{10}$ and $5d_5$ levels). The metre is a length equal to 1,650,763·73 wavelengths of this radiation.

Spectral lines in the infrared, and of any shorter wavelength, may be recorded photographically. Glass prisms, however, are not transparent to radiations much lower than the limit of the visible range which is about $3{\cdot}6 \times 10^{-5}$ cm, but quartz will transmit wavelengths down to about $2{\cdot}0 \times 10^{-5}$ cm below which absorption occurs both in quartz and in air. To record lines of still shorter wavelength and, indeed, into the extreme ultraviolet, gratings are used.

Gratings are made by finely ruling parallel, equidistant grooves with a diamond point either on a glass plate to form a transmission grating or on a polished metal to make a reflection grating. There may be 2,000 to 10,000 grooves per cm width of the grating. The principle of the transmission grating is illustrated in Fig. 2.5. Parallel monochromatic light is seen coming from the left. The thick lines represent an end-on view of the lines or grooves ruled on the grating, the light passing through the spaces between them. By Huygens' principle each parallel transparent slit becomes a source of wavelets which spread outwards in the manner denoted by the concentric, equidistant circles in the diagram. Wavefronts, in which reinforcement of the wavelets occurs, are found by drawing tangents as shown. Parallel beams of light normal to these wavefronts are thus projected on the right-hand side of the grating in particular directions, and each direction, respectively, is said to have an order of interference denoted as zero, first, second and so on. This is made clear in the figure. If the wavelength of the light is changed, the direction of all but the zero order is changed also, as the broken lines show for the first order. Thus different wavelengths in the incoming light, if this is not monochromatic, may be separated by the grating.

In the case of reflection gratings, lines are ruled on the polished mirror surface of speculum metal, an alloy of copper and tin, and the incident light is then diffracted from the untouched surfaces between the rulings. If the mirror surface is made concave then waves of different frequencies in the incident light are not only separated but may also be directly focused. ROWLAND [2] made the first concave gratings on a spherical surface with ruled lines equally spaced on a chord. When light of any wavelength, from a slit parallel to the rulings, falls on the grating it is brought to a focus on a circle, called the Rowland circle, if the slit also lies on this circle, and where the diameter of the Rowland circle is equal to the radius of curvature of the grating. Figure 2.6 shows diagrammatically a Paschen spectrograph employing these principles. The light to be analysed enters the dark room containing the instrument from a slit *S* and is diffracted from a concave reflection grating *G*. The spectral lines are focused on to a curved photographic plate or film *P* placed with its active surface on the Rowland circle. The approximate position of the spectrum in the second order of interference is shown in the figure. A large diameter of circle in such an instrument may be used, for example 20 feet, so that wavelengths may be compared with great accuracy.

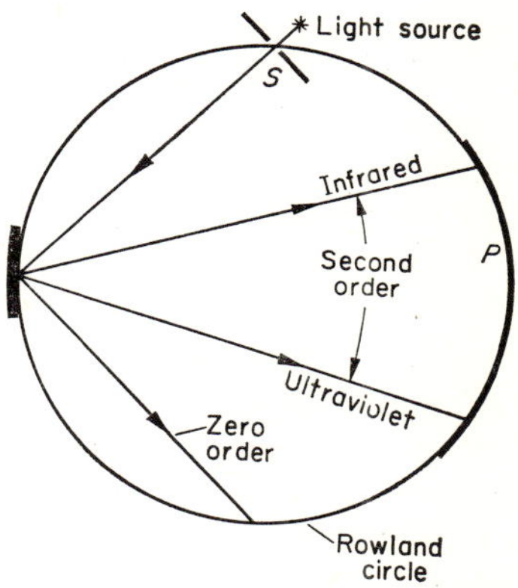

FIG. 2.6. The Paschen spectrograph.

2.2 The Bohr Theory of the Atom

In 1901 PLANCK [3] formulated a law of radiation which, for the first time, successfully accounted for the observed radiation of light from hot bodies. He found it necessary to postulate that light, radiated at any frequency ν, was not a perfectly continuous stream but was in discrete packets or quanta each of which possessed a definite amount of energy $h\nu$ where h was a constant, now known as Planck's constant. These individual quanta are also sometimes called photons and their existence is undisputed in modern physics.

When light quanta in the ultraviolet range are absorbed by a clean metal surface it is found that the metal emits electrons. Visible light also has this effect for a few elements. Observation, however, shows that the velocity of the emitted electrons depends only upon the frequency of the light and not upon its intensity, although increasing the intensity increases the number of photoelectrons. This photoelectric effect was explained by EINSTEIN [4] in 1905 in terms of the equation $h\nu = W + \frac{1}{2}mv^2$. The equation refers to each individual event where a single quantum or photon of energy $h\nu$ is absorbed by the metal and an electron is ejected with kinetic energy $\frac{1}{2}mv^2$. W is the energy required to drag an electron away from the atoms at the metal surface. It is called the work function and its value differs for different elements. Extensive experimental observations based on the photoelectric effect and Einstein's equation enabled Millikan [5] subsequently to determine accurately the value of Planck's constant. It is now given as

$$h = 6{\cdot}6252\ (\pm 0{\cdot}0002) \times 10^{-34}\ \text{joule second.}$$

Experiments carried out during the period from about 1909 to 1913 on the scattering of α-particles from metal foils led Rutherford to certain conclusions about the constitution of atoms. α-particles are positive ions obtained by removing the two electrons from helium atoms. RUTHERFORD [6] proposed that the whole of the positive charge and nearly the whole of the mass of an atom were concentrated in a central region, now called the nucleus, less than 10^{-12} cm in diameter and that, in atomic units, the positive charge was about half the atomic weight.

At this time spectral lines had been extensively studied but the key to their origin had not been found. However, in 1913, BOHR [7] made an outstanding advance with a theory of the hydrogen atom which embraced the ideas mentioned in the preceeding paragraphs. The theory not only accounted accurately and quantitatively for the lines then known in the hydrogen spectrum, but also predicted further series of lines, since found, and furthermore provided a model for the structure of atoms of all elements.

In the Bohr model of the hydrogen atom nearly the whole mass of the atom is concentrated in a central nucleus which has also one unit of positive charge. There is also one extra-nuclear electron bearing one unit of negative charge and this moves around the nucleus, assumed to be at rest, in a circular orbit. The mass of the hydrogen nucleus, which is in fact a single proton, is

$$m_p = 1{\cdot}6725 \times 10^{-27}\ \text{kg},$$

the mass of the electron is $m = 9{\cdot}1066 \times 10^{-31}$ kg, and one unit of charge $e = 1{\cdot}602 \times 10^{-19}$ coulomb. The ratio of the mass of the hydrogen atom to that of the electron is therefore 1837·5.

This system of two oppositely-charged particles, one of which is in motion, has potential and kinetic energy and the total energy depends upon the radius of the orbit in which the electron moves. Bohr made the arbitrary assumption that certain orbits only were allowed, these orbits being determined by the proposition that the angular momentum of an orbital electron had to be an integral multiple of $h/2\pi$ where h is Planck's constant. When the electron is in the innermost allowed orbit in the hydrogen atom the total energy is a minimum and the angular momentum is one unit. The atom is then said to be in the normal or ground state. If the electron jumps outwards to the next orbit, and this requires that the atom shall absorb a certain quantity of energy, the angular momentum of the electron is two units. In the next allowed orbit outwards the electron has three units of angular momentum and so on. The atom, with the electron in an outer orbit, is said to be excited and if the electron is completely removed the atom is ionized. In the Bohr theory no energy is absorbed or radiated so long as the electron remains in one orbit. The atom is then said to be in a stationary state. Energy, however, is emitted, when an electron drops inwards from one allowed orbit to another, equal in amount to the difference of the total energies of the atom in the two states. This quantity of energy $W_2 - W_1$ is in the form of a packet of radiation of frequency ν such that $W_2 - W_1 = h\nu$. A spectral line corresponding to this frequency is the result of a great number of quanta emitted for equal transitions in a great many excited atoms.

An analysis of the hydrogen atom may now be made on the suppositions that Newtonian mechanics and the Coulomb law of force between charges are valid in the atom, but with restrictions on the allowed orbital angular momenta.

A more indefinite physical picture of the atom is envisaged in more elegant recent work involving wave mechanics, in which these suppositions are not necessary, but this work will be referred to later.

The nucleus of the hydrogen atom has positive charge e. However, let us consider the rather more general case of a hydrogen-like atom in which the nucleus has positive charge Ze, where Z is the atomic number, and where there is still only one electron present. Thus for hydrogen $Z = 1$. For helium $Z = 2$ and the neutral atom has two electrons. The atom, however, becomes hydrogen-like if it is ionized by the complete removal of one electron. For lithium $Z = 3$ with three electrons in the neutral atom. However, if the lithium atom is doubly-ionized by the removal of two electrons it also becomes hydrogen-like.

The total electrostatic flux emanating from a nucleus having charge Ze is also Ze and the electric flux density at a distance x from the nucleus is $Ze/4\pi x^2$. The electric field strength at this point is then $E = Ze/4\pi\epsilon_0 x^2$ where ϵ_0 is the absolute permittivity of free space. $\epsilon_0 = (\mu_0 c^2)^{-1} = 8{\cdot}8542 \times 10^{-12}$ farad m^{-1} where the absolute permeability of free space is, by definition, $\mu_0 = 4\pi \times 10^{-7}$ henry m^{-1}. The force outwards on an electron having negative charge e is then

$$F = E(-e) = \frac{-Ze^2}{4\pi\epsilon_0 x^2} \tag{2.1}$$

the negative sign meaning that the electron is attracted towards the nucleus. The work done by the electron when it moves to an infinite distance from a point x from the centre is the potential energy of the system of two charges.

$$\therefore \text{ Potential energy} = \int_x^\infty F\,dx = \int_x^\infty \frac{-Ze^2}{4\pi\epsilon_0 x^2}\,dx$$

$$= \frac{-Ze^2}{4\pi\epsilon_0 x} \tag{2.2}$$

The negative sign indicates that work must be done on the electron to move it outwards.

If now the electron of mass m is moving with velocity v in an allowed circular orbit of radius r the centrifugal force is mv^2/r. This is balanced by the electrostatic attraction inwards. Whence, from equation (2.1), putting $x = r$,

$$\frac{Ze^2}{4\pi\epsilon_0 r^2} = \frac{mv^2}{r} \tag{2.3}$$

The kinetic energy of the electron is $mv^2/2$. Whence, from equation (2.3),

$$\text{kinetic energy} = \frac{Ze^2}{8\pi\epsilon_0 r} \tag{2.4}$$

The total energy of the system is the sum of the potential and kinetic energies. Therefore, from equations (2.2) and (2.4), with $x = r$,

$$\text{total energy } W = \frac{-Ze^2}{4\pi\epsilon_0 r} + \frac{Ze^2}{8\pi\epsilon_0 r}$$

$$\therefore W = \frac{-Ze^2}{8\pi\epsilon_0 r} \tag{2.5}$$

The angular momentum of the electron is mvr and, according to Bohr, this must be an integral multiple of $h/2\pi$.

$$\therefore mvr = \frac{nh}{2\pi} \tag{2.6}$$

where $n = 1, 2, 3 \ldots$. Eliminating v between equations (2.3) and (2.6) gives

$$r = \frac{n^2\epsilon_0 h^2}{\pi m Z e^2} \tag{2.7}$$

Equation (2.7) gives the radius of each allowed orbit in terms of known constants and the integer n. For the smallest orbit of the hydrogen atom $Z = 1$ and $n = 1$, and substituting numerical values in equation (2.7) gives

$$r_1 = 0{\cdot}5292 \times 10^{-10}\text{ m} = 0{\cdot}5292 \text{ ångström} = a_0$$

where a_0 is the radius of the first Bohr orbit.

For $n = 2$, $r_2 = 4r_1$ and indeed $r_n = n^2 r_1$.

Eliminating r between equations (2.3) and (2.6) gives

$$v = \frac{Ze^2}{2\epsilon_0 nh} \tag{2.8}$$

and the frequency of the orbit, that is the number of revolutions of the electron every second, is given by

$$\nu = \frac{v}{2\pi r} = \frac{mZ^2e^4}{4\epsilon_0^2 n^3 h^3} \tag{2.9}$$

For the hydrogen atom with $n = 1$

$$v_1 = \frac{e^2}{2\epsilon_0 h} = 0{\cdot}02187 \times 10^8 \text{ m.s}^{-1} = \frac{c}{137{\cdot}1}$$

The velocity of the electron in the first orbit of the Bohr hydrogen atom is therefore 1/137 of the velocity of light. From equation (2.8) it will be seen that v decreases as n increases but increases as the atomic number Z increases. The frequency of the first orbit for hydrogen is, from equation (2.9),

$$\nu_1 = \frac{me^4}{4\epsilon_0^2 h^3} = 6{\cdot}576 \times 10^{15}\ \text{s}^{-1}$$

The orbital frequency is inversely proportional to n^3 and directly proportional to Z^2 as equation (2.9) shows. The relative sizes of the inner allowed orbits for hydrogen, ionized helium and doubly-ionized lithium are shown in Fig. 2.7 together with the electron velocities and orbital frequencies.

The electron in an excited atom will continue only briefly in an allowed orbit. The atom will revert in an interval of time of the order of 10^{-8} second

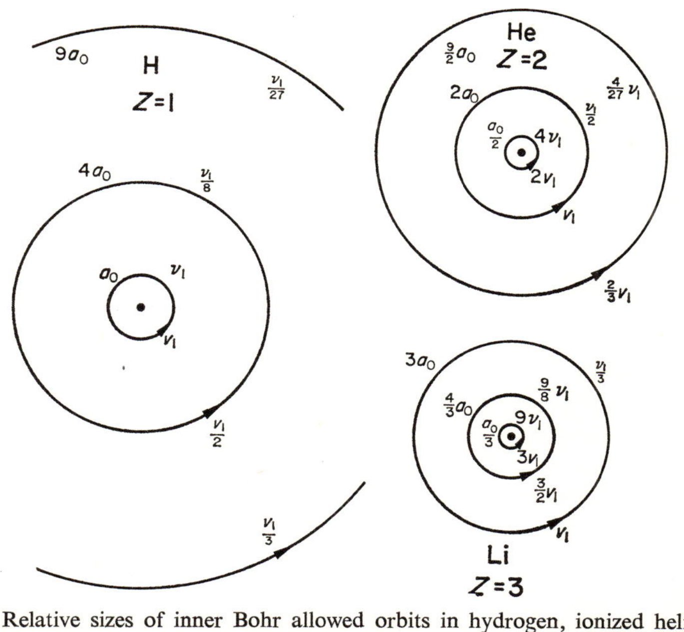

FIG. 2.7. Relative sizes of inner Bohr allowed orbits in hydrogen, ionized helium and doubly-ionized lithium. For the smallest orbit in hydrogen: radius $a_0 = 0{\cdot}5292 \times 10^{-10}$ m $= 0{\cdot}5292$ Å, electron velocity $v_1 = 0{\cdot}02187 \times 10^8$ m s^{-1} $= c/137{\cdot}1$, orbital frequency $\nu_1 = 6{\cdot}576 \times 10^{15}$ s^{-1}.

to a state of lower energy. The total energy of the atom for the electron in any orbit may be obtained by substituting for r from equation (2.7) in equation (2.5). Whence

$$\text{total energy } W = \frac{-me^4Z^2}{8\epsilon_0^2h^2n^2} \tag{2.10}$$

Putting numerical values for the hydrogen atom in equation (2.10) gives, for $n = 1$ which is the ground state, $W_1 = -218{\cdot}0 \times 10^{-20}$ joule. When $n = 2$, $W_2 = -54{\cdot}5 \times 10^{-20}$ joule, and so on for the other orbits, the total energy rising from negative values to zero when $n = \infty$.

The change in energy when an electron jumps inward from an orbit n_2 to an orbit n_1 is given by

$$W_{n_2} - W_{n_1} = \frac{-me^4Z^2}{8\epsilon_0^2h^2}\left(\frac{1}{n_2^2} - \frac{1}{n_1^2}\right)$$

and $W_{n_2} - W_{n_1} = h\nu$ where ν is the frequency of the radiation emitted. We also have the wave number $\nu' = \nu/c = 1/\lambda$ and it follows that

$$\nu' = \frac{me^4Z^2}{8c\epsilon_0^2h^3}\left(\frac{1}{n_1^2} - \frac{1}{n_2^2}\right)$$

or

$$\frac{1}{\lambda} = RZ^2\left(\frac{1}{n_1^2} - \frac{1}{n_2^2}\right) \tag{2.11}$$

where

$$R = \frac{me^4}{8c\epsilon_0^2h^3} \tag{2.12}$$

R is called the Rydberg constant.

The quantity

$$T = \frac{RZ^2}{n^2} = \frac{-W}{hc} \tag{2.13}$$

is called the term or term value of the electron when in orbit n. It represents the total energy of the system, the term value being in wave number units and with the negative sign discarded. Then for a jump from orbit n_2 to orbit n_1, $\nu' = T_{n_1} - T_{n_2}$.

Equation (2.11) is a theoretical relationship capable of direct comparison with observation. The equation predicts, for example, that in the case of hydrogen, electrons dropping inward from orbit $n = 2$ to orbit $n = 1$ should produce a spectral line for which $1/\lambda = R(1 - \frac{1}{4})$, and if the jump is from $n = 3$ to $n = 1$, $1/\lambda = R(1 - \frac{1}{9})$ and so on for any initial and final n values. If the atom is ionized helium, for which $Z = 2$, the corresponding values of $1/\lambda$ are four times as much.

Certain lines of the hydrogen and ionized helium spectra, however, should have the same wavelengths, for example, if the jump is from $n = 3$ to $n = 2$ in hydrogen, and from $n = 6$ to $n = 4$ in helium, then in both cases $1/\lambda = R(\frac{1}{4} - \frac{1}{9})$. As already described wavelengths may be measured and compared with great accuracy. It is found in fact that the lines in the hydrogen and helium spectra which should be of the same wavelength are not exactly so, there being a small discrepancy of about 0·04%.

This small difference serves in a remarkable way to confirm the general validity of the Bohr theory. The value of the Rydberg constant given in equation (2.12) is based on the assumption of an electron orbiting about a stationary central nucleus. It may, however, be shown in mechanics that such a pair of particles would each orbit about a common intermediate centre, the radii of the two orbits being in inverse ratio to the particle masses. The orbital radius of the nucleus is obviously small compared with that of the electron, even in the hydrogen atom, and theoretically becomes zero when the nuclear mass is infinite. To correct for the motion of the nucleus it is necessary to divide the energy of the atom calculated above, and the value of R in equation (2.12), by the factor $1 + m/M$ where M is the nuclear mass. For hydrogen $m/M = 1/1836{\cdot}5 = 0{\cdot}0005445$ and for helium $m/M = 0{\cdot}0001361$. The value of R given in equation (2.12) corresponds to a stationary nucleus, or $M = \infty$, and may be denoted as R_∞. For hydrogen R_1 is then approximately 0·054% less than R_∞ and for helium R_4 is 0·014% less, and these calculated differences account almost exactly for the measured difference of about 0·04% between the hydrogen and helium lines.

The Rydberg constant can be determined by measuring wavelengths for different elements and applying equation (2.11). Hence R_∞ may be deduced with considerable accuracy. Its value is found to be $R_\infty = 10{,}973{,}731\ \text{m}^{-1}$ and this agrees with the value calculated from equation (2.12) to within experimental error thus remarkably supporting the Bohr theory.

The total energy of the atom when the electron is in any orbit is, from equation (2.13), given by $W = -hcRZ^2/n^2$. The total energy may therefore be calculated for any value of n in terms of known constants, and energy-level diagrams constructed. The energy-level diagram for hydrogen ($Z = 1$) is shown in Fig. 2.8 where, when $n = 1$, $W_1 = -217{\cdot}9 \times 10^{-20}$ joule. When $n = \infty$, $W = 0$. The corresponding term values $T = RZ^2/n^2$ are also shown.

The first series of spectral lines to be observed for hydrogen was in the visible range—the Balmer series. This spectrum is shown on the right-hand side of Fig. 2.8 (and also in Fig. 2.3). The lines are due to electron jumps from orbits $n = 3, 4, 5, \ldots$, respectively, to orbit $n = 2$ as the diagram indicates. The wavelengths of the lines may be calculated from equation (2.11) and observation and calculation are in excellent agreement. On the left of Fig. 2.8 is shown the spectrum obtained when the electrons jump to the innermost orbit $n = 1$. This is the Lyman series and the lines involve larger energy differences and hence higher frequencies and shorter wavelengths. This series is in fact in the ultraviolet range (see Fig. 2.2). Other observed series for hydrogen are the Paschen, Brackett, Pfund and Humphreys series for which the final orbital quantum numbers n are 3, 4, 5 and 6, respectively. The six series are illustrated diagrammatically in Fig. 2.9, in which, however, the orbits are not drawn to scale.

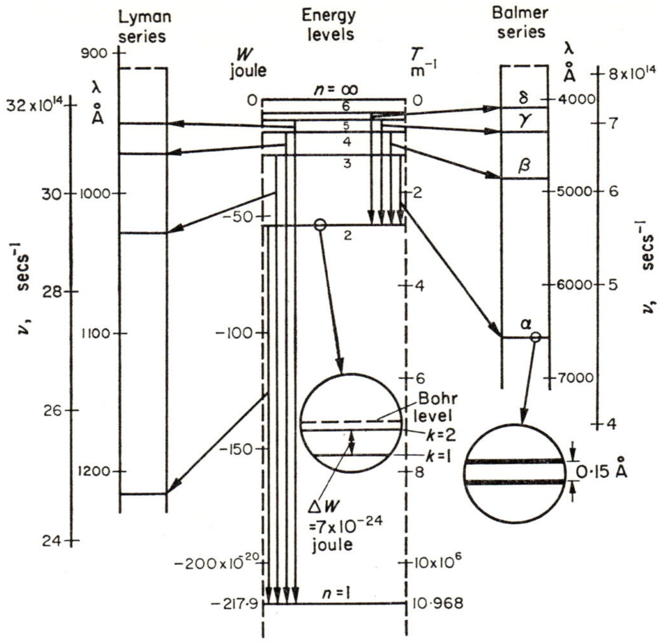

FIG. 2.8. Energy-level diagram for hydrogen showing the transitions giving rise to the lines in the Lyman and Balmer series. Also illustrating line splitting and Sommerfeld energy sub-levels.

It may be noted that the frequency of a line due to a particular transition is not the same as the orbital frequency of either of the orbits involved but is an intermediate value. For example, the orbital frequencies for hydrogen when n is 1 and 2 are $65{\cdot}76 \times 10^{14}$ and $8{\cdot}22 \times 10^{14}$ s^{-1}, respectively, whereas

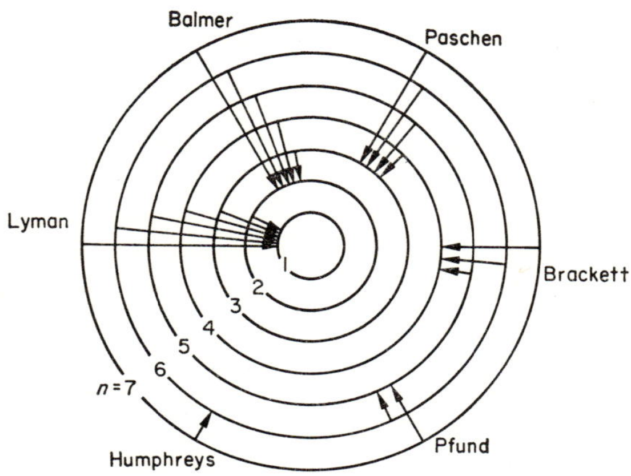

FIG. 2.9. Hydrogen orbits (not to scale) showing series transitions.

the frequency of the line in the Lyman series corresponding to the transition from $n = 2$ to $n = 1$ is $24{\cdot}66 \times 100^{14}\ s^{-1}$.

An electron moving in an orbit is equivalent to a circuital electric current $i = ev/2\pi r$, and this in turn, considered as a magnetic shell, has a magnetic moment $\mu = \mu_0 iA$ where $A = \pi r^2$ is the area of the circuit.

$$\therefore \mu = \frac{\mu_0 evr}{2} \tag{2.14}$$

The angular momentum of any orbit n is

$$p = mvr = \frac{nh}{2\pi} \tag{2.15}$$

Eliminating vr between equations (2.14) and (2.15) gives

$$\mu = \frac{\mu_0 e}{2m} \cdot \frac{nh}{2\pi} \tag{2.16}$$

$$\therefore \frac{\mu}{p} = \frac{\mu_0 e}{2m} = 1{\cdot}1053 \times 10^5 \tag{2.17}$$

Equation (2.17) shows that the ratio of magnetic moment to angular momentum of an electron in any Bohr orbit is a fixed quantity independent of n. From equation (2.16) the magnetic moment of an electron in the innermost orbit of the Bohr hydrogen atom is obtained by putting $n = 1$, and substituting the appropriate numerical values. This magnetic moment is called the Bohr magneton and has the value $1\mu_B = 1{\cdot}1653 \times 10^{-29}$ weber m.

2.3 Limitations and Extension of the Bohr Theory

The Bohr theory as outlined in the foregoing pages, in spite of great successes, was found to be inadequate in explaining all the experimental data available. On detailed examination it was observed that many spectral lines had a fine structure. An example of this is shown in Fig. 2.8 where the α line of the Balmer series of hydrogen, corresponding to radiation of wavelength $\lambda = 6563$ Å, is seen to consist of two closely-spaced lines with an interval between them corresponding to $\Delta\lambda = 0{\cdot}15$ Å approximately. The implication is that the principal energy levels need to be further subdivided.

SOMMERFELD [8] proposed that for any principal energy level n, there were n possible states for the electron in a hydrogen-like atom, and each state could be denoted by a second quantum number k where k had integral values from 1 to n. Electrons in these states moved in elliptical orbits with the nucleus at one focus. For a given value of n all the ellipses had equal major axes and the ratio of minor to major axis for any ellipse was k/n. The allowed orbits for the case of hydrogen with $n = 1$, 2 and 3, respectively, are shown in Fig. 2.10. The orbit for $k = n$ became a circle of the same radius and angular momentum as in the Bohr atom, but the angular momentum of an electron in any other orbit was $kh/2\pi$.

Sommerfeld showed that the total energy of an electron of principal quantum

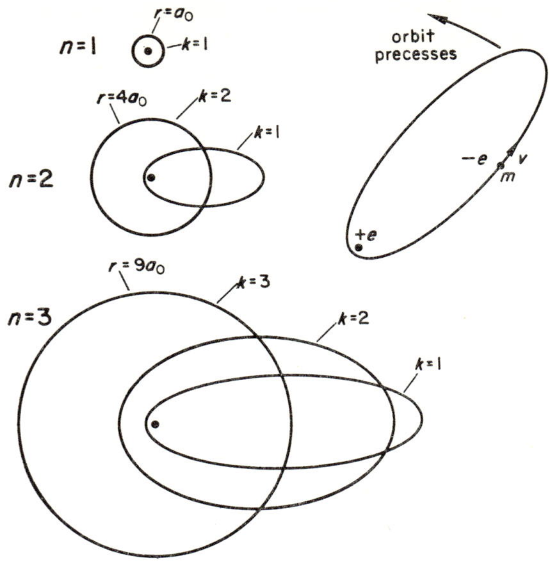

FIG. 2.10. Sommerfeld orbits for hydrogen, drawn to scale, showing also precession of an orbit.

number n depended slightly upon the value of k. The orbital velocities are high, for example, $v = c/137{\cdot}1$ for the first Bohr orbit of hydrogen. These velocities are high enough to make a significant difference to the mass of the electron for, according to the theory of relativity, the relation between mass and velocity is given by $m = m_0\{1-(v^2/c^2)\}^{-\frac{1}{2}}$, where m_0 is the mass at a low velocity, or the rest mass. Moreover in an elliptical orbit the velocity of an electron varies round the orbit and is greatest when the electron passes nearest to the nucleus, so that the mass also changes during its circuit. As a result of these relativity effects an orbit as a whole will precess, although comparatively slowly, around the nucleus, as illustrated in Fig. 2.10, and a correction to the calculated Bohr energy is required. The energy levels now involve both n and k, and when Z is small, may be determined from the formula

$$W_{n,k} = W_n\left\{1 + \left(\frac{Z}{137{\cdot}1n}\right)^2\left(\frac{n}{k} - \frac{3}{4}\right)\right\} \tag{2.18}$$

where W_n is the negative Bohr energy for orbit n as previously given. Taking the case, for example, of hydrogen and $n = 2$, the very small energy interval between the states where $k = 1$ and $k = 2$ is easily calculated from equation (2.18) to be

$$\Delta W = W_2 \times 1{\cdot}33 \times 10^{-5}$$
$$= -54{\cdot}48 \times 10^{-20} \times 1{\cdot}33 \times 10^{-5} = -7{\cdot}2 \times 10^{-24}\ \text{joule}$$

These sub-levels, both lying below the Bohr energy level, are shown in Fig. 2.8.

The splitting of the α line in the Balmer series, also shown in the figure, may now be accounted for quantitatively by transitions from certain close sub-levels in $n = 3$ to these two sub-levels in $n = 2$.

The Sommerfeld extension of the Bohr theory met with a good deal of success but still failed to account for all the experimental facts, and has now given way to the new quantum theory. The Bohr atom always involved an illogical combination of classical mechanics and arbitrary quantum restrictions. This inconsistency is removed in the wave mechanics of SCHRÖDINGER [9] and in the system of quantum mechanics independently developed by HEISENBERG [10] and DIRAC [11]. Although these mathematical systems differ in their formulation they lead to the same correct results except for minor differences. Unfortunately the beautiful mechanical model of the atom, which can be envisaged in the Bohr–Sommerfeld theory, must now be discarded at least so far as its precise detail is concerned.

It has become apparent in science that events on an atomic scale cannot be described with accuracy at a particular instant in the usual terms of velocity, position, energy and the like, terms which may, on the other hand, be simultaneously applied in classical mechanics to massive bodies. In particular, no physical means can be conceived whereby the velocity and position of an orbital electron, at any instant, might be measured, which did not at the same time disturb the quantities being investigated. The indeterminacy involved in such an attempt is enunciated in Heisenberg's Uncertainty Principle. In the case of the orbital electron the principle may be expressed thus: If the uncertainties in the angular momentum and in the angular position are Δp and $\Delta\theta$, respectively, then the product $\Delta p . \Delta\theta$ cannot be less than Planck's constant h.

An electron in an atom may be considered in relation to the system of polar coordinates shown in Fig. 2.11 where the nucleus is at the origin O. For an electron in a particular energy state no attempt is made in the new mechanics to pin-point its position. The electron may exist at any point in the space surrounding the nucleus. The probability of finding the electron in any small volume ΔV near P is then derived mathematically in terms of the product of three separate functions of r, θ and ϕ, respectively. To provide some physical picture of the electron state it is convenient to imagine the charge e of the electron to be distributed as a cloud of charge around the nucleus its density varying in a manner which is specified by the probability functions.

The simplest case to consider is that of hydrogen in the ground state. The charge cloud is a spherically-symmetrical ball around the nucleus. The distribution of charge within this spherical volume may be denoted by either of the curves shown in Fig. 2.11. Here ρ is the charge per unit volume and $U(r)$ is defined so that $U(r)dr$ gives the fraction of the total electronic charge contained within a spherical shell of radius r and wall thickness dr. In this case $\rho 4\pi r^2 \, dr = U(r)dr$ or $\rho = U(r)/4\pi r^2$. The quantity $U(r)$ does not therefore

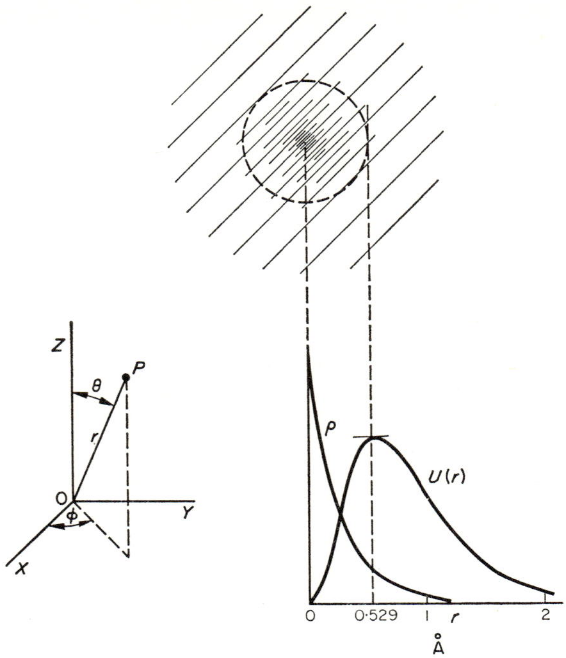

FIG. 2.11. Quantum-mechanical picture of the hydrogen atom in the ground state. The upper diagram represents the negative cloud of charge, spherically symmetrical, with the highest density at the centre.

indicate the charge density at a radius r but the probability of finding the electron at this distance from the nucleus. In this case for the hydrogen atom the $U(r)$ curve has a maximum value when $r = 0{\cdot}529$ Å $= a_0$ indicating that the electron is more often to be found at a distance from the nucleus equal to the radius of the first Bohr orbit a_0 than in any other position. It will be noted, however, that the charge density is greatest, not at this radius, but theoretically at the centre of the atom. Other shapes of the electron cloud occur for the electron in other states.

While it is not now possible to say that the electron is moving in any definite flat orbit as in the Bohr–Sommerfeld atom, the electron still possesses total energy as before, and moreover angular momentum and magnetic moment are associated with it just as if it had orbital motion about the nucleus. Furthermore, detailed spectrum analysis has made it necessary to introduce an additional component of angular momentum and of magnetic moment to every electron. The electron is supposed to be spinning upon its own axis like a gyroscope and this spin is an inherent, unalterable property of every electron and leads to half units of angular momentum. The conception of electron spin is of the greatest importance in magnetism.

It may be noted at this point that for the hydrogen atom in the ground state, shown in Fig. 2.11, the electron is considered to have no orbital components but spin components of angular momentum and magnetic moment only.

The quantum-mechanical approach, in which electron clouds are imagined, gives results in agreement with spectroscopic data. It is, however, an advantage, while making full use of these results, to retain an orbital picture of the atom and to represent the orbital and spin components of angular momentum and of magnetic moment by vectors which can be summed according to certain rules. It must, however, be clearly understood that this vector-orbital model is only a mental aid. It is not a precise mechanical model of the atom and must be used with some discretion.

2.4 Larmor Precession

The precession of an orbit about an axis through the nucleus has already been mentioned in connection with the Sommerfeld orbits. The concept of precession is continually encountered in atomic theory, and the Larmor precession which occurs for an orbit in the presence of a steady magnetic field will now be considered [12].

Take first an orbit of radius r and an electron circulating with velocity v_0. The angular momentum $p = mv_0r$ may be represented by a vector normal to the plane of the orbit in the direction given in Fig. 2.12(a) for the direction of motion of the electron shown. The equivalent current in the orbit is $i = ev_0/2\pi r$. Following the usual convention for the direction of current flow, i circulates in the opposite direction to the negative charge e. The magnetic

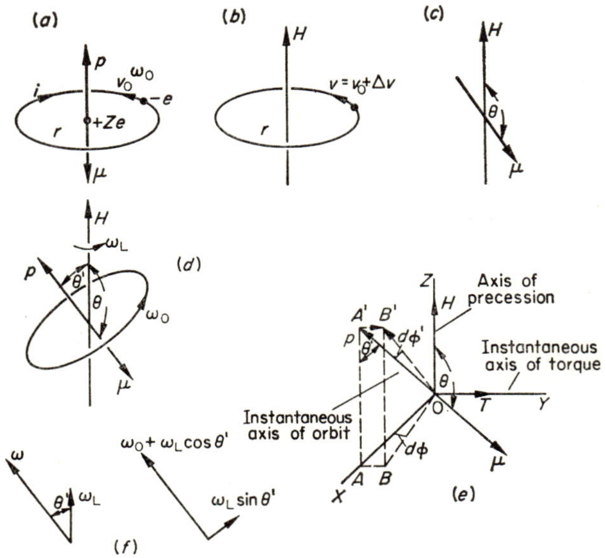

FIG. 2.12. Precession of an orbit in a magnetic field.

moment of the orbit is then, from equation (2.14), $\mu = \mu_0 e v_0 r/2$ and this moment may be represented by a vector in the opposite direction to p as shown in the figure. Clearly if the direction of rotation of the electron in the orbit is reversed, the direction of the vectors representing p and μ are also reversed. We also note from equation (2.17) that for any orbit $\mu/p = \mu_0 e/2m$.

Consider now the hypothetical case of a field H applied normal to the plane of the orbit, as shown in Fig. 2.12(*b*). It may be shown from classical electrodynamics that the radius of the orbit is unaltered but that the velocity of the electron changes. If, for present purposes, we accept without proof that the radius remains constant, the change in velocity and kinetic energy are easily deduced. In Fig. 2.12(*b*) if v is the velocity with the field H the inward force on the electron due to H is $\mu_0 H i 2\pi r = \mu_0 H e v$. As in Section 2.3, the electrostatic attraction to the nucleus is given by $Ze^2/4\pi\epsilon_0 r^2$. Equating the sum of these forces to the centrifugal force gives

$$\frac{mv^2}{r} = \frac{Ze^2}{4\pi\epsilon_0 r^2} + \mu_0 H e v \tag{2.19}$$

The kinetic energy of the electron may therefore be expressed, from equation (2.19), as

$$\text{kinetic energy} = \tfrac{1}{2}mv^2 = \frac{Ze^2}{8\pi\epsilon_0 r} + \frac{\mu_0 H e v r}{2}$$

Let ΔW be the additional kinetic energy arising from the presence of the field H, then $\Delta W = \mu_0 H e v r/2$. If $v = v_0 + \Delta v$ and the increment in velocity Δv due to H is relatively small, then approximately

$$\Delta W = \frac{\mu_0 H e v_0 r}{2} = H\mu$$

In Fig. 2.12(*c*) is shown a magnet of moment μ inclined at an angle θ to a field H. As shown in Section 1·1 the work done in bringing the magnet from infinity, that is the potential energy of the magnet in the field, is given by $-H\mu\cos\theta$. In Fig. 2.12(*b*) the potential energy of the orbit in the field, where the orbital current is equivalent to a magnet of moment μ and $\theta = 180°$, is thus $H\mu$. It is therefore clear that the potential energy is in fact stored in the form of an increment of kinetic energy of the electron. If the direction of rotation in the orbit is reversed, the vector μ is reversed. Then $\theta = 0°$ and the potential energy is $-H\mu$. But $v = -v_0 + \Delta v$ in this case and so $\Delta W = -H\mu$ also. Hence the negative potential energy is accounted for by a fall in the electron velocity and a reduction of the kinetic energy of the electron in the orbit.

We may also write

$$\Delta W = \tfrac{1}{2}m(v_0 + \Delta v)^2 - \tfrac{1}{2}mv_0^2$$

$$\therefore \Delta W = m.\Delta v.v_0 \quad \text{if } \Delta v \ll v_0$$

Whence
$$m.\Delta v.v_0 = \frac{\mu_0 H e v_0 r}{2} \tag{2.20}$$

If the increment of angular velocity induced by the field H on the electron is ω_L, then from equation (2.20),

$$\omega_L = \frac{\Delta v}{r} = H\frac{\mu_0 e}{2m} = H\frac{\mu}{p}$$

It will be seen later that, for the model of the atom we shall adopt, only discrete values of the angle θ between an applied field H and the normal to the plane of an orbit are allowed and $\theta = 0°$ is not one of them. It is therefore of more general interest to consider the case of an orbit inclined to the direction of H as in Fig. 2.12(*d*). Consider an instant when the vectors representing p and μ are in the XOZ plane shown in Fig. 2.12(*e*). Let OA' represent p. With H in direction OZ the torque acting on the orbit, regarded as equivalent to a magnet of moment μ, is $T = H\mu \sin\theta$. The axis of the torque T is OY and the vector representing T may be set along OY as shown. If in time dt there is a small rotation of the orbit about OZ so that p moves from OA' to OB' through an angle $d\phi'$, then $A'B'$ represents the change in angular momentum in the direction $A'B'$. The change of angular momentum about OZ is therefore $p\,d\phi' = p\sin\theta\,d\phi$ where $d\phi$ is the angular rotation in the plane XOY. Now the torque T is equal to the rate of change of angular momentum which is $p\sin\theta\,d\phi/dt$. Whence

$$H\mu \sin\theta = p\sin\theta\frac{d\phi}{dt} \tag{2.21}$$

Now $d\phi/dt$ represents the angular velocity of precession ω_L of the orbit about the direction of H. Therefore, from equation (2.21),

$$\omega_L = H\frac{\mu}{p} \tag{2.22}$$

This is the Larmor precession and the increment of angular velocity of the electron in an orbit, the plane of which is assumed normal to H, as obtained in the preceding paragraph, is seen to be the special case where $\theta = 180°$. It will be noted that ω_L is independent of θ and the sense of the rotation is always, for any value of θ, to produce an increment of magnetic moment in opposition to the direction of H. This is the origin of diamagnetism, and accounts for the negative susceptibility, or a relative permeability slightly less than unity, observed in diamagnetic substances.

There will be an increment of kinetic energy arising from the precessional motion. The electron has an orbital angular velocity $\omega_0 = v_0/r$, and a precession ω_L, and the angle between the axes of these rotations is $\theta' = 180° - \theta$. Resolving ω_L, as shown in Fig. 2.12(*f*), and taking the resultant orbital angular velocity to be $\omega_0 + \omega_L\cos\theta'$ and the angular velocity about a diameter of the orbit to be $\omega_L\sin\theta'$ then the kinetic energy becomes

$$\tfrac{1}{2}I(\omega_0 + \omega_L\cos\theta')^2 + \tfrac{1}{4}I(\omega_L\sin\theta')^2$$

Where $I = mr^2$ and $I\omega_0 = p$. If $\omega_L \ll \omega_0$ the increment of kinetic energy due to the precession is given approximately by

$$\Delta W = I\omega_0\omega_L\cos\theta' = pH(\mu/p)(-\cos\theta) = -H\mu\cos\theta$$

Thus, as before, the potential energy of the orbit in the field H, may be regarded as being stored as an increment of kinetic energy of the electron, this positive or negative component of energy arising from the precessional motion.

2.5 Vector Model with a Single Electron

We will consider a single electron in different energy states in a hydrogen-like atom according to the modern theory.

The total energy of the electron is specified by a total or principal quantum number n and this energy has the same value as in the Bohr atom, that is the value given by equation (2.10). This total energy is, however, subject to increments or corrections, as it was with the Sommerfeld orbits, depending upon the precise state of the electron at this energy level.

The orbital angular momentum which is to be attributed to the electron is denoted by a total orbital angular momentum number l. For any value of n, l may take the n values 0, 1, 2 . . . $(n - 1)$. The orbital angular momentum of an electron is then given by $p_l = l^*(h/2\pi)$ where $l^* = \{l(l + 1)\}^{\frac{1}{2}}$. If μ_l is the orbital magnetic moment, then from equation (2.17),

$$\mu_l = p_l\,(\mu_0 e/2m) = l^*(\mu_0 eh/2m2\pi) = l^*\mu_B$$

where μ_B is one Bohr magneton and $\mu_B = 1{\cdot}1653 \times 10^{-29}$ weber m.

Now every electron is also regarded as spinning on its own axis. Spin is a fixed property of the electron and the spin angular momentum is given by $p_s = s^*(h/2\pi)$ where $s^* = \{s(s + l)\}^{\frac{1}{2}}$ and $s = \frac{1}{2}$. Therefore $s^* = \frac{1}{2}\sqrt{3}$ and has only one value. The experimental evidence requires the ratio of spin magnetic moment to spin angular momentum to be double the value for the orbital motion. Hence, if μ_s is the magnetic moment due to spin,

$$\mu_s/p_s = 2(\mu_0 e/2m)$$

or $\mu_s = 2s^*\mu_B$. $\therefore$ $\mu_s = \sqrt{3}\mu_B$. The intrinsic magnetic moment of a spinning electron is therefore $\sqrt{3}$ Bohr magnetons.

For any value of l, except $l = 0$, the electron will now have both orbital and spin motions. The electron may be considered to be spinning in a magnetic field produced by the orbital motion. To find this magnetic field at the electron, the electron is imagined to be at rest and the nuclear charge Ze moving round it. If r is the radius of the orbit and v the velocity, the equivalent current $i = Zev/2\pi r$ and the field H at the electron due to this circular current is $H = i/2r = Zev/4\pi r^2$. Putting $mvr = p_l$ and substituting for v, $H = p_l\,(Ze/4\pi mr^3)$ or

$$H = l^*\frac{h}{2\pi}\frac{Ze}{4\pi mr^3} \tag{2.23}$$

The spin axis of the electron might now be expected to precess about the direction of H with an angular velocity, from equation (2.22),

$$\omega = H(\mu_s/p_s) = H.2(\mu_0 e/2m)$$

However, it has been shown that, because of relativity effects in this case, the precessional velocity is halved and becomes $\omega = H(\mu_0 e/2m)$. The spinning electron may also be considered equivalent to a circular current producing a field in which the orbital axis will precess. The final result is that both the spin and orbital axes precess about a common axis which is the direction of the resultant angular momentum of p_l and p_s. The precessional velocity is the same for both, and in the same direction, and is given by $\omega = H(\mu_0 e/2m)$, where H has the value given in equation (2.23).

Figure 2.13 illustrates the vector model for this spin–orbit interaction. The orbital and spin angular momenta are represented by $p_l = l^*(h/2\pi)$ and $p_s = s^*(h/2\pi)$, respectively, summing vectorially to give a resultant of $p_j = j^*(h/2\pi)$, about the direction of which the vectors of angular momentum precess.

It is of interest to consider the order of magnitude of H and ω in this case. For hydrogen, with $Z = 1$, taking for example $n = 2$ and $l = 1$, we may put $r = 4a_0$. If the appropriate numerical values are put in equation (2.23), we obtain $H = 2{\cdot}20 \times 10^5$ A.m^{-1} and $\omega = H(\mu_0 e/2m) = 2{\cdot}43 \times 10^{10}$ s^{-1} or the precessional frequency $\nu = \omega/2\pi$ $= 3{\cdot}87 \times 10^9$ s^{-1}. The precessional frequency is therefore small compared with the orbital frequency, since for an electron in the Bohr atom with $n = 2$ the orbital frequency is $8{\cdot}22 \times 10^{14}$ s^{-1} (see Fig. 2.7).

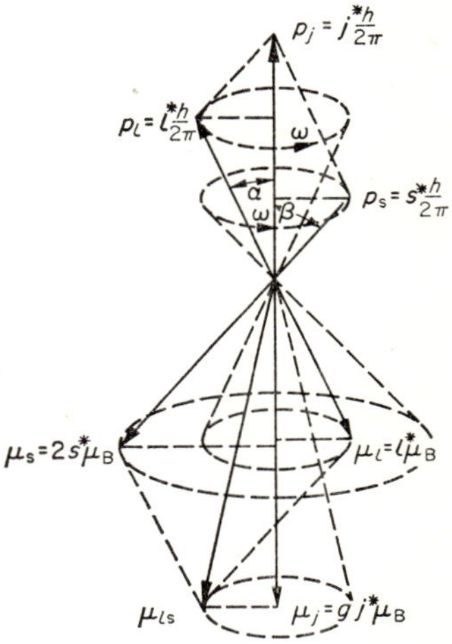

FIG. 2.13. Vector model for a bound electron with spin–orbit interaction.

The spin and orbital motions now interact or are coupled in such a way that the resultant total angular momentum is quantized, or has discrete values, given by $p_j = j^*(h/2\pi)$ where $j^* = \{j(j+1)\}^{\frac{1}{2}}$, and where $j = l \pm \frac{1}{2}$ starting from $+\frac{1}{2}$. j is the total angular momentum quantum number.

Electrons are said to be in s, p, d, f states when $l = 0, 1, 2, 3$, respectively, independently of the value of n. Thus an electron in an s state, for which $l = 0$ and $j = \frac{1}{2}$, has only one value of j^* and p_j. For any other state j^* and p_j have two possible values. Values are tabulated in Table 2.1 and the symbol used to denote the state is shown in the last column. For an electron at a particular principal or total energy level, n would be replaced by its appropriate numerical value, thus for example, $2s_{\frac{1}{2}}$ or $3d_{\frac{5}{2}}$. Vector diagrams representing the states are shown in Fig. 2.14.

Now precession of the axes of the spin and orbit leads to increments of kinetic energy, as already discussed in connection with the Larmor precession, and the total energy of the electron in any state is given by

$$W_{nj} = W_n\left\{1 + \left(\frac{Z}{137{\cdot}1n}\right)^2\left(\frac{n}{j+\frac{1}{2}} - \frac{3}{4}\right)\right\} \tag{2.24}$$

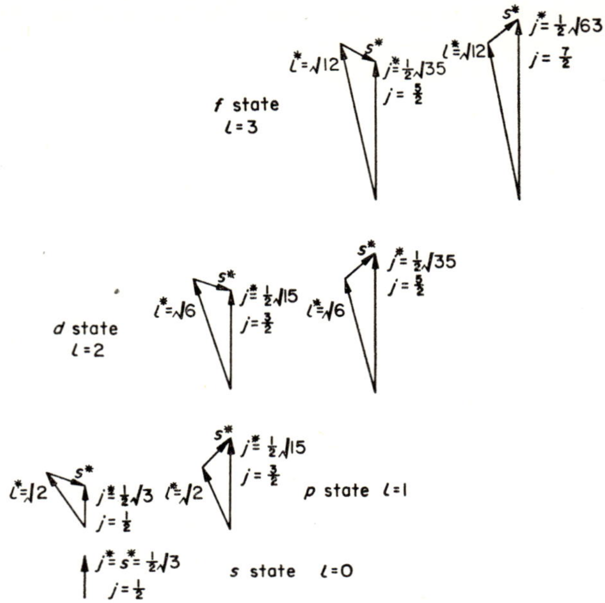

FIG. 2.14. Vector diagrams of electrons in *s*, *p*, *d* and *f* states.

TABLE 2.1

Electron	l	l^*	j	j^*	g	Possible Values of n	Symbol
f	3	$\sqrt{12}$	$\frac{7}{2}$	$\frac{1}{2}\sqrt{63}$	$\frac{8}{7}$	4, 5, 6 . . .	$nf_{\frac{7}{2}}$
			$\frac{5}{2}$	$\frac{1}{2}\sqrt{35}$	$\frac{6}{7}$		$nf_{\frac{5}{2}}$
d	2	$\sqrt{6}$	$\frac{5}{2}$	$\frac{1}{2}\sqrt{35}$	$\frac{6}{5}$	3, 4, 5 . . .	$nd_{\frac{5}{2}}$
			$\frac{3}{2}$	$\frac{1}{2}\sqrt{15}$	$\frac{4}{5}$		$nd_{\frac{3}{2}}$
p	1	$\sqrt{2}$	$\frac{3}{2}$	$\frac{1}{2}\sqrt{15}$	$\frac{4}{3}$	2, 3, 4 . . .	$np_{\frac{3}{2}}$
			$\frac{1}{2}$	$\frac{1}{2}\sqrt{3}$	$\frac{2}{3}$		$np_{\frac{1}{2}}$
s	0	0	$\frac{1}{2}$	$\frac{1}{2}\sqrt{3}$	2	1, 2, 3 . . .	$ns_{\frac{1}{2}}$

where W_n is the negative Bohr energy, $W_n = -hcRZ^2/n^2$. Equation (2.24) differs from equation (2.18) for the Bohr–Sommerfeld atom only by the substitution of $j + \frac{1}{2}$ for k in the older formula. For any value of n the intervals between the energy sub-levels for the allowed values of l are small and become smaller as n increases.

The student must try not to be confused by the established practice in this subject of using the same letter symbols for different purposes. For example, s is used to denote an electron state as well as for spin, p for a state and for angular momentum, and later, m as a magnetic quantum number and for the mass of the electron.

We now require to find the resultant magnetic moment of an electron having spin and orbital motions. Referring to Fig. 2.13, the orbital magnetic moment $\mu_l = l^*\mu_B$ will be represented by a vector in the opposite direction to p_l. The spin moment is $\mu_s = 2s^*\mu_B$ and the resultant of these two vectors μ_{ls} is clearly not in line with the resultant p_j, but must precess about the direction of the vector representing p_j, as shown in the figure. There is no steady component of magnetic moment normal to the axis of precession, but the resultant moment along the axis is made up of the resolved component of μ_{ls}, or the sum of the resolved components of μ_l and μ_s. It will be seen from the figure that

$$j^* = l^* \cos \alpha + s^* \cos \beta$$

$$l^* \sin \alpha = s^* \sin \beta$$

and if the resultant magnetic moment is $\mu_j = gj^*\mu_B$

$$gj^* = l^* \cos \alpha + 2s^* \cos \beta$$

Eliminating α and β from these three equations gives

$$g = 1 + \frac{(j^*)^2 + (s^*)^2 - (l^*)^2}{2(j^*)^2} \tag{2.25}$$

g is called the Landé factor and we therefore have for a single electron in any state,

$$p_j = j^*\frac{h}{2\pi}, \quad \mu_j = gj^*\mu_B \quad \text{and} \quad \frac{\mu_j}{p_j} = g\frac{\mu_0 e}{2m}$$

For any s, p, d, f state, having respectively the l values 0, 1, 2, 3, and always with $s = \frac{1}{2}$, g may readily be evaluated. Values of g are given in Table 2.1. When $l = 0$ and the electron has spin only $g = 2$, while in the hypothetical case of orbital motion with no spin g would be unity.

We now consider the hydrogen-like atom with a single electron in an excited state in the presence of an external steady magnetic field. In Fig. 2.15(a) the vectors p_j and μ_j represent the total angular momentum and magnetic moment with spin–orbit interaction and the vectors are inclined to the direction of the magnetic field H. Precession of the vectors about H will occur with an angular velocity $\omega = H(\mu_j/p_j) = Hg(\mu_0 e/2m)$. ω will be much less than the precessional velocity of p_l and p_s about the axis of p_j if H is much less than the value given by equation (2.23). We may then refer to the field H in Fig. 2.15(a) as a weak field. The fine structure of spectral lines, obtained under these conditions, indicates that p_j can take up only certain angular positions relative to the field such that the resolved components of $p_j = j^*(h/2\pi)$ in

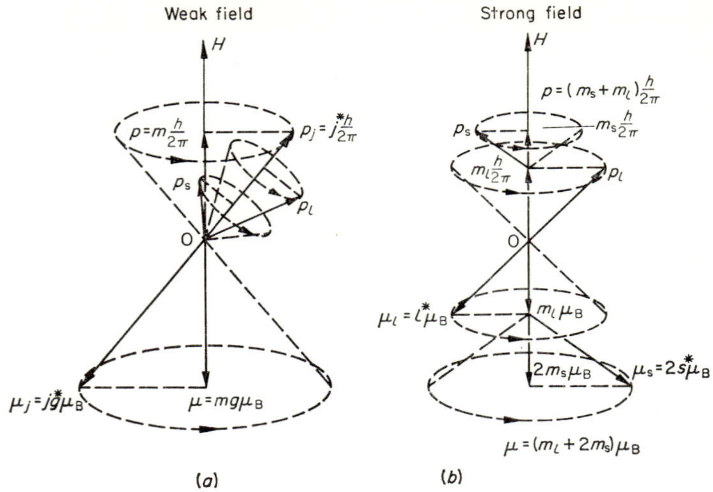

FIG. 2.15. Precession in a weak field (Zeeman effect) and in a strong field (Paschen–Back effect).

the field direction are $p = m(h/2\pi)$, where m is a magnetic quantum number. The allowed values of m are $\pm\frac{1}{2}, \pm\frac{3}{2} \ldots \pm j$. The resolved values of μ_j are $\mu = gm\mu_B$.

If the electron is in an s state, then $l = 0$ and $j = \frac{1}{2}$ and $m = \pm\frac{1}{2}$. Thus if an electron has spin but no orbital motion the components of angular momentum in the field direction are $\pm\frac{1}{2}\dfrac{h}{2\pi}$ and the components of magnetic moment in the field direction are $\pm\frac{1}{2}g\mu_B = \pm 1\mu_B$ since $g = 2$ in this case. Thus if an electron spinning in a magnetic field has no orbital motion, it can have resolved components of magnetic moment of plus or minus one Bohr magneton only. This result is of the greatest importance in ferromagnetism.

An electron in a p state has two possible values of j^* corresponding to $j = \frac{1}{2}$ and $\frac{3}{2}$. It will be apparent that when $j = \frac{1}{2}$ the resolved values p and μ

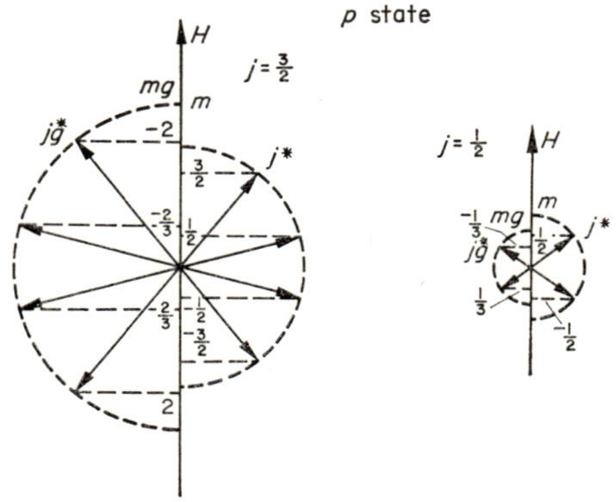

FIG. 2.16. Magnetic quantum numbers m and Landé splitting factors mg for a p state.

each have two allowed values and when $j = \frac{3}{2}$ there are four, since then $m = \pm\frac{1}{2}$ and $\pm\frac{3}{2}$. For a d state there are $4 + 6 = 10$ values of the resolved vectors and for an f state $6 + 8 = 14$ values. The diagrams in Fig. 2.16 show, for example, the values of m and of mg for the case of the p state.

The potential energy of the system consisting of μ_j in the field H in Fig. 2.15(a) is given by $\Delta W = Hmg\mu_B$. The s state will therefore have superimposed increments of energy in a magnetic field H of

$$\Delta W = \pm\tfrac{1}{2}.2.H\mu_B = \pm H\mu_B$$

that is the s energy levels will be split into two. p levels are split into six, d levels into ten and f levels into fourteen. The quantity mg is known as the

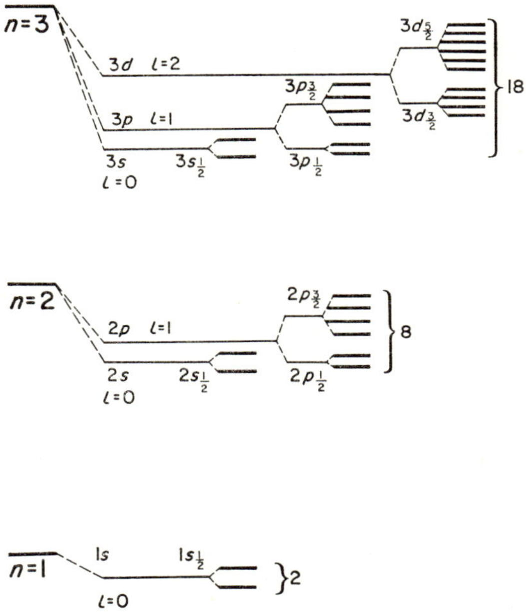

FIG. 2.17. Energy levels in a hydrogen-like atom with a single electron in a magnetic field—not to scale.

Landé Splitting Factor. The energy levels for $n = 1$, 2 and 3 are illustrated in Fig. 2.17, which, however, is not drawn to scale. It will be seen that the number of allowed energy levels for a single electron in a hydrogen-like atom, when this is in a magnetic field, are 2 for $n = 1$, 8 for $n = 2$, 18 for $n = 3$, and in fact $2n^2$ for any value of n. The line splitting which occurs in a magnetic field as a result of this subdivision of the energy levels is known as the ZEEMAN effect [13].

In Fig. 2.15(a) H is a weak field and the angular velocity of precession of p_j about H is less than the angular velocity of precession of p_s and p_l about p_j. However, if H is made a very strong field the angular velocity $\omega = H(\mu_j/p_j)$ about the field direction could become greater than the spin–orbit precessional velocity. In these circumstances the spin–orbit interaction, if H is strong enough, is completely broken down and the spin and orbit axes each precess independently about the direction of H, as shown diagrammatically in Fig. 2.15(*b*). This is the PASCHEN–BACK effect [14]. Spin and orbit components each have their own separate magnetic quantum numbers m_s and m_l. The only possible values for spin are $m_s = \pm\frac{1}{2}$, but m_l may have the values 0, $\pm 1, \ldots \pm l$. For given values of m_s and m_l it will be clear that $p = (m_l + m_s)\, h/2\pi$ and $\mu = (m_l + 2m_s)\mu_B$. It turns out that all the allowed values of the magnetic quantum number m, for the weak field condition, and for any given state of the electron, that is for any value of l, are obtainable by summing m_l and m_s in the various $2(2l + 1)$ possible combinations. Thus we may write $m = m_l + m_s$.

The precise state of an electron in the hydrogen-like atom may therefore be specified by the use of four quantum numbers, n, l, m_l and m_s where m_l and m_s are quantum numbers obtaining in the presence of a strong magnetic field. It will be found in the next section that these four quantum numbers are important when considering atoms with more than one extra-nuclear electron.

2.6 Electronic Structure of Atoms

Any atom of atomic number Z will have Z extra-nuclear electrons making up the electronic configuration of the atom. Each electron will in general have spin and orbital components of angular momentum and magnetic moment, and these components may be combined or coupled together, and quantized, in a quite complicated way to give a resultant angular momentum and a resultant magnetic moment to the atom as a whole. The atom may be in its normal or ground state, that is in a stable state of lowest energy, or it may be excited in which case one or more electrons will be in outer states of higher energy. In reverting to the normal state electromagnetic energy is radiated at characteristic frequencies as previously mentioned producing spectral lines which may be examined by means of a spectroscope.

From such spectroscopic studies it has been possible to build up the complete electronic structure of all the elements in the periodic table, making use of principles and rules which have been evolved to fit the experimental data. These structures are tabulated in Appendix 1 for all the elements from hydrogen, $Z = 1$, to uranium, $Z = 92$.

We have seen in Section 2.5, that for a hydrogen-like atom having a single bound electron, the electron may exist in an energy state specified by the quantum numbers, n determining the principal energy level, $l = 0, 1, \ldots (n - 1)$ specifying the orbit, s always equal to $\frac{1}{2}$ denoting the spin, and finally the quantum number $j = l \pm \frac{1}{2}$ starting from $+\frac{1}{2}$, which determines the vector sum of the coupled orbital and spin angular momenta. Four quantum numbers, n, l, s and j therefore specify the particular state of the electron.

The state of the electron may be given by four numbers in an alternative way. If the atom is in a weak magnetic field the resultant angular momentum $j^*(h/2\pi)$, corresponding to the particular state, can have only the values resolved in the field direction of $m(h/2\pi)$ where $m = \pm\frac{1}{2}, \pm\frac{3}{2} \ldots \pm j$. But if now the magnetic field is made so strong that the *ls* coupling is broken down the orbital and spin momenta are separately quantized giving the resolved components $m_l(h/2\pi)$ and $m_s(h/2\pi)$ where $m_l = 0, \pm 1 \ldots \pm l$ and $m_s = \pm\frac{1}{2}$. We suppose that the resultant resolved component of angular momentum remains unchanged as H is reduced from a high value until it is vanishingly small, so that $m = m_l + m_s$. The state of the electron may then be specified by the four quantum numbers n, l, m_l and m_s.

If we now consider any atom with Z extra-nuclear electrons we may imagine an applied magnetic field hypothetically so strong that all couplings within the atom are completely broken down so that every electron, each having four quantum numbers n, l, m_l and m_s, would have its orbital and spin momenta separately and individually quantized or resolved in the field direction. Under these circumstances the resolved angular momentum of the atom as a whole would be $M(h/2\pi)$ where M is a magnetic quantum number and $M = \Sigma\, m_l + \Sigma\, m_s$ for all the electrons. M would remain unaltered as the field was reduced to zero.

The PAULI Exclusion Principle [15], enunciated with this imaginary Complete Paschen–Back Effect in view, states that no two electrons in one atom can have the same four quantum numbers n, l, m_l and m_s. Application of the principle limits the number of possible electron states in an atom and accounts for the form of the periodic table in Appendix 1.

The normal state for an atom of hydrogen is for one electron with $n = 1$. Since l may have only the values $0, 1 \ldots (n - 1)$ and m_l the values $0, \pm 1 \ldots \pm l$, l and m_l are each zero and the electron is in an *s* state, but m_s may be $\pm\frac{1}{2}$. If we now consider helium, $Z = 2$, the Pauli Exclusion Principle will allow a second electron to be accommodated at the same principal energy level $n = 1$, for the values of n, l, m_l and m_s for the two electrons, respectively, will be $1, 0, 0, +\frac{1}{2}$ and $1, 0, 0, -\frac{1}{2}$. A third electron at this level is, however, not permitted. Therefore for helium in the normal state there are two *s* electrons having $n = 1$ and this energy level is then completely full. Electrons having the same value of n are said to belong to a shell. These are known as *K, L, M, N* . . . shells for $n = 1, 2, 3, 4 \ldots$, respectively. The *K* shell is therefore complete with two electrons and the magnetic quantum number for the shell has only the one value $M = \Sigma\, m_s = 0$. The completed shell has in fact two electrons without orbital motion ($l = 0$) and having antiparallel or oppositely-directed spins. Figure 2.18(*a*) shows this condition, diagrammatically, in an imaginary very strong magnetic field about the direction of which the spin vectors s^* are precessing. The *K*, or 1*s*, shell therefore has zero angular momentum and magnetic moment, and the atom is also chemically inert.

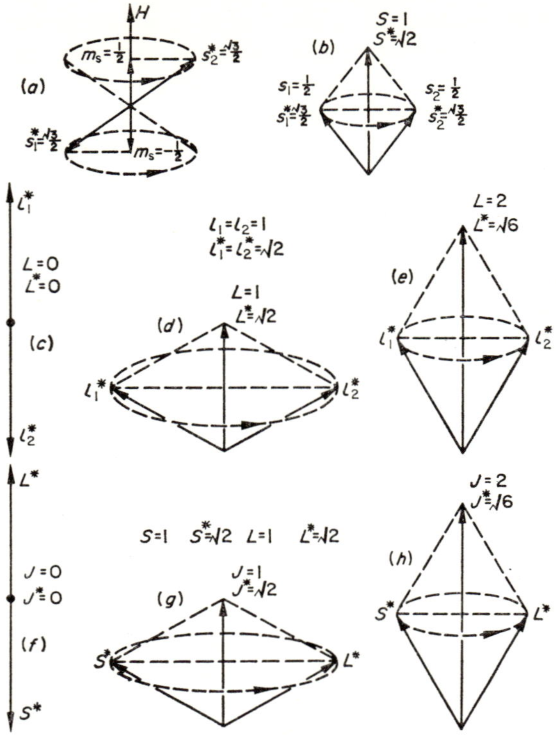

FIG. 2.18. *ss*, *ll* and *LS* coupling between two equivalent *p* electrons.

When $Z = 2$, further electrons are added and, the *K* shell being full, these electrons must go into the *L* shell where $n = 2$. When $n = 2$, l may have the values 0 and 1. When $l = 0$ we can have only two electrons each in an *s* state as before. When $l = 1$ we may have $m_l = 0, \pm 1$. When $n = 2$ and $l = 1$ the only combinations of m_l and m_s which comply with the Pauli principle are, for m_l and m_s respectively, $+1, +\frac{1}{2}$: $0, +\frac{1}{2}$: $-1, +\frac{1}{2}$: $+1, -\frac{1}{2}$: $0, -\frac{1}{2}$: $-1, -\frac{1}{2}$. Therefore six electrons are required to fill completely the sub-level where $l = 1$, that is six electrons in *p* states. The two sub-levels in the *L* shell, $n = 2$, may be referred to, respectively, as 2*s* and 2*p* sub-shells. The *L* shell is therefore completely filled with eight electrons, the atom then having a total of ten electrons giving $Z = 10$ or neon, another inert element.

In the *M* shell, $n = 3$ and l may be 0, 1 and 2 and so there are three sub-shells, 3*s*, 3*p* and 3*d*: and so on for $n > 3$. It is easily deduced that the total number of electrons to fill any main shell is $2n^2$ and the number to fill any sub-shell is $2(2l + 1)$. These numbers are the same as the numbers of levels for an electron in an excited hydrogen atom in a magnetic field shown in Fig. 2.17.

Experimental evidence confirms that the successive atoms in the periodic table build up in a regular manner until potassium, $Z = 19$, is reached, and then as seen in Appendix 1 an electron goes into the $4s$ sub-shell before the $3d$ sub-shell starts to fill. Many other subsequent anomalies occur in the table.

2.7 The Normal States of Free Atoms

The electronic structures outlined in Appendix 1 may be used to deduce the state of lowest energy or normal state of a free atom.

We have seen that there is an ls coupling between the orbital and spin motions of a single electron in an excited atom. Similarly in an atom which contains a number of electrons there are, as already mentioned, couplings between the spin and orbital motions of all the electrons. In what follows we shall consider only what is called LS coupling or RUSSELL–SAUNDERS coupling [16], a system which appears to be the only one of importance in magnetism although it is not the only system.

Consider first a p sub-shell containing two electrons and thus incompletely occupied. Electrons in a particular sub-shell having the same n and l values are known as equivalent electrons. The spin and orbital quantum numbers for the two electrons are $s_1 = \frac{1}{2}$, $l_1 = 1$ and $s_2 = \frac{1}{2}$, $l_2 = 1$.

In Russell–Saunders coupling there is first a strong interaction between the spins. This is considered to be electrostatic rather than magnetic in character and may be referred to as ss coupling. The individual electrons each have a spin angular momentum of s^* units but when coupled the resultant is S^* units where $S^* = \{S(S+1)\}^{\frac{1}{2}}$ and, for two electrons, S may have only the values 0 and 1, i.e. $S = s_1 \pm s_2$. The coupling when $S = 1$ is shown in Fig. 2.18(*b*) in which the electron spin vectors s_1^* and s_2^* precess at a high angular velocity about the resultant S^*.

There is similarly a very strong electrostatic coupling between the orbital motions or ll coupling. The orbital vectors l_1^* and l_2^* precess about a resultant L^* as shown in Fig. 2.18(*d*) and (*e*). $L^* = \{L(L+1)\}^{\frac{1}{2}}$ where in this case L may have only the values 0, 1 and 2. These three conditions are shown at (*c*), (*d*) and (*e*) in the figure.

L^* and S^* now combine to give a resultant J^*, each precessing about the direction of J^*. This is LS coupling and is magnetic in character in just the same way as the spin–orbital coupling already discussed for a single electron. Now $J^* = \{J(J+1)\}^{\frac{1}{2}}$ and the values of J for the case considered of two electrons are limited to 0, 1, 2 and 3.

If now the atom is situated in a relatively weak magnetic field H then J^* will precess about H such that the components of angular momentum resolved in the field direction are $M(h/2\pi)$ where M may have the values $0, \pm 1 \ldots \pm J$.

It will be apparent that two p electrons may be coupled together in many ways, for if s_1 and s_2 combine to give $S = 0$, then for $S = 0$ we have possible couplings of l_1 and l_2 to give $L = 0$, 1 or 2. If, however, s_1 and s_2 couple to give $S = 1$, this spin resultant may be combined with an orbital resultant also having $L = 0$, 1 or 2. And for each LS coupling there will be one or more values of J and J^*. All the conceivable combinations for two p electrons are listed in Table 2.2.

However, when these combinations are examined it is found that the Pauli Exclusion Principle disallows some of them. The student must be referred to other textbooks for a detailed discussion of the operation of the Pauli principle to the coupling between two or more equivalent electrons. (See especially H. E. WHITE: *Introduction to Atomic Spectra*.) All the allowed combinations are indicated in the table. It will be seen, for example, that an allowed condition is when $S = 1$ and $L = 1$ and there are then three possible values for J and for J^*. These three conditions are shown by the vector diagrams in Fig. 2.18 at (*f*), (*g*) and (*h*).

Each vectorial combination of spin and orbital angular momenta corresponds to an energy level in the atom known as a term. The terms are denoted by the capital letters $S, P, D, F\ldots$ when the values of L are 0, 1, 2, 3 . . ., respectively. Spectroscopists use a prefixed superscript to the capital letter, and this prefixed figure is in fact equal to $2S + 1$. The actual value of J for a particular J level is shown by a subscript. Thus $^3P_2$ is a term for which $L = 1$, $2S + 1 = 3$ or $S = 1$ and $J = 2$.

From Table 2.2 it is seen that the allowed states of an atom with two equivalent p electrons are $^1S_0$, $^1D_2$, $^3P_0$, $^3P_1$ and $^3P_2$. These represent different energy levels and the state of lowest energy will be the normal state for the atom. When there is LS coupling the state of lowest energy will be determined by the use of HUND's Rules [17]. To apply the rules in this case, consider all the terms in Table 2.2. First select the terms of highest S. These are the ^3P and ^3D terms. Of the terms of highest S select those of highest L. The ^3D terms have the higher L but these terms are forbidden by the Pauli principle. The ^3P terms remain. Then according to the rules, for a sub-shell more than half full the lowest term will have $J = L + S$, but if the sub-shell is less than half full, as it is in this particular case with two p electrons, then the lowest term has $J = L - S$. For ^3P terms $L = 1$, $S = 1$ and therefore $J = 0$ gives the lowest term, which accordingly will be $^3P_0$. It happens in this particular case with $J = 0$ that the atom has zero angular momentum and zero magnetic moment when it is in the normal state, and this is the condition shown diagrammatically in Fig. 2.18(*f*).

In a similar way the normal state for an atom with more than two equivalent p electrons may be determined, and this may be further extended to the d and f sub-shells. The normal states for from one to six equivalent p electrons and for from one to ten equivalent d electrons are given in Table 2.3. From the symbol denoting the normal state the values of S, L and J may be written down. It may be noted that S and J, and hence also M, have whole-number values for an even number of electrons and half-integer values for an odd number.

It will be seen from the successive values of S in the table, increasing at first in units of one half, that electrons are added to a sub-shell with parallel

TABLE 2.2

S	S^*	L	L^*	J	J^*	g	M	Term	Remarks
		0	0	0	0	1	0	$^1S_0$	Allowed
		1	$\sqrt{2}$	1	$\sqrt{2}$	1	1 0 -1	$^1P_1$	Not allowed
		2	$\sqrt{6}$	2	$\sqrt{6}$	1	2 1 0 -1 -2	$^1D_2$	Allowed
0	0	0	0	1	$\sqrt{2}$	2	1 0 -1	$^3S_1$	Not allowed
				0	0		0	$^3P_0$	Allowed: Normal state
		1	$\sqrt{2}$	1	$\sqrt{2}$	$\frac{3}{2}$	1 0 -1	$^3P_1$	Allowed
				2	$\sqrt{6}$	$\frac{3}{2}$	2 1 0 -1 -2	$^3P_2$	Allowed
1	$\sqrt{2}$			1	$\sqrt{2}$	$\frac{1}{2}$	1 0 -1	$^3D_1$	Not allowed
		2	$\sqrt{6}$	2	$\sqrt{6}$	$\frac{7}{6}$	2 1 0 -1 -2	$^3D_2$	Not allowed
				3	$\sqrt{12}$	$\frac{4}{3}$	3 2 1 0 -1 -2 -3	$^3D_3$	Not allowed

or positive spins until the sub-shell is half full, and then subsequently with antiparallel or negative spins. When a sub-shell is half full the orbital component denoted by L is zero. When the sub-shell is completely full, S, L and J are all zero so that a completed sub-shell has zero angular momentum and magnetic moment.

The normal states shown in Table 2.3 for the p and d sub-shells are the same as the normal states of free atoms which have an incompleted p or d sub-shell all other sub-shells being full. For example, referring to Appendix 1, the atoms of carbon $Z = 6$, silicon $Z = 14$, germanium $Z = 32$, tin $Z = 50$, and lead $Z = 80$ all have an incompleted p shell with two electrons and the normal state in each case is $^3P_0$ as in Table 2.3.

TABLE 2.3

Sub-shell	Number of Electrons	Normal State S	L	J	Notation
	1	$\frac{1}{2}$	1	$\frac{1}{2}$	$^2P_{\frac{1}{2}}$
	2	1	1	0	$^3P_0$
p	3	$\frac{3}{2}$	0	$\frac{3}{2}$	$^4S_{\frac{3}{2}}$
	4	1	1	2	$^3P_2$
	5	$\frac{1}{2}$	1	$\frac{3}{2}$	$^2P_{\frac{3}{2}}$
	6	0	0	0	$^1S_0$
	1	$\frac{1}{2}$	2	$\frac{3}{2}$	$^2D_{\frac{3}{2}}$
	2	1	3	2	$^3F_2$
	3	$\frac{3}{2}$	3	$\frac{3}{2}$	$^4F_{\frac{3}{2}}$
	4	2	2	0	$^5D_0$
d	5	$\frac{5}{2}$	0	$\frac{5}{2}$	$^6S_{\frac{5}{2}}$
	6	2	2	4	$^5D_4$
	7	$\frac{3}{2}$	3	$\frac{9}{2}$	$^4F_{\frac{9}{2}}$
	8	1	3	4	$^3F_4$
	9	$\frac{1}{2}$	2	$\frac{5}{2}$	$^2D_{\frac{5}{2}}$
	10	0	0	0	$^1S_0$

2.8 Mechanical and Magnetic Moments of a Free Atom

It is of some interest now to determine the angular momentum and magnetic moment of a free atom in the normal state. Let us take the case of an atom of iron.

An atom of iron has 26 electrons with six equivalent electrons in an incompleted $3d$ sub-shell. The normal state is $^5D_4$ as may be seen from Appendix 1 or Table 2.3, and $S = 2$, $L = 2$ and $J = 4$. The vector model drawn to scale is shown in Fig. 2.19(a). The angular momentum of the atom is $J^*(h/2\pi)$ where $J^* = \{J(J+1)\}^{\frac{1}{2}} = \sqrt{20}$. In the presence of a weak magnetic field H precession of the vector J^* about the direction of H occurs such that the

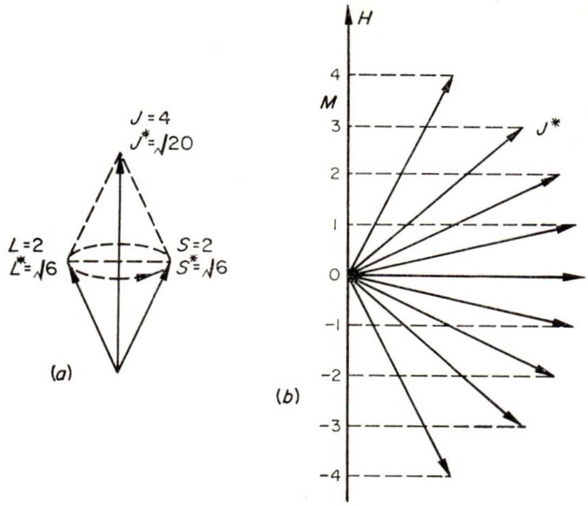

FIG. 2.19. Vector diagrams for a free iron atom.

resolved values of J^* are $M = 0, \pm 1 \ldots \pm J$. J^* can therefore take up $2J + 1$ angular positions with respect to the direction of H, or 9 possible positions in this case, as shown in Fig. 2.19(*b*).

In Fig. 2.20 the components of angular momentum and of magnetic moment are shown for precession of the atom in a weak magnetic field, the example selected being for the particular case of $M = 3$. The *LS* coupling, and Zeeman precession about the field H, are precisely similar to the *ls* coupling and precession for a single electron shown in Figs. 2.13 and 2.15(*a*) and described in Section 2.5 The vectors $L^*(h/2\pi)$ and $S^*(h/2\pi)$, representing the resultant orbital and spin angular momenta of the atom, respectively, precess at a relatively high angular velocity about the axis of the resultant

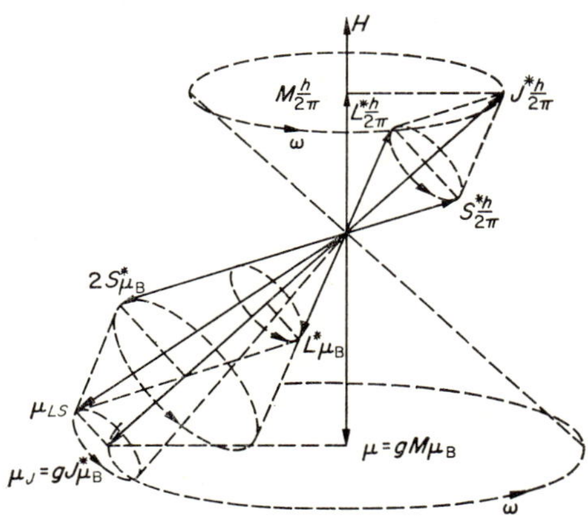

FIG. 2.20. Vector model of free iron atom in the normal state showing precession in a weak magnetic field for $M = 3$: $S = 2$, $L = 2$, $J = 4$, $g = 3/2$.

$J^*(h/2\pi)$, which in turn precesses with a lower angular velocity ω about the direction of H to give a resolved component $M(h/2\pi)$ as shown in the figure.

The components of magnetic moment are represented by the vectors $L^*\mu_B$ and $2S^*\mu_B$ precessing fast about the axis of J^* with a resolved component $\mu_J = gJ^*\mu_B$ as shown. As in the case of ls coupling with a single electron g is the Landé factor, g may be derived as in Section 2.5 and is given by

$$g = 1 + \frac{(J^*)^2 + (S^*)^2 - (L^*)^2}{2(J^*)^2} \tag{2.26}$$

in which capital letters now take the place of the lower-case letters of equation (2.25). μ_J precesses with angular velocity ω about the direction of H to give a steady resolved component $\mu = gM\mu_B$ in the field direction. In this particular case of the free iron atom $S^* = L^*$ and, on substituting the appropriate values in equation (2.26), $g = \frac{3}{2}$. The precessional angular velocity

$$\omega = H(\mu_J/p_J) = Hg(\mu_0 e/2m)$$

as in the case of ls coupling, and the Zeeman energy associated with the precession is $\mu H = gM\mu_B H$.

It should be emphasized in concluding this chapter that the foregoing discussion and the atomic structures given in Appendix 1 apply to single free atoms. The outer electronic configurations, the atomic magnetic moments and their response in an externally-applied magnetic field will in general be modified when the atoms come together in the solid state.

PROBLEMS

2.1 Calculate the following, in each case, for a hydrogen atom on the Bohr theory when $n = 1$, $n = 2$, $n = 10$:

(a) the orbit radius;

(b) the total energy of the atom; hence write down the kinetic energy of the electron and the potential energy;

(*c*) the velocity of the electron and the current in the orbit;

(*d*) the magnetic field, on classical theory, at the centre of the orbit;

(*e*) the magnetic moment of the orbit.

(*Note:* the magnetic field at the centre of a circular orbit may be obtained by using Biot and Savart's Law.)

2.2 Calculate the revolution frequency of the electron in a hydrogen atom, on the Bohr theory, when $n = 1$ and when $n = 2$, and calculate the frequency of the emitted radiation for a transition from $n = 2$ to $n = 1$.

2.3 Repeat the calculations of Problem 2.2 for a triply-ionized lithium atom.

2.4 Repeat the calculations of Problem 2.2 for a hydrogen atom when $n = 50$ and $n = 51$, and again when $n = 500$ and $n = 501$, and compare the revolution frequencies with the frequency of the emitted radiation.

2.5 The normal state of a free nickel atom is given in spectroscopic notation by $^3F_4$. Find S, L, J and the Landé factor g for the atom, and calculate the precessional angular velocity for the atom in an applied field $H = 4000$ A/cm.

2.6. Describe in spectroscopic notation the normal state of the trivalent iron ion Fe^{3+}, and determine the Landé factor g.

SUGGESTED READING

LONGHURST, R. S. *Geometrical and Physical Optics*, Longmans Green, London (1957).

WHITE, H. E. *Introduction to Atomic Spectra*, McGraw-Hill, New York (1934).

HUME-ROTHERY, W. *Atomic Theory for Students of Metallurgy*, 3rd edn., Institute of Metals, London (1960).

TOLANSKY, S. *Introduction to Atomic Physics*, 3rd edn., Longmans Green, London (1949).

BATES, L. F. *Modern Magnetism*, 4th edn., Cambridge University Press, London (1961).

REFERENCES

[1] LONGHURST, R. S. *Geometrical and Physical Optics*, p. 172, Longmans Green, London (1957).

[2] ROWLAND, H. A. 'Manufacture and theory of gratings for optical purposes', *Phil. Mag.* **13,** 469 (1882).

[3] PLANCK, M. 'Ueber die Elementarquanta der Materie und der Elektrizität', *Annln Phys.* **4,** 564 (1901).

[4] EINSTEIN, A. 'A heuristic standpoint concerning the production and transformation of light', *Annln Phys.* **17,** 132 (1905).

[5] MILLIKAN, R. A. 'A direct photoelectric determination of Planck's h', *Phys. Rev.* **7,** 355 (1916).

[6] RUTHERFORD, L. M. 'Scattering of α and β particles by matter and the structure of the atom', *Phil. Mag.* **21,** 669 (1911).

[7] BOHR, N. 'Constitution of atoms and molecules', *Phil. Mag.* **26,** 1, 476, 857 (1913).

[8] SOMMERFELD, A. 'Zur Quantentheorie der Spektrallinien', *Annln Phys.* **51,** 1, 125 (1916).

[9] SCHRÖDINGER, E. 'Quantisierung als Eigenwert Problem', *Annln Phys.* **79,** 361, 489 (1926); **80,** 437 (1926); **81,** 109 (1926).

[10] BORN, M., JORDAN, P. and HEISENBERG, W. 'Zur Quantenmechanik', *Z. Phys.* **35,** 557 (1926).

[11] DIRAC, P. A. M. 'The quantum theory of the electron', *Proc. R. Soc.* A, **117,** 610 (1928); A, **118,** 351 (1928).

[12] LARMOR, J. 'Aether and matter', *Phil. Mag.* **44,** 503 (1897).

[13] ZEEMAN, P. 'Influence of magnetism on the nature of the light emitted by a substance', *Phil. Mag.* **5,** 43, 226 (1897).

[14] PASCHEN, F. and BACK, E. 'Normale und anomale Zeemaneffekte', *Annln Phys.* **39,** 897 (1912); **40,** 960 (1913).

[15] PAULI, W. 'Über den Zusammenhang des Abschlusses der Elektronengruppen im Atom mit der Komplexstruktur der Specktren', *Z. Phys.* **31,** 765 (1925).

[16] RUSSELL, H. N. and SAUNDERS, F. A. 'New regularities in the spectra of the alkaline earths', *Astrophys.* **61,** 38 (1925).

[17] HUND, F. *Linienspektren und periodisches System der Elemente*, p. 124, Springer, Berlin (1927).

Chapter 3

CRYSTALS

3.1 Interatomic Cohesive Forces

When atoms of like or unlike elements approach each other, interatomic forces come into play. Thus atoms may be bound together to form molecules or solid crystals. Intermolecular forces also may result in the formation of molecular crystals. The cohesive forces involved are of a number of different kinds which may overlap and exist simultaneously although one particular type of force may be predominant.

When two or more atoms come together to form a molecule, a new electronic configuration is involved in which the inner electrons remain associated with their respective nuclei but the outer electrons may be said to belong to the molecule as a whole. As with a free atom the electrons are now in electronic quantum states, to which the Exclusion Principle applies, so that no two electrons in the molecule, taking spin into account, can be in the same state. The energy levels may be studied from molecular spectra. A complicated system of molecular quantum numbers is involved.

Similarly a crystal may be regarded as one very large molecule. Electrons are present in it in discrete energy levels, and the X-ray spectra of crystals provide a possible means of exploring the levels.

Three principal types of bonding between atoms will be briefly considered.

3.1.1 *The Covalent Bond*

A covalent bond, also referred to as a homopolar or valency bond or linkage, is formed between two atoms by the sharing of two electrons. In the bond the shared electrons are of opposite spin but are otherwise in the same quantum state.

In the upper part of Fig. 3.1(*a*) two hydrogen atoms with their respective electrons, denoted by the dot and cross, are shown approaching each other. The electrons are of opposite spin. There are electrostatic forces between the charges on the electrons and the nuclei, and the energy of the system will depend on the separation between the nuclei. This particular problem is capable of exact solution in wave mechanics. It is found that, as the nuclei approach, the energy of the system decreases to a minimum when the distance between the nuclei is 0·74 Å (0·74 $\times$ 10^{-8} cm). For a closer approach

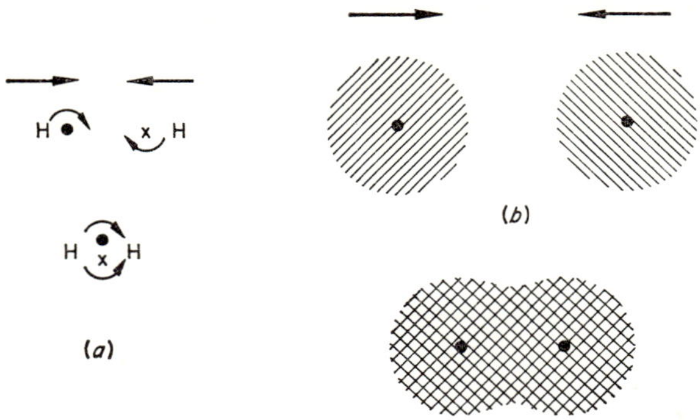

FIG. 3.1. Formation of a hydrogen molecule by covalent bonding.

the energy rises again because repulsion between the positive nuclei becomes the predominant force. A stable molecule, with oppositely-spinning electrons, is thus produced, since work would have to be done to increase or decrease the optimum distance between the nuclei.

Figure 3.1(*b*) is a semi-pictorial representation of the atoms and the molecule. As mentioned in Chapter 2, all that can be known of the position of the electron in the atom is the probability of finding the electron at any point in the neighbourhood of the nucleus, and the resultant effect of the electron in motion is as though it were smeared out into a cloud of negative charge. For the hydrogen atom, having in the normal state a single *s* electron, the density of charge is greatest very near the nucleus, and falls off rapidly with increasing distance, but with spherical symmetry as indicated in Fig. 2.11. The two hydrogen atoms are shown diagrammatically in the upper part of Fig. 3.1(*b*). When the molecule is formed, the electron clouds are superimposed, the charge density still being greatest near the nuclei but also high between them with a 'dumb-bell' kind of distribution. The cloud density becomes more spherical and more attenuated as the distance from the nuclei increases.

Because of the opposing spins of the electrons, the magnetic moment of the molecule in an applied field is zero. The electrons do not belong to either nucleus but there is a free exchange. The forces occurring in the covalent bond are in fact called forces of exchange and are important not only in cohesion but also in ferromagnetism.

An electron in an *s* state, considered as a cloud of charge, has spherical symmetry and so can form a covalent bond in any direction. A *p* sub-shell containing its full complement of six electrons also has a spherical charge distribution. The shell is incapable of forming a covalent bond because the electrons are already paired. If, however, the sub-shell is one electron short, the charge distribution is no longer spherical, and moreover a covalent bond with another atom is possible. If more than one electron is absent from the *p* sub-shell more than one bond with one or more atoms can occur by electron sharing and these bonds are directional in space.

If a nucleus is considered at the origin of Cartesian coordinates OX, OY, OZ, and fields exist along all three axes, the charge clouds for three p electrons, p_X, p_Y, p_Z, may be represented by the three dumb-bell shapes shown in Fig. 3.2. The figure is purely diagrammatic since the charge density dies away gradually at the outer surfaces but does, in fact, come to zero at the origin. Accurate calculations of the charge densities are possible in wave mechanics for electrons in different states in excited hydrogen or helium atoms, though not for atoms of higher atomic number. In the Schrödinger theory the effect of orbital motions only are included. In the Dirac theory, spin is also taken into account but this modifies the distributions only a little.

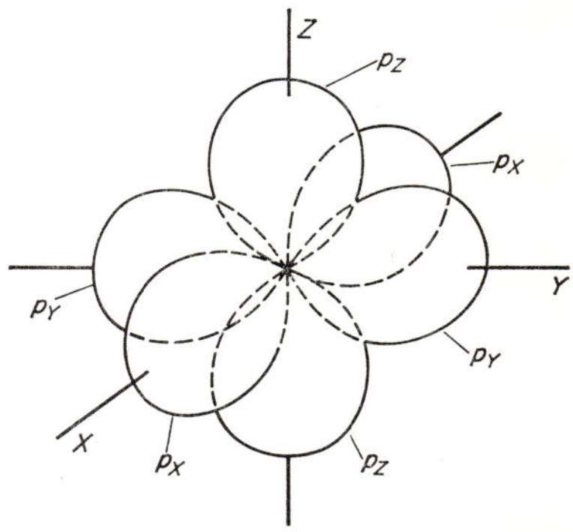

FIG. 3.2. Charge clouds for electrons in p states.

In Fig. 3.2 the total distribution, summed for the three electrons, is again spherical. If a fourth electron is added, for example to complete the p_X pair, then the p_X charge density is doubled, and similarly for the y and z axes. The covalent bonds which may be formed by pairing with electrons from neighbouring atoms are directed in space along the coordinate axes.

As an example, an atom of oxygen has in its electronic structure one pair of 2s electrons and four 2p electrons (see Appendix 1). The structure is therefore two electrons short of the group of eight electrons required to complete the L main shell, and it may form two covalent bonds. A molecule of water may thus be represented as in Fig. 3.3, the stable group of eight electrons being formed by electron sharing with two hydrogen atoms. The angle between the bonds, expected to be 90°, is found to be 104° 31′, no doubt due to the asymmetery of the molecule and repulsion between the hydrogen nuclei.

Carbon in the form of diamond is a case of a covalent crystal. The carbon atom has four electrons in the L main shell. In the crystal each atom makes up a group of eight by sharing one electron from each of four near neighbours by four covalent bonds. The crystal structure is shown in Fig. 3.4 in relation to a cubic lattice. Each atom has four near neighbours arranged on the corners of a tetrahedron. The covalent bonds are shown for some of the atoms by the thicker lines and the angle between the bonds is now 109°.

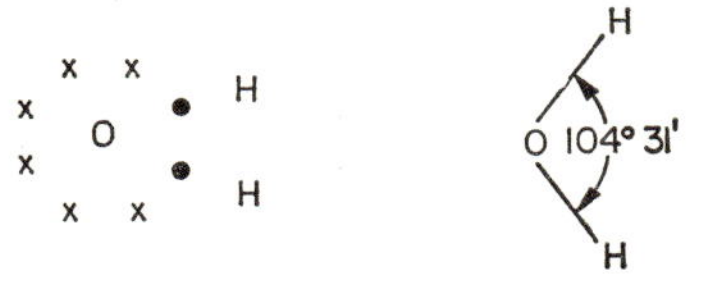

FIG. 3.3. Water molecule.

For carbon the unit cube, or unit cell, in the figure, has a length of side, or lattice constant, $a = 3{\cdot}561$ Å. The shortest distance between atom centres is $\sqrt{3}a/4 = 1{\cdot}542$ Å and there are eight atoms per cell. Assuming the atoms to be hard spheres in contact, the fraction of the total volume occupied by the atoms is $\pi\sqrt{3}/16 = 0{\cdot}34$, so that the diamond structure is relatively empty.

Silicon, germanium and grey tin crystallize on the same structure as diamond (compare the electronic configurations of these elements in Appendix 1) and their lattice constants are 5·431, 5·657 and 6·46 Å, respectively. It may be mentioned here that in these materials all the valency electrons take part in the bonding. At very low temperatures the electrons are not free to move through the crystals, which are therefore insulators. With increasing temperature, however, electrons may acquire sufficient energy from the thermal vibrations of the atoms to become detached and mobile in an electric field, at the same time leaving behind an electron vacancy or positive hole. The materials now have some electrical conductivity, and in the pure state are known as intrinsic semiconductors.

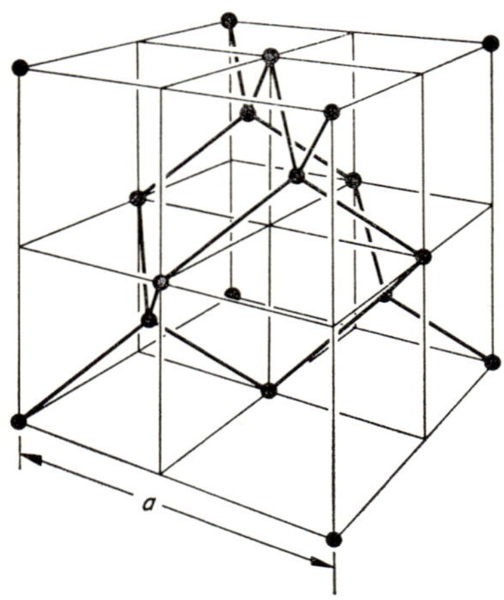

FIG. 3.4. Crystal structure of diamond, silicon, germanium and grey tin showing covalent bonds.

3.1.2 *The Ionic Bond*

In ionic crystals cohesion is due to electrostatic attraction, or Coulomb forces, between oppositely-charged ions. The binding force between ions is known as an ionic, heteropolar or electrovalent bond. Sodium chloride is a typical ionic crystal.

An atom of sodium, $Z = 11$, has its K and L shells full and one $3s$ electron. Chlorine, $Z = 17$, has the K and L shells full but with two $3s$ electrons and five $3p$ electrons. The $3p$ sub-shell is therefore one electron short of the six required to fill it, or one short of a group of eight in the M shell. When atoms of sodium and chlorine come together the $3s$ sodium electron is transferred to the $3p$ sub-shell of the chlorine atom. Both atoms now have stable outer groups of eight electrons, but the sodium has become a positive ion, and the chlorine a negative ion, each with one unit of charge. The ions approach each other by the electrostatic force between the opposite charges until the attraction is counterbalanced by repulsion between the positive nuclei.

In spite of what has been said about diffuse electron clouds the equilibrium achieved between the ions, in effect, gives them definite radii which remain approximately unaltered for the same ions in other chemical compounds. In fact the distance between the nuclei for ions in contact at low temperatures is very closely equal to the sum of their radii. The radius of the sodium positive ion is 0·98 Å and of the chlorine negative ion is 1·81 Å while the separation between the nuclei in a sodium chloride crystal, when extrapolated to 0°K, is $a/2 = 2{\cdot}79$ Å.

Figure 3.5 shows the crystal structure of sodium chloride. The chlorine ions are arranged in what is called a face-centred cubic system, that is, as drawn in the figure, the chlorine ions are situated at the corners of a cubic lattice of lattice constant a with an ion at the centre of each cube face. The sodium ions are also on an identical interpenetrating face-centred cubic lattice in the manner shown. The plan view shows the ions, in their correct relative sizes, in contact in the cube-edge directions.

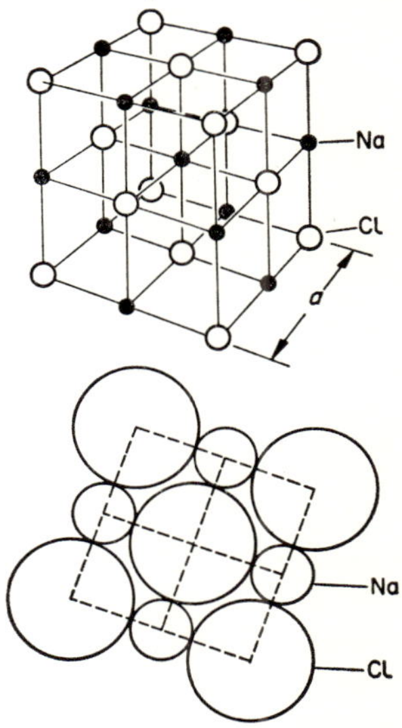

FIG. 3.5. Crystal structure of sodium chloride showing, in the lower figure, the relative sizes of the ions.

The iron oxide Fe_3O_4, magnetite or lodestone, is the basis of a useful group of magnetic materials known as ferrites, and this oxide also forms ionic crystals on a cubic lattice. The structure, which is not quite as simple as that of sodium chloride, will be discussed further in Chapter 9.

In ionic crystals in general the electrons are tightly bound and the crystals have no electrical conductivity which can be associated with the free movement of the valency electrons, although at high temperatures conductivity due to the movement of ions may occur.

3.1.3 *The Metallic Bond*

Most metals crystallize on one of three characteristic types of structure which are, respectively, face-centred cubic, close-packed hexagonal and body-centred cubic. The three structures are represented in Fig. 3.6. The atoms are shown as spheres in contact which is a good pictorial method of depicting a crystal. As already mentioned the atoms or ions are not in fact hard spheres although, because of the repulsive forces between the outer shells of electrons, they may, for present purposes, be considered to behave as such.

Face-centred crystals have atoms at each cube corner and at the centre of each cube face of the crystal lattice. A typical example, which might represent the structure of nickel at room temperature, is shown at *A*. The lattice constant is a and the interatomic distance which is the diameter of the spheres representing the atoms is $a/2^{\frac{1}{2}}$. There are four whole atoms per unit cell and therefore the packing factor, that is the fractional volume of the crystal occupied by the atoms is easily shown to be $\pi\sqrt{2}/6 = 0{\cdot}74$.

The close-packed hexagonal structure shown in the figure at *B* is formed by placing seven spheres in contact in the bottom layer, then laying spheres

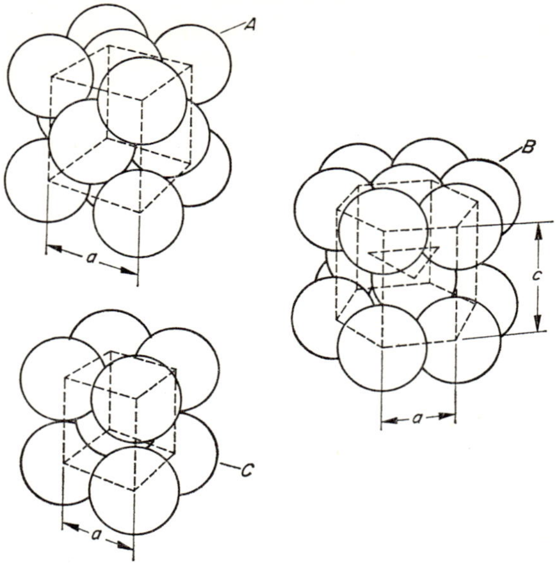

FIG. 3.6. Typical structures of metal crystals. *A*—face-centred cubic: *B*—close-packed hexagonal: C—body-centred cubic.

in contact in the second layer, so that each sphere rests on three below, and adding a third layer, in contact with the second, with the spheres immediately above those in the first layer. Cobalt at room temperature has this structure. The interatomic distance in this case equals the lattice constant a and the axial ratio $c/a = (8/3)^{\frac{1}{2}} = 1{\cdot}633$. There are six whole spheres in the unit volume shown by the broken lines in the figure, and the packing factor is again $\pi\sqrt{2}/6 = 0{\cdot}74$, exactly the same as for the face-centred cubic arrangement.

The body-centred cubic structure shown at *C* has atoms at the corners and one atom at the centre of the cube. The corner atoms are not in contact with each other but are all touching the central atom. This is the room-temperature structure of iron. In this case the closest interatomic distance is $a\sqrt{3}/2$. There are two atoms in the unit cell and the packing factor is $\pi\sqrt{3}/8 = 0{\cdot}68$.

The arrangements *A* and *B* both give the closest possible packing of spheres, or the minimum interstitial volume, in spite of the geometrical difference between the face-centred and hexagonal structures. As a matter of interest, spheres packed at random in a container are found experimentally to occupy 15 to 20% more space. Thus the directional forces between atoms in *A* and *B* lead to the highest possible packing factor. On the other hand, the directional bonds between the atoms in the body-centred structure cause the packing factor to be about 8% lower.

Some room-temperature constants for nickel, cobalt and iron crystals are given in Table 3.1, from which it will be seen that the interatomic distances, or the diameters of the spheres in contact taken to represent the atoms, are for these three elements in three different structures nearly the same. The observed axial ratio c/a for cobalt is seen to be slightly lower than the theoretical value for close-packed spheres in the hexagonal structure. In fact a slightly lower ratio occurs for most metal crystals of this type, and indicates a close packing of spheroids rather than spheres.

TABLE 3.1

Element	Structure	Lattice Constant a Å	Nearest Interatomic Distance Å	Observed Axial Ratio c/a
Ni	f.c.c.	3·517	2·487	
Co	c.p.hex.	2·507	2·499	1·624
Fe	b.c.c.	2·861	2·478	

The electronic structures of atoms which form metal crystals are not in general suitable for the formation of either covalent or ionic bonds in which the electrons are closely bound to particular atoms. In metals there are free electrons, which give rise to high electrical conductivity, and the crystals are made up of a regular geometrical pattern of positive ions in a mobile swarm of valency electrons. The whole assemblage of ions and electrons obviously must form a stable system of lowest energy, but the problem of the cohesive forces involved is too complex to be solved rigorously. Cohesion, however, may be said to be mainly due to Coulomb forces of attraction between the positive ions and the negative charge of the valency electrons.

There will be a field of force near any positive ion tending to hold a neighbouring valency electron. However, in a crystal the fields of force of the ions in the structure will overlap, so that energy barriers to the movement of an electron through the crystal become very small. The crystal may be regarded, as already mentioned, as one very large molecule. The inner electrons of the atoms remain associated with their particular nuclei but the outer electrons belong to the crystal as a whole. As with a free atom or a molecule the electrons occupy electronic quantum states. There are obviously in any metal crystal an immense number of states with very small energy differences between them, for the Exclusion Principle applies and only two electrons, one of positive and one of negative spin, can occupy any one state.

3.2 Energy Bands in Crystals

The energy levels in free atoms have already been discussed. The optical spectra obtained when electrons drop inwards from outer to inner shells of excited atoms consist of sharp lines. When, however, the atoms come together to form a solid crystal, the electronic configuration of the atoms, so far as the outermost electrons are concerned, is altered. The experimental evidence for this comes from soft X-ray spectroscopy on solid crystals. It is found that, when the atoms of the crystal are excited so that transitions may occur in which outer electrons drop inwards to inner shells of the atoms, instead of sharp spectral lines, broad bands of varying intensity are obtained. The experimental results lead to the conclusion that the outer valency electrons of atoms in a crystal may occupy a wide range of energy levels which are very close together. The energy differences between these various levels and the nearly fixed energy level of an inner shell therefore cover a range of values. The quantum frequencies of the associated electromagnetic radiations therefore also cover a range and thus cause the spectral lines to appear as a band. Measurement of the intensity of the emission over the width of the band enables conclusions to be reached of the distribution of the outer electrons in the valency energy band or zone of the crystal.

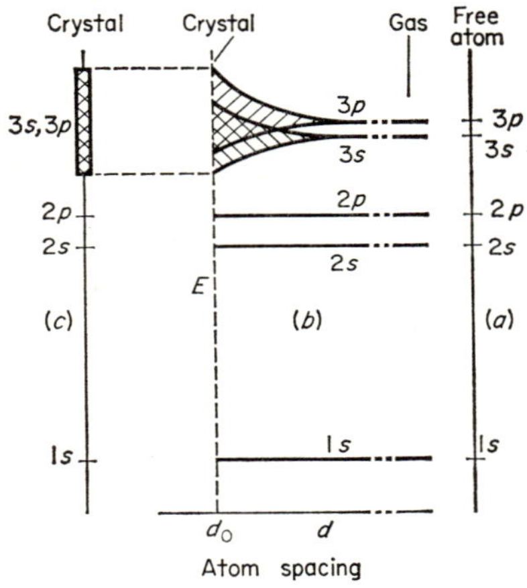

FIG. 3.7. Comparison of allowed energy levels in a free atom and in a crystal of sodium.

As an example the case of sodium is illustrated in Fig. 3.7. The sub-shell energy levels for a free atom are as shown at (*a*), although these are not drawn to scale. For a sodium crystal in the normal state the 1*s*, 2*s* and 2*p* sub-shells are full. There is one electron in 3*s* and none in 3*p*. These are the conditions to be expected for atoms when the element is in an attenuated gaseous condition, and this would be the case in Fig. 3.7(*b*) when d, the distance between the atoms, was great. We assume now that the atoms come together until when $d = d_0$ a crystal is formed. The energy of an electron in, for example, a 2*s* sub-shell of an atom is determined mainly by its own positive nucleus and is hardly changed by neighbouring atoms even when $d = d_0$. The energy of the valency electrons, however, is increasingly affected as d decreases. It is now considered that the Exclusion Principle also applies so that only two electrons in the crystal, of positive and negative spin respectively, may occupy any one energy level. The effect is that the allowed energy levels spread out into a band as shown in Fig. 3.7(*b*). The allowed 3*p* level, which is unoccupied in the free atom in the normal state, also spreads into a band, and, as shown in the figure, the 3*s* and 3*p* bands overlap.

The energy levels for solid sodium may therefore be represented as shown in the figure at (*c*). In the 1*s*, 2*s* and 2*p* sub-shells the spread in energy levels which occur on crystallization is so small that these sub-shells remain virtually at single levels. The allowed levels for the valency electrons are, however, spread out as shown diagrammatically, and the width of this band is a characteristic of solid sodium and depends on the interatomic spacing, but the number of discrete levels in the band depends upon the number of atoms, that is, upon the size of the crystal.

If one imagines the crystal bereft of all its valency electrons, which are then returned one by one, the electrons will occupy the lowest energy levels in succession, with two electrons, of opposite spin, to each level until all the valency electrons, that is in this case one electron per atom, have been restored. However, when this occurs, the allowed energy band is not full, since the higher levels are unoccupied. This is a characteristic not only of sodium, but of the metals in general. Metals have high electrical conductivity because electrons in the valency band may be excited to states of higher energy by the application of an electric field. This is possible because the higher allowed energy levels are existing in the energy band and may be occupied by electrons accelerated in a particular direction. There is then a slight average

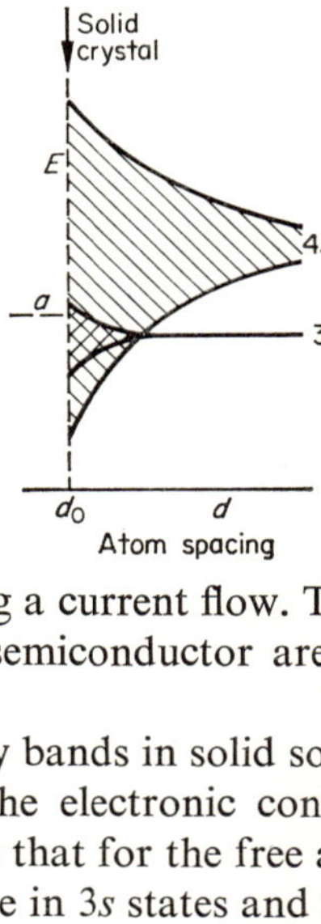

FIG. 3.8. Showing spread and overlapping of allowed energy levels of nickel as the atoms approach to form a crystal. Levels in the crystal are filled to *a*.

drift of electrons in this direction constituting a current flow. The conditions which make a material an insulator or a semiconductor are described in Section 3.4.

One result of the overlapping of the energy bands in solid sodium, as with other elements in the solid state, is that the electronic configuration for atoms in the crystal is no longer the same as that for the free atom. In solid sodium a fraction of the valency electrons are in 3*s* states and the remainder in 3*d* states. This overlapping of energy bands has an important effect in determining the average atomic magnetic moments in ferromagnetic crystals, the observed values being different from the moments for free atoms.

Figure 3.8 shows diagrammatically how the energy bands for nickel spread as the atoms approach to form a crystal. The atomic spacing in nickel is represented by $d = d_0$. It will be seen that the 3*d* band and the 4*s* band, which together contain all the electrons of the atom which are in incompleted sub-shells, overlap. The energy levels in these bands are now occupied to the point *a* in the figure, neither band in the crystal being completely full. Reference to Appendix 1 shows that the free atom of nickel has eight electrons in the $3d$ sub-shell and two electrons in $4s$. In the crystal, however, the average number of electrons per atom are 9·4 in $3d$ states and 0·6 in $4s$ states. It should be noted that to fill completely the $3d$ band of levels the number of electrons per atom required would be ten, which is the number required to fill the $3d$ sub-shell in a free atom. Two electrons per atom would be required to fill the $4s$ band. The electronic configurations for the free atom, and on the average for the atom in the crystal, are therefore respectively,

$$(1s)^2(2s)^2(2p)^6(3s)^2(3p)^6(3d)^8(4s)^2 \quad \text{and} \quad (1s)^2(2s)^2(2p)^6(3s)^2(3p)^6(3d)^{9\cdot4}(4s)^{0\cdot6}$$

where the superscripts give the number of electrons in each state, and this again emphasizes the point that these configurations are not the same in the two cases.

3.3 The Free-Electron Theory of Metals

In the free-electron theory of metals the valency electrons, consisting in the simplest case of one electron per atom, may move within the crystal-like gas particles in a box. The electric charges of the electrons and ions are neglected and there is therefore no component of potential energy associated with them. The electron energies are purely kinetic but are subject to quantum restrictions which arise as a result of the Heisenberg Uncertainty Principle and the Pauli Exclusion Principle. At the surfaces of the box or crystal there is, however, a sharp rise of potential and to overcome this potential barrier, and escape from the crystal, electrons would require considerable energy, so that under normal circumstances the electrons are confined to the crystal.

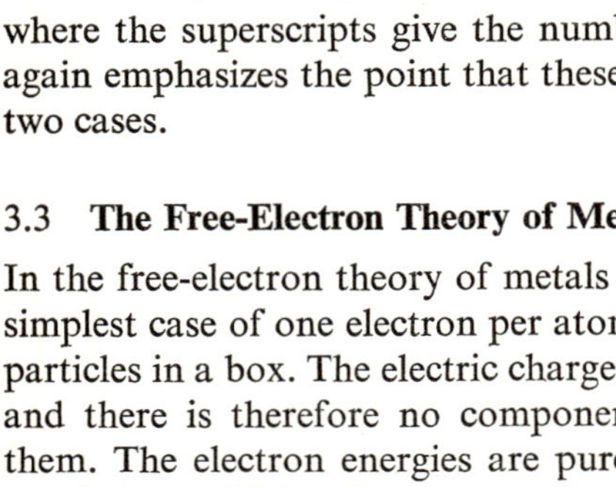

FIG. 3.9. Momentum diagram in the free-electron theory of metals.

Now there is no conceivable method of measuring the simultaneous position and velocity of an electron. The more accurately one is known the less accurately is the other known. The Heisenberg Principle states that if Δx is the uncertainty in the position and Δp the uncertainty of the momentum of a particle then the product $\Delta x\, \Delta p$ has a minimum value equal to Planck's constant h.

It can be shown that the following argument will apply to a crystal of any shape. However, consider the crystal represented by a cubical box of side L with coordinate axes OX, OY, OZ parallel to edges of the box. Let p be the momentum of an electron in the box and let p_X, p_Y, p_Z be its components in directions parallel to OX, OY and OZ, respectively. A *Momentum Diagram* as shown in Fig. 3.9 may be constructed in which the coordinate axes, parallel to OX, OY and OZ, now represent momentum. The point P in this diagram has the coordinates p_X, p_Y, p_Z

and the length OP is equal to p. The direction of OP represents the direction of motion of the electron. Then $p^2 = p_X^2 + p_Y^2 + p_Z^2$ but the diagram of course gives no indication of the position of the electron in the box.

Now p cannot be known with certainty. If it were, then by the Uncertainty Principle its position would be infinitely uncertain. But in fact the position of the electron in the box, with respect for example to the x-axis, is known with an uncertainty not greater than L. If we put $\Delta x \Delta p = h$, with $\Delta x = L$, the minimum uncertainty in the momentum p_X is $\Delta p_X = h/L$. The same applies to the components p_Y and p_Z, so that the point P, representing the momentum of an electron, cannot be plotted more precisely in the momentum diagram in Fig. 3.9 than as being associated with a volume of *momentum space* equal to $(h/L)^3 = h^3/V$ where V is the volume of the crystal.

Each elementary volume h^3/V in the momentum diagram is identified with a quantum state and, applying the Pauli Principle, each quantum state can be occupied by not more than two electrons of opposite spin. In the diagram, therefore, the number of allowed states in a shell of thickness dp and radius p is given by $4\pi p^2\, dp/(h^3/V)$ and since the energy of an electron is $E = mv^2/2 = p^2/2m$ the energy of the states increases as the radius p in the diagram increases. If there is a total of N free electrons in the crystal, there will be $N/2$ states and the total energy of the electrons will be a minimum when they occupy states completely filling a spherical volume of the momentum diagram. This is so for, if an electron moves from a state inside the sphere to a state outside, the energy will be increased.

The condition of lowest energy occurs at absolute zero of temperature and the electron gas possesses in fact substantial kinetic energy at 0°K. If the temperature is raised, a redistribution of the electrons in the highest energy states occurs, and this redistribution may be determined by the use of the Fermi–Dirac statistics.

Considering the crystal at 0°K there will be $N/2$ allowed states in a volume of the momentum diagram equal to $4\pi p_0^3/3$, where p_0 is the momentum of electrons of highest energy. If E_F is the energy of an electron in the highest occupied level, $E_F = p_0^2/2m$. Each state occupies a volume h^3/V and therefore

$$Nh^3/2V = 4\pi p_0^3/3 = (4\pi/3)(2mE_F)^{\frac{3}{2}} \tag{3.1}$$

If $n_0 = N/V$ is the total number of electrons per unit volume we then have

$$E_F = (h^2/2m)(3n_0/8\pi)^{\frac{2}{3}} \tag{3.2}$$

E_F is called the Fermi energy and is the energy of an electron in the highest occupied level at 0°K.

If n is the number of electrons per unit volume completely occupying states up to momentum p in the momentum diagram then, using equation (3.1),

$$n = (8\pi/3h^3)(2mE)^{\frac{3}{2}}$$

If dn is the number of electrons occupying states between the energy levels E and $E + dE$ we have, by differentiating,

$$dn/dE = (8\pi/3h^3)(2m)^{\frac{3}{2}}(3\sqrt{E}/2)$$

and
$$n_0 = (8\pi/3h^3)(2m)^{\frac{3}{2}}E_F^{\frac{3}{2}}$$

$$\therefore dn/dE = 3n_0\sqrt{E}/2E_F^{\frac{3}{2}} \tag{3.3}$$

The relation between dn/dE and E, a parabola as plotted in Fig. 3.10, gives the distribution of the electrons in relation to energy at 0°K. Thus the hatched area represents the number of electrons in states having energies lying between E and $E + dE$. There is a sharp cut-off in the distribution when $E = E_F$, the Fermi energy, and the total area beneath this line represents n_0, the total number of electrons per unit volume in the crystal.

The total kinetic energy at 0°K of the n_0 electrons and the mean energy of an electron are readily calculated. The total kinetic energy is given by

$$U_0 = \int_0^{E_F} E\, dn = \int_0^{E_F} (3n_0/2)(E/E_F)^{\frac{3}{2}}\, dE = 3n_0E_F/5$$

and the mean kinetic energy per electron $E_m = 3E_F/5$. Take for example the case of the metal sodium which has one free electron per atom. At room temperature sodium crystallizes on the body-centred cubic system with a lattice constant $a = 4{\cdot}2820$ Å and therefore

$$n_0 = 2/(4{\cdot}2820 \times 10^{-10})^3 = 2{\cdot}55 \times 10^{28}/m^3$$

The Fermi energy is therefore

$$E_F = (h^2/2m)(3n_0/8\pi)^{\frac{2}{3}}$$
$$= (6{\cdot}625 \times 10^{-34})^2(3 \times 2{\cdot}55 \times 10^{28}/8\pi)^{\frac{2}{3}}/2 \times 9{\cdot}107 \times 10^{-31}$$
$$= 5{\cdot}0 \times 10^{-19} \text{ joule}$$

It is convenient to express this in terms of the electronvolt (eV), an electronvolt being the energy acquired by an electron when accelerated through a potential difference of one volt. 1 eV $= 1{\cdot}602 \times 10^{-19}$ joule. Therefore $E_F = 5/1{\cdot}602 = 3{\cdot}1$ eV and the mean kinetic energy of the electrons is $E_m = 3E_F/5 = 1{\cdot}9$ eV.

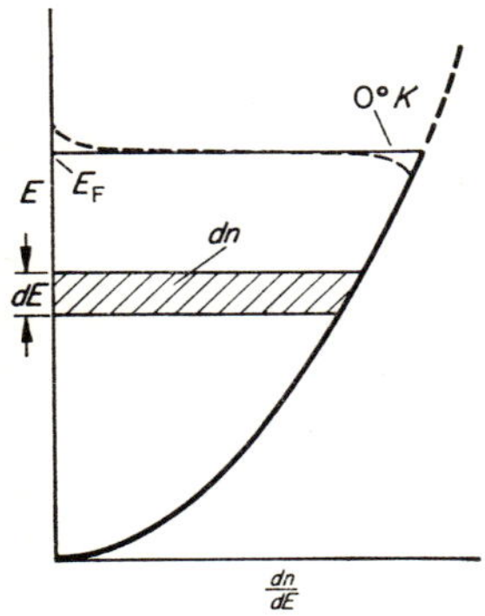

FIG. 3.10. Distribution of electrons in relation to energy in the free-electron theory.

If in this case the electrons in the box had been treated in accordance with the classical kinetic theory of gases then, instead of the large energies just found, the electrons at absolute zero would have been at rest and have had zero energy. At temperature T the mean kinetic energy of the particles on classical theory is $3kT/2$ where k is Boltzmann's constant. A mean energy of $E_m = 1{\cdot}9$ eV would therefore require the gas to be at an absolute temperature of $T = 2E_m/3k$, that is

$$T = 2 \times 1{\cdot}9 \times 1{\cdot}602 \times 10^{-19}/3 \times 1{\cdot}380 \times 10^{-23} = 15{,}000°\text{K}$$

The distribution of the electrons over an energy range from zero up to

E_F in Fig. 3.10 corresponds to absolute zero of temperature, and it will be apparent that the energies of the electrons are enormously greater than normal thermal energies, so that an increase in temperature in the crystal of any practical amount has only a small effect upon the distribution of the electrons in the allowed quantum states. It is shown in the Fermi–Dirac statistics that an increase of temperature affects only electrons near the Fermi level, and causes some electrons below to be transferred to states above the level. This is illustrated by the broken line in the figure. It is, however, adequate for many purposes to neglect this small redistribution and to suppose that there is a sharp cut-off at the Fermi level.

3.4 The Zone Theory

In the free-electron theory just discussed, in which the electrons are assumed to have a constant potential energy and only the kinetic energies are considered, a parabolic distribution of allowed levels is found. These levels continue upward with increasing energy and with no upper limit. The actual condition in a crystal, however, is that there are bands of allowed levels, as shown in Figs. 3.7 and 3.8, with upper and lower limits, although in these two cases there are overlapping bands. The existence of an upper limit may be deduced mathematically by the use of wave mechanics and the Schrödinger wave equation. The detailed mathematical aspects are beyond the scope of this book but the general picture may be presented as follows.

It is found experimentally that if a beam of electrons impinges upon a thin metal foil, a proportion of the electrons pass through the metal and on striking a photographic film form a series of rings. The patterns observed are the same as those to be expected from an incident pencil of X-rays if the wavelength of the X-rays had a value given by $\lambda = h/mv$, where m is the mass and v the velocity of the electrons in the electron beam. An electron thus behaves as a particle under some circumstances and as a wave packet under others. These experimental results are in conformity with the physical theories of moving electrons which involve wave mechanics.

A metal crystal consists of a regular geometrical arrangement of positive ions and a valency electron will therefore have, not only kinetic energy, but also potential energy in the electrostatic field of the ions. For an electron, moving for example parallel to a row of ions, the electron potential energy will vary in a periodic manner, whereas we have previously assumed the potential energy to be a constant.

The kinetic energy of an electron is $E_K = mv^2/2$ and its momentum $p = mv$. Treated as a wave, the electron wavelength $\lambda = h/mv = h/p$. Whence $E_K = p^2/2m = h^2/2m\lambda^2$. It is usual to define the wave number of an electron as $k = 2\pi/\lambda$ and then the kinetic energy of the electron may be written $E_K = h^2k^2/8\pi^2m$.

In the wave-mechanical theory, the total energy of an electron in a crystal is taken into account, and this includes a potential energy component.

Whereas the relation between the kinetic energy E_K and the wave number k, as given above, is a parabola, it is found that the total energy E of the electron departs from the parabolic form as k increases until, when k reaches a critical value k_c, there is a sudden discontinuity in the relation between E and k. It will, of course, be understood that since the Exclusion Principle is operative there are discrete increments in allowed energy levels, but these increments are so small that the relation between E and k is taken to be a continuous function, but when k reaches the value k_c there is a discontinuous jump in energy of a quite different order of magnitude.

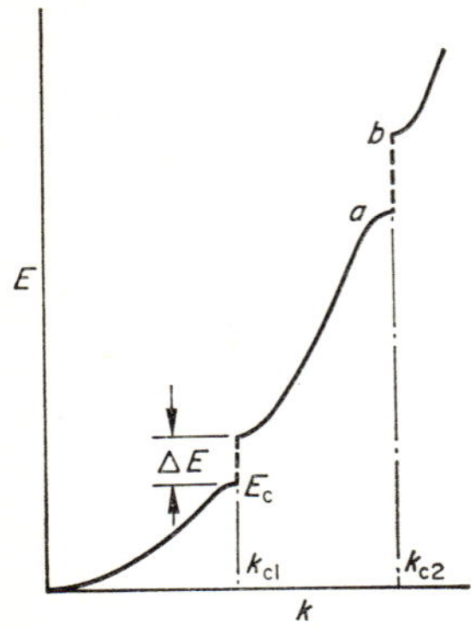

FIG. 3.11. Showing allowed energy bands for electrons moving in the same direction in a crystal.

The wave number k is related to the electron wavelength and velocity as indicated above. The E–k relation, as k approaches the critical value, and the value of k_c itself, however, depend also on the direction of motion of the electron with respect to the crystal axes as will be described, but Fig. 3.11 shows a relation between E and k for electrons all moving in the same direction in a crystal. When k reaches the value denoted by k_{c1} in the figure the energy $E = E_c$. There is a sudden jump to $E = E_c + \Delta E$ at this point following which the smooth relation between E and k is resumed. There are no allowed energy levels which may be occupied by electrons in the energy gap from E_c to $E_c + \Delta E$, or again in the gap denoted by ab in the figure, where a second critical value k_{c2} occurs in the wave number in this example.

Since electrons in a crystal are here considered as wave packets, with an assigned wavelength dependent on the energy, their behaviour may once again be regarded as analogous to that of X-rays in crystallography. Figure 3.12 shows an X-ray beam impinging on parallel planes of atoms in a crystal of simple cubic structure. The waves are diffracted from the atoms in all directions and interfere with or reinforce one another depending upon the wavelength, the atom spacing and the angle of incidence of the beam. The rays are reinforced and reflected, after penetrating only a small distance into the crystal, if the Bragg relation $n'\lambda = 2a \cos \theta$ is satisfied, where n' is the order of the reflection, a is the distance between the parallel planes of atoms in the crystal and θ is the angle between the incident beam and a normal

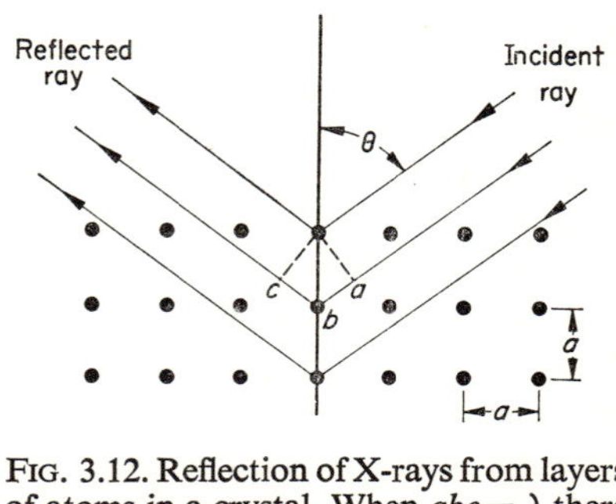

FIG. 3.12. Reflection of X-rays from layers of atoms in a crystal. When $abc = \lambda$ there is a first-order reflection.

to the planes. We need consider only a first-order reflection, that is $n' = 1$, and then $k = 2\pi/\lambda = \pi/a \cos\theta$.

For moving electrons, considered as waves inside the crystal, Bragg reflections will similarly occur for a particular direction θ when k has the critical value $k_c = \pi/a \cos\theta$. When the value k_c is reached, and the energy is E_c, the electron can no longer move freely through the crystal in the direction θ but may be considered to suffer repeated internal reflections.

In Fig. 3.13(*a*) we view, in two dimensions, a crystal of simple cubic structure with the coordinate axes *OY*, *OZ* parallel to cube edges as shown. The coordinate axes are reproduced in Fig. 3.13(*b*) but now distances measured along them in this figure represent wave number k. Thus the length OP' corresponds with the wave number of an electron and the direction OP'

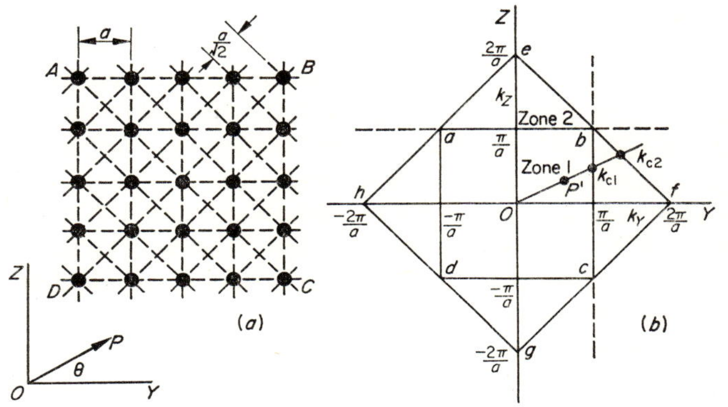

FIG. 3.13. Outlines of the first and second Brillouin Zones in one plane in a crystal of simple cubic structure. The wave number of an electron moving in the direction *OP* in real space may be represented in magnitude and direction by OP' in k space.

with the direction of motion in relation to the crystal axes. For electrons moving along *OP* in Fig. 3.13(*a*) at an angle θ to *OY*, reflection from atom planes normal to the paper and parallel to *BC* will occur for a critical value of the wave number $k_c = \pi/a \cos\theta$. Thus when $\theta = 0°$, $k_c = \pi/a$ and when $\theta = 45°$, $k_c = \sqrt{2}\pi/a$. In fact the locus of k_c for these planes is the straight line *bc* produced in both directions. Similarly when reflections from planes parallel to *AB* are considered θ must now be taken in relation to the direction *OZ*, which is normal to these particular planes, and the locus of k_c is then the straight line *ab* in the figure. When the directions of motion of the electrons are reversed then the loci are the lines *ad* and *cd*. Thus the square *abcd*, intersecting the axes at $\pm\pi/a$, gives the critical values of the wave number k_c for any direction of motion of an electron in planes parallel to *YOZ*.

If, on the other hand, planes of atoms perpendicular to the paper and parallel to the diagonals *AC* or *BD* had been considered, then critical values of k_c would be determined from $k_c = \pi/a' \cos\theta'$ where now a' is the distance between the planes and θ' is the angle between the direction of motion of an electron and a normal to the planes. In this case $a' = a/2^{\frac{1}{2}}$. For the OY or OZ directions $\theta' = 45°$ and then $k_c = 2\pi/a$. For electrons moving normally to the planes, that is for $\theta' = 0°$, $k_c = \sqrt{2}\pi/a$. The locus of k_c for these planes is therefore the square *efgh* in Fig. 3.13(*b*).

The *E–k* relation shown in Fig. 3.11 may be considered to correspond to some direction in Fig. 3.13(*b*) such as *OP'*, and the critical wave number values k_{c1} and k_{c2} are shown in both figures.

If the analysis is extended into three dimensions in space and we take, for example, electrons moving in a direction equally inclined to all three coordinate axes *OX*, *OY*, *OZ*, then for planes such as *AB* or *BC*, $\cos\theta = 1/3^{\frac{1}{2}}$ and $k_c = \sqrt{3}\pi/a$. If planes parallel to *AC* or *BD* are considered then $\cos\theta' = \sqrt{\frac{2}{3}}$, $a' = a/2^{\frac{1}{2}}$ and again $k_c = \sqrt{3}\pi/a$.

It will be clear that, in three dimensions, the values of k_c for planes parallel to cube faces of the crystal fall on the six surfaces of a cube of side $2\pi/a$, the cube occupying a volume in what is called *k* space. For the second set of planes considered, the values of k_c fall on the triangular surfaces of six pyramids of vertical height π/a based on the first cube.

These volumes in *k* space are known as Brillouin Zones. It must be understood that they do not represent volumes in real space. Directions from the origin in *k* space correspond to directions of motion of an electron in the real crystal while the distance from the origin represents the wave number of the electron. Diagrams in *k* space may be compared with the momentum-space diagram in Fig. 3.9 in which momentum *p* was plotted instead of wave number *k*.

Zones 1 and 2 for the simple cubic crystal structure are shown in Fig. 3.14. Brillouin zones may similarly be constructed for face-centred cubic, hexagonal or body-centred cubic structures, although these are rather more complicated in shape than those shown for the simple cubic structure.

Further zones are produced for any crystal structure by considering other sets of planes of atoms, and it turns out that successive zones are each of

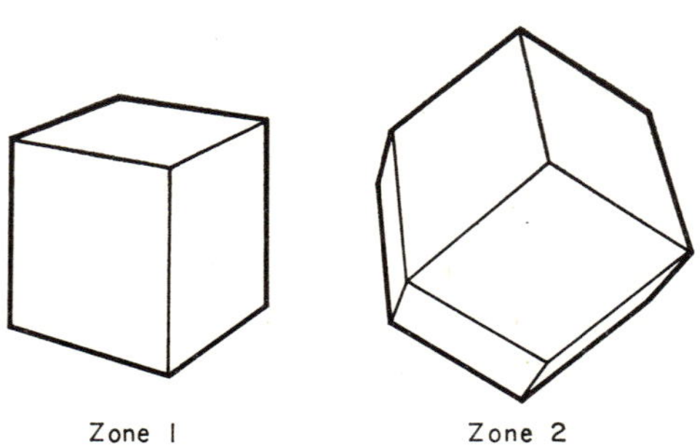

FIG. 3.14. Volumes in *k* space of the first and second Brillouin Zones for a crystal of simple cubic structure.

the same volume in k space. However, usually only the first two zones are required to be known.

We may now consider the energy bands in a crystal which correspond with the Brillouin zones. For the first zone in Fig. 3.14 the critical value of the wave number for electrons moving in any cube-edge direction is $k_c = \pi/a$, and for the two diagonal directions previously considered k_c has the values $\sqrt{2}\pi/a$ and $\sqrt{3}\pi/a$, respectively. The E–k relation for electrons moving, respectively, in these three directions may therefore be represented by the curves 0a, 0b and 0c in Fig. 3.15(a). Electrons having wave number values less than π/a may move in any direction in the crystal. The number of electrons dn per unit volume in an energy increment dE, between E and $E + dE$, may be plotted as dn/dE in relation to E as shown in Fig. 3.15(b). The lower part of the curve 0a' is parabolic and is the same as that plotted in Fig. 3.10 for the free-electron theory. However, when k approaches the first critical

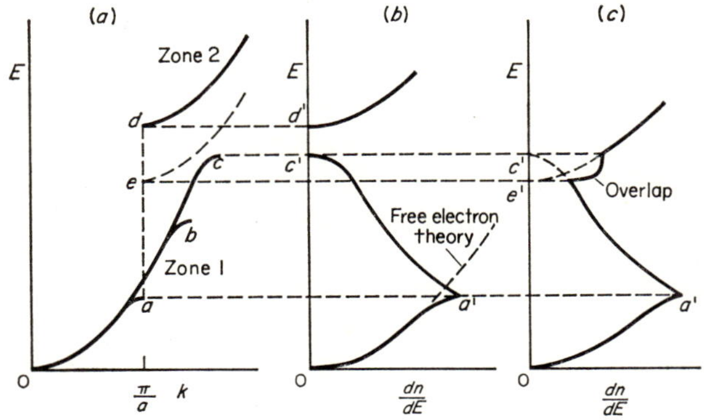

FIG. 3.15. Relation between energy-wave number curves and energy-electron distribution bands in a crystal.

value π/a, the theory shows that the allowed energy levels tend to crowd together causing the upper part of the curve 0a' to depart from the parabolic form as shown. When the points a and a' are reached further electrons may no longer move in the cube-edge directions. Electrons may still move in other directions but the number of allowed levels falls rapidly with increasing k and increasing E. The last allowed levels are for electrons moving in long diagonal directions and these vanish, in this example, when $k = \sqrt{3}\pi/a$ at the points c and c'. The energy band for Zone 1 is therefore 0c' and the area 0$a'c'$ represents the total number of electrons per unit volume to fill the band completely. The number of discrete energy levels in the band is, of course, half the number of electrons required to fill them, since two electrons, of positive and negative spin, respectively, may occupy each level.

A second energy band for Zone 2 may similarly be determined. In Figs. 3.15(a) and (b) the second zone is shown to start at d, and the energy band at d', such that there is an energy gap $c'd'$ between the bands. On the other hand if the E–k relation for Zone 2 had started at the point e in Fig. 3.15(a) and followed the broken curve, the second energy band would have started at e' in Fig. 3.15(c), and the first and second energy bands would then have overlapped as shown.

If the number of valency electrons in the crystal are such as just to fill completely the first band, and an energy gap exists between the bands, then at 0°K the crystal will be an insulator. The number of electrons moving in a particular direction, in which an electric field is applied, cannot be increased if all the allowed energy levels are already full. There can therefore be no conduction current.

If, however, the gap between the energy bands is small, so that with increasing temperature electrons in the upper levels of the lower band acquire sufficient additional energy to jump the gap into the lower levels of the upper band, then the crystal ceases to be a perfect insulator and becomes a semiconductor. Vacant energy levels are now left in the lower or valency band so that a movement of electrons in this band may also occur. It is customary to think of the vacant levels as positive 'holes' which themselves move in the crystal. It may be shown mathematically that these positive holes have a lower mobility due to an apparent mass which is greater than that of the 'free' electrons which have entered the upper or conduction band. This leads, in the field of semiconductors, to interesting consequences which will not be pursued further here.

On the other hand when the levels in the lower band are only partly filled by electrons, or when the bands overlap and the upper band is only partly occupied, the crystal will exhibit the electrical conductivity characteristic of the metals. In this case we are also concerned with the effect of an externally-applied magnetic field. The energy levels which formerly were occupied each by a pair of electrons of opposite spin are now split, as in the Zeeman effect. As a result an excess of positive to negative spins will occur, and the material has a paramagnetic susceptibility, which disappears when the magnetic field is removed. In ferromagnetism an excess of positive to negative spins occurs even in the absence of an external magnetic field. These phenomena will be discussed further in the following chapters, although it may be mentioned here that rigid calculations of the effects in actual crystals do not appear to be possible because of the mathematical complexities of the problems involved.

PROBLEMS

3.1. Calculate the packing factor for an arrangement of hard spheres in the following configurations:

(a) simple cubic, (b) body-centred cubic, (c) face-centred cubic, (d) close-packed hexagonal.

3.2. Determine the crystal directions of closest atomic packing in each of the structures of Problem 3.1.

SUGGESTED READING

HUME-ROTHERY, W. *Atomic Theory for Students of Metallurgy*, 3rd edn., Institute of Metals, London (1960).

RAYNOR, G. V. *Introduction to the Electron Theory of Metals*, Institute of Metals, London (1947).

Chapter 4

DIAMAGNETISM AND PARAMAGNETISM

4.1 Diamagnetism

It was shown in Chapter 2 that, in the electronic structure of an atom, any completely filled sub-shell has a resultant magnetic moment which is zero. If all the sub-shells in an atom are full, as for example in the inert gases such as helium, neon or argon, the atomic magnetic moment is also zero. However, in any atom, whether the sub-shells are full or not, there will be a small induced magnetic moment when a magnetic field is applied, because each electron orbit in the atom will precess about the direction of the applied field. This precession, the Larmor precession, has been discussed in Section 2.4 and it was shown there that an orbit will precess in a field H with an angular velocity $\omega_L = H\mu/p = H\mu_0 e/2m$, where μ and p are, respectively, the magnetic moment and angular momentum due to the orbital motion of the electron. ω_L is dependent only on the magnetic field strength H and on the mass and charge of the electron, and so is independent of the radius of the orbit and of the angle between the direction of H and the plane of the orbit. The precessional component ω_L of the rotation of the electron, producing in effect a circuital current about the axis of the precession, is also for any orbit, always in a direction to produce a component of induced magnetic moment in opposition to H. Since every extra-nuclear electron in a substance is acted upon in this way, the effect exists in all materials and is known as diamagnetism. When considered in relation to ferromagnetism the effect is comparatively very feeble. However, if $-M$ is the induced intensity of magnetization, that is the induced magnetic moment per unit volume, in a material when the applied field is H then the absolute diamagnetic susceptibility $\kappa_D = -M/H$.

4.1.1 *Diamagnetism due to a Single Orbit*

Consider first the diamagnetism associated with the single circular orbit of radius r precessing, in Fig. 4.1(*a*), in a field H directed along the axis OZ. The angular momentum of the electron, moving with angular velocity ω_0 in the direction shown, may be represented by the vector p normal to the plane of the orbit and rotating around OZ on the surface of a cone of semi-vertical angle θ' with a precessional angular velocity ω_L. The orbital magnetic moment will then be represented as shown in Fig. 4.1(*b*) by a precessing vector μ in the opposite direction to p, that is at an angle to the direction of H of $\theta = \pi - \theta'$.

Consider a moment in time when a diameter of the orbit is along the axis XOX', and the radius OA perpendicular to XOX' is in the plane YOZ: then $\angle AOY = \theta'$. Let P be a point on the orbit having coordinates x, y, z such that $\angle AOP = \alpha$. Then it is easy to show that $x = r \sin \alpha$, $y = r \cos \theta' \cos \alpha$ and $z = r \sin \theta' \cos \alpha$. As the orbit precesses P will rotate about OZ at a distance r_1 where $r_1^2 = x^2 + y^2 = r^2(\sin^2 \alpha + \cos^2 \theta' \cos^2 \alpha)$. The angular velocity ω_0 is high compared with ω_L and we may consider the orbit as a ring of charge.

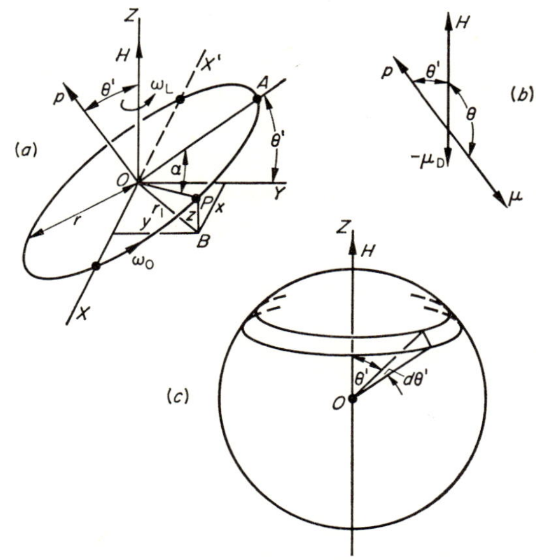

FIG. 4.1. Illustrating the classical theory of diamagnetism.

The element of charge associated with a length $r\,d\alpha$ of the orbit at P will be $e\,d\alpha/2\pi$. This element rotates about OZ with angular velocity ω_L in a circle of radius r_1 and so corresponds to a current $di = (1/T_L)(e/2\pi)d\alpha$ where $1/T_L = f_L = \omega_L/2\pi \; \therefore di = (\omega_L/4\pi^2)e\,d\alpha$. The magnetic moment due to the current di in a circuit of area πr_1^2 is $d\mu_D = -\mu_0 \pi r_1^2\, di$ where the minus sign means that the moment is opposite to H in direction. Substituting for r_1 and di and putting $\omega_L = H\mu_0 e/2m$ we have

$$d\mu_D = -(\mu_0^2/4\pi^2)(e^2r^2H/2m)(\sin^2 \alpha + \cos^2 \theta' \cos^2 \alpha)d\alpha$$

The total magnetic moment due to the precession of the whole orbit is $\mu_D = \int_{\alpha=0}^{\alpha=2\pi} d\mu_D$. Carrying out the integration and putting in the limits gives

$$\mu_D = -(\mu_0^2 e^2 r^2 H/4m)(\cos^2 \theta' + \sin^2 \theta'/2) \tag{4.1}$$

The potential energy due to the orbital moment μ in the field H is $-H\mu \cos\theta$. Since μ_D increases linearly as H increases the potential energy resulting from the diamagnetic moment μ_D in opposition to the field is the positive quantity $H\mu_D/2$. The potential energy due to the diamagnetism is therefore

$$W_D = (\mu_0^2 e^2 r^2 H^2/8m)(\cos^2\theta' + \sin^2\theta'/2) \qquad (4.2)$$

In Section 2.4 on the Larmor precession it was shown that the total kinetic energy of the precessing electron orbit was

$$W_T = (I/2)(\omega_0 + \omega_L \cos\theta')^2 + (I/4)(\omega_L \sin\theta')^2$$

where $I = mr^2$. The kinetic energy associated directly with the precession, that is $W_T - I\omega_0^2/2$, is then

$$W_P = I\omega_0\omega_L \cos\theta' + I(\omega_L \cos\theta')^2/2 + I(\omega_L \sin\theta')^2/4$$

It was shown in Section 2.4 that, in this equation, the first and much the largest term $I\omega_0\omega_L \cos\theta'$ is equal to $-H\mu\cos\theta$, the potential energy due to the orbital moment μ in the field H, and it is now clear that on substituting for I and ω_L in the second two terms, the sum of these two components of kinetic energy is exactly equal to the potential energy W_D in equation (4.2). On this basis then we may say that the magnetic potential energy associated with μ and μ_D in the field H is stored in the system in the form of the precessional components of kinetic energy.

4.1.2 *Diamagnetism with Many Orbits*

The diamagnetic moment due to the precession of a single orbit of radius r, in which the momentum vector is inclined at an angle θ' to the direction of the applied field, is given in equation (4.1). We now wish to consider the total diamagnetic moment when a very large number of orbits are involved of different radii and covering uniformly all values of θ'.

We first think of orbits of radius r and all values of θ'. The tips of the angular momentum vectors, in this case, may then be considered to be uniformly distributed on the surface of a sphere of unit radius, corresponding to a solid angle 4π, as in Fig. 4.1(*c*), and all will precess about OZ which is the direction of H. If there are b vectors per unit area of the surface of the sphere, that is per unit solid angle, the number of vectors in the increment $d\theta'$ in the figure is $2\pi b \sin\theta'\, d\theta'$ and the diamagnetic moment due to each of these orbits is given by μ_D in equation (4.1). The total number of orbits is $4\pi b$ and so the average value of the orbital magnetic moments for the whole sphere is given by

$$\mu_{av} = -(\mu_0^2 e^2 r^2 H/4m)\left\{\int_0^\pi 2\pi b \sin\theta' (\cos^2\theta' + \sin^2\theta'/2)d\theta'\right\}/(4\pi b)$$

$$= -(\mu_0^2 e^2 r^2 H/4m)(2/3)$$

Whence $\mu_{av} = -\mu_0^2 e^2 r^2 H/6m$ (4.3)

If now there are further orbits of other values of r then the average value for all the orbits is obtained from equation (4.3) by substituting the root mean square value $\bar{r}$ for all orbits in place of r.

For a material in bulk containing n atoms per unit volume each atom having Z orbits, where Z is the atomic number of the atom, then the diamagnetic moment M per unit volume, divided by the field strength H, or in other words the absolute diamagnetic susceptibility, is given by

$$\kappa_{\mathrm{D}} = M/H = -nZ(\mu_0 e\bar{r})^2/6m \tag{4.4}$$

Equation (4.4) is the classical formula for the diamagnetic volume susceptibility. The same formula is, however, obtainable in quantum theory in which orbital motions are no longer envisaged. The quantity $\bar{r}$ now has a different significance and is dependent on the charge distribution within the atom. When many electrons are present, calculation of $\bar{r}$ is a matter of great difficulty in quantum mechanics and involves the use of approximate solutions.

The subject of diamagnetism will not be pursued further except to say again that the small, negative diamagnetic effect is present in all materials even though it may be overwhelmed by positive paramagnetic and ferromagnetic effects.

4.2 Paramagnetism

Diamagnetic substances are those in which the atoms or molecules have a zero magnetic moment. They are characterized by a negative susceptibility, or a relative magnetic permeability very slightly less than unity. The phenomenon of paramagnetism, on the other hand, occurs in materials where the atoms or molecules have a magnetic moment and, in this case, there is a positive susceptibility, or a relative permeability slightly greater than unity.

When a magnetic field H is applied to a paramagnetic substance each elementary magnet will acquire potential energy $-H\mu \cos \theta$ where μ is its magnetic moment and θ the angle between the magnetic axis and the direction of H. It will, however, be remembered that the effect of the field is to produce a precession and not, of itself, a change in the angle θ, but the potential energy of the system in the field will decrease if, by some means, the magnetic axes turn into line with H. However, the particles also have thermal motions and the thermal kinetic energies are, except at very low temperatures, very much greater than the magnetic potential energies for any practical value of H. Thus any lining-up of the axes in the presence of the field, which may in fact occur as a result of collisions, is very strongly conditioned by the comparatively violent thermal agitation.

The problem of finding the resultant intensity of magnetization in a given field is a statistical one, and the first satisfactory quantitative treatment of it was due to Langevin [1].

4.2.1 *Classical Theory of Paramagnetism*

We consider a perfect gas at temperature T consisting of molecules of mass

m and magnetic moment μ. It is assumed that there is no magnetic interaction between the particles.

In the presence of a field H the molecules have thermal energy and magnetic potential energy determined by the value of H and the angle of inclination θ of the magnetic axes to the direction of the field.

The case is analogous to that of a gas in a long vertical column where the molecules have thermal energy with potential energy due to their weight and height above a given datum plane in a gravitational field. In the column the density of the gas will be greatest at the bottom where the potential energy of the gas particles is least. The particles are thus not uniformly distributed up

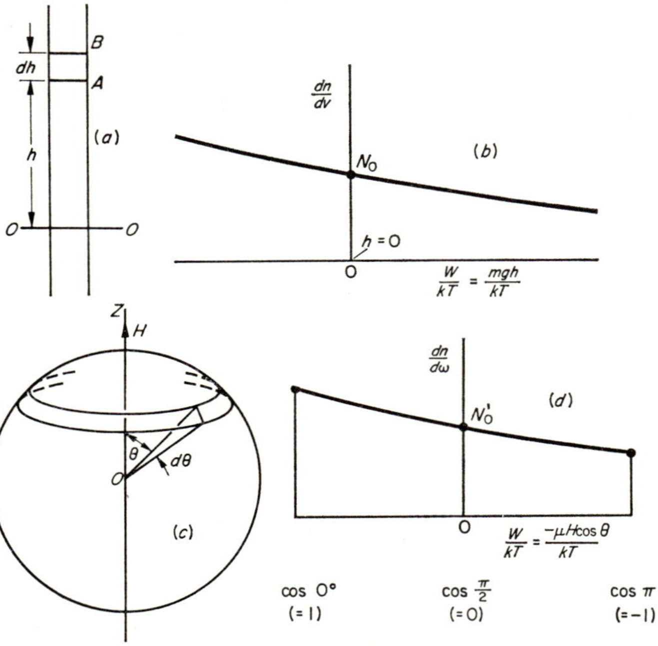

FIG. 4.2. Classical theory of paramagnetism.

the column but will tend to concentrate in positions of lower potential energy. Similarly in the paramagnetic gas in a magnetic field the molecular magnets will tend to take up angular positions of lower magnetic potential energy, and this involves a turning of the magnets to give a resultant component of magnetization in the field direction. The problem of the column is readily solved from first principles, by using the laws of a perfect gas, and we shall do this first and then apply the result to the magnetic case.

Figure 4.2(a) represents a very long vertical gas column of unit cross-section. In the earth's atmosphere the temperature of the gas would vary with height but in the present problem we suppose the gas in the column is everywhere at temperature T. Let the gas pressure at A at a height h above a datum level O be P. Then at A for a perfect gas $PV = RT$ where V is the volume per gramme mole of the gas and $R = 8{\cdot}317$ J g mole^{-1} °K^{-1} is the universal constant. Now the number of molecules in a gramme mole is Avogadro's number $N = 6{\cdot}0249 \times 10^{23}$ and $R/N = k$ where $k = 1{\cdot}3804 \times 10^{-23}$ J/°K is Boltzmann's constant. Whence $P = NkT/V = N_1kT$ where N_1 is the number of molecules per unit volume of the gas at A. For an increment of height dh, corresponding to point B, the increment of pressure is

$$-dP = -kT\,dN_1$$

and the difference in pressure between A and B is due to the weight of the molecules in a volume dh. Therefore if the molecules are of mass m, and g is the gravitational constant

$$-kT\,dN_1 = N_1 mg\,dh$$

The negative sign appears because the number of molecules N_1 per unit volume decreases as h increases.

Whence $$dN_1/N_1 = -(mg/kT)dh$$

or $$\log N_1 = -mgh/kT + A$$

If, for the datum position where $h = 0$, $N_1 = N_0$ the constant $A = \log N_0$

and $$N_1/N_0 = \exp(-mgh/kT)$$

or $$N_1/N_0 = \exp(-W/kT)$$

where $W = mgh$ is the potential energy of a molecule at height h.

If the number of molecules in a volume dv at height h is dn then $N_1 = dn/dv$ and

$$dn/dv = N_0 \exp(-W/kT) \tag{4.5}$$

where N_0 is the number of molecules per unit volume at the datum position where $h = 0$.

Figure 4.2(b) shows the relation between dn/dv, which is proportional to the density of the gas, and the potential energy of the molecules. The density increases as the potential energy, or height above the datum position, decreases, while the number of molecules of potential energy W is directly proportional to $\exp(-W/kT)$.

Reverting now to a given volume of a paramagnetic gas having a total of n molecules in a steady field H at temperature T, the potential energy of each molecule in the field is $W = -\mu H \cos\theta$. The potential energy of the particles depends on θ, the angular position of the axes, instead of upon the height above a datum plane in the gravitational case.

Referring to Fig. 4.2(c), let there be dn molecules whose axes are contained within a solid angle $d\omega$ in an element $d\theta$ at an angle θ to H. Thus $dn/d\omega$ is the number of molecules per unit solid angle which have potential energy $-\mu H \cos\theta$ and equation (4.5) becomes

$$dn/d\omega = N_0^1 \exp(-W/kT)$$

where $W = -\mu H \cos\theta$ and N_0^1 is the number of molecules per unit solid angle having energy $W = 0$, that is when $\cos\theta = 0$ or $\theta = \pi/2$.

The relation between the number of molecules per unit solid angle $dn/d\omega$ and the potential energy W, or the value of $\cos\theta$, is shown in Fig. 4.2(d) which may be compared with Fig. 4.2(b).

Putting $\mu H/kT = a$, and $d\omega = 2\pi \sin\theta\, d\theta$ then

$$dn = N_0' e^{a\cos\theta}\, 2\pi \sin\theta\, d\theta$$

The total number of molecules in the entire gas is therefore

$$\begin{aligned} n &= 2\pi N_0' \int_0^\pi e^{a\cos\theta} \sin\theta\, d\theta \\ &= (4\pi N_0'/a)\sinh a \end{aligned}$$

Now the element of magnetic moment, resolved in the direction of H, due o the molecular magnets pointing inside the solid angle $d\omega$ is

$$\begin{aligned} dM' &= \mu \cos\theta\, dn \\ &= 2\pi N_0' \mu\, e^{a\cos\theta} \sin\theta \cos\theta\, d\theta \end{aligned}$$

∴ the total magnetic moment

$$\begin{aligned} M' &= 2\pi N_0' \mu \int_0^\pi e^{a\cos\theta} \sin\theta \cos\theta\, d\theta \\ &= (4\pi N_0' \mu/a^2)(a\cosh a - \sinh a) \end{aligned}$$

∴ the mean magnetic moment per molecule in the direction of H is

$$\bar{\mu} = M'/n = \mu(\coth a - 1/a)$$

or

$$\bar{\mu}/\mu = \coth a - 1/a$$

This may be written, for a constant volume of gas, as

$$M_T/M_0 = \coth a - 1/a = L(a) \tag{4.6}$$

where M_T is the intensity of magnetization in the field H at temperature T, and M_0 is the saturation value of M_T.

The quantity $\coth a - 1/a = L(a)$ is known as the Langevin function. This function is the classical solution which assumes that the magnetic axes of the molecular magnets may take any angular relation with respect to the direction of the magnetic field. However, if the quantum restriction is imposed that there can be only certain discrete orientations of the magnetic axes then another formula is obtained. The quantum-mechanical solution is due to Brillouin [2], and the Langevin function is a special case of it.

4.2.2 *Quantum Theory of Paramagnetism*

We consider a gas of atoms which each have a magnetic moment and to which a steady magnetic field H is applied. The atoms are supposed to be in their normal or ground state and may be specified by quantum numbers J, L, S as described in Sections 2.7 and 2.8. The magnetic moment of each atom is now $\mu_J = gJ^*\mu_B$ where g is the Landé factor as given in equation (2.26). In the presence of a field H of any practical amount the axes of the magnetic moments μ_J will precess about the field direction such that the resolved components of μ_J in the field direction are $gM\mu_B$, where M is the magnetic quantum number having values $M = 0, \pm 1 \ldots \pm J$ if J is whole number, and $M = \pm 1/2, \pm 3/2 \ldots \pm J$ if J is half-integer.

If, for example, the normal state of the atom is represented, as discussed in Section 2.6, by the notation $^2P_{\frac{3}{2}}$ the quantum numbers are $J = 3/2$, $L = 1$, $S = 1/2$. Then $J^* = (1/2)\sqrt{15}$, $g = 4/3$ and $M = \pm 1/2$ and $\pm 3/2$. In this case $\mu_J = (2/3)\sqrt{15}\mu_B$ and its resolved values in the field direction are $gM\mu_B = \pm(2/3)\mu_B$ and $\pm 2\mu_B$. This particular example is illustrated in

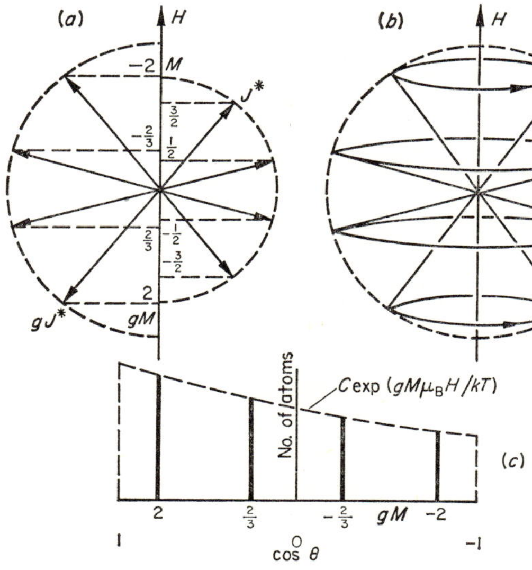

FIG. 4.3. Paramagnetism for atoms of normal state $^2P_{\frac{3}{2}}$.

Fig. 4.3(*a*). It will be remembered that the vectors representing magnetic moment, that is $gJ^*\mu_B$ and $gM\mu_B$ are opposite in direction from those for the angular momentum of the atom derived from J^* and M. Figure 4.3(*b*) represents the conditions in the paramagnetic gas when the quantum restrictions are applied. The vectors for the magnetic moments of the atoms will precess about the H direction. In this particular case the vectors will be in four groups, each group precessing on the surface of a cone as indicated in the figure. To find the resultant intensity of magnetization in the field direction it then becomes necessary to determine the fraction of the total number of atoms in each group.

Comparison of the paramagnetic gas in a field H with the column of gas in a gravitational field shows that there will be equilibrium between the thermal and potential energies in either case when, as given in equation (4.5), the number of atoms with potential energy W is proportional to $\exp(-W/kT)$. In the paramagnetic gas $W = -gM\mu_B H$ so that

$$\exp(-W/kT) = \exp(gM\mu_B H/kT) = e^{Mx}$$

where, for convenience, we put $x = g\mu_B H/kT$ and M has a series of values from $+J$ to $-J$ as mentioned in the preceding paragraph.

In the particular case shown in Fig. 4.3 gM has four allowed values and the number of atoms in each state are represented by the thick vertical lines in Fig. 4.3(*c*).

The total number of atoms in the gas is given by $n = \Sigma\, Ce^{Mx}$ where C is a constant, and the total magnetic moment M' in the field direction is given by $M' = \Sigma\, gM\mu_B Ce^{Mx}$. Bearing in mind the precessional motions the magnetic moment perpendicular to the field direction is zero.

The mean magnetic moment per atom $\bar{\mu}$ is therefore

$$\bar{\mu} = M'/n = g\mu_B\, \Sigma\, Me^{Mx}/\Sigma\, e^{Mx}$$

Now $\quad d(\log \Sigma\, e^{Mx})/dx = \Sigma\, Me^{Mx}/\Sigma\, e^{Mx}$

$$\therefore\ \bar{\mu} = g\mu_B d(\log \Sigma\, e^{Mx})/dx$$

But

$$\begin{aligned}\Sigma\, e^{Mx} &= e^{Jx} + e^{(J-1)x} + \ldots + e^{-Jx}\\ &= (e^{(J+1)x} - e^{-Jx})/(e^x - 1)\\ &= \sinh(J + \tfrac{1}{2})x/\sinh x/2\end{aligned}$$

Differentiating this expression we have

$$d(\log \Sigma\, e^{Mx})/dx = \left(\frac{2J+1}{2}\right)\coth\left(\frac{2J+1}{2}\right)x - \frac{1}{2}\coth(x/2)$$

Put $a = Jx$, and then

$$\bar{\mu} = gJ\mu_B\left\{\left(\frac{2J+1}{2J}\right)\coth\left(\frac{2J+1}{2J}\right)a - (1/2J)\coth(a/2J)\right\} \qquad (4.7)$$

where $a = Jx = gJ\mu_B H/kT$

The quantity

$$B(a) = \left(\frac{2J+1}{2J}\right)\coth\left(\frac{2J+1}{2J}\right)a - (1/2J)\coth(a/2J) \qquad (4.8)$$

is called the Brillouin function

$$\therefore\ \bar{\mu} = gJ\mu_B B(a)$$

Under saturation conditions when $H \to \infty$, or $T \to 0, a \to \infty$. Then we may put $\coth a = 1$, $B(a) = 1$ and $\bar{\mu} = gJ\mu_B$. This corresponds with the condition where $M = J$ and all the atomic magnets have taken up the allowed angular position nearest to the field direction. We may thus write for a constant volume

$$M_T/M_0 = B(a) \qquad (4.9)$$

where M_T is the intensity of magnetization in the field H at temperature T and M_0 is the saturation value.

In the classical theory the atomic magnetic axes may assume any angular position with respect to the field. This would correspond in the quantum case with the hypothetical condition of J very large and with $2J + 1$ allowed positions. If $J = \infty$ equation (4.8) becomes $B(a) = \coth a - 1/a$ which is the Langevin function $L(a)$ given in equation (4.6).

If $J = 0$ then $B(a) = 0$, as may be expected since the gas atoms would then have zero magnetic moment and the gas would not be paramagnetic, but a limiting case is when $J = 1/2$. In this case the number of allowed positions for the magnetic axes is $2J + 1 = 2$, and $M = \pm 1/2$. Two angular positions only of the magnetic axes would be allowed in this event with resolved moments in the field direction of $gM\mu_B = \pm(g/2)\mu_B$. A particular example of $J = 1/2$ would be for an atom having a single uncompensated s electron, and then $L = 0$, $S = 1/2$ and $g = 2$. The carriers of the magnetic moment in the atoms would be single spinning electrons which could precess only in parallel or antiparallel positions so that the resolved moments were $\pm 1\mu_B$. This condition is of particular interest in relation to ferro-, ferri- and antiferromagnetism and will be considered again later.

If $J = 1/2$

$$B(a) = 2 \coth 2a - \coth a = \tanh a$$

$$\therefore\ M_T/M_0 = \tanh a \tag{4.10}$$

The relations between M_T/M_0 and a, given generally in equation (4.9),

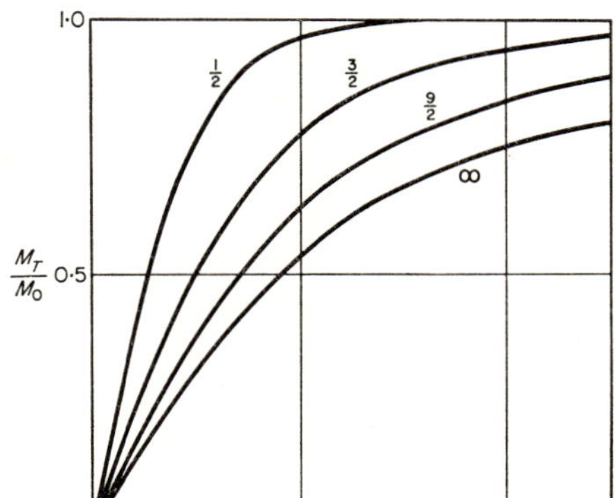

FIG. 4.4. Brillouin function for $J = 1/2, 3/2, 9/2$ and ∞.

are plotted in Fig. 4.4 for several values of J and, since $a = gJ\mu_B H/kT$, give in quantitative terms the theoretical magnetization curves of paramagnetic gases.

To establish the general order of magnitude of the field H in these curves take, for example, the case of $J = 3/2$, $g = 4/3$. To reach 70% of saturation in this case the value of a is 1·65. Whence at 20°C,

$$\begin{aligned} H &= akT/gJ\mu_B \\ &= 1{\cdot}65 \times 1{\cdot}38 \times 10^{-23} \times 293/2 \times 1{\cdot}165 \times 10^{-29}\ \text{A/m} \\ &= 2{\cdot}86 \times 10^6\ \text{A/cm} \end{aligned}$$

Such high values of H are unattainable in the laboratory and therefore only a small part of the magnetization curves near the origin can be investigated at room temperatures. However, for temperatures within a few degrees of absolute zero, much lower values of H are required to reach saturation, and under these conditions the whole of the magnetization curves has been determined on some materials as mentioned in Section 4.2.5, and shown in Fig. 4.7.

It may also be noted here that a sample of soft polycrystalline iron may be brought to 70% of saturation with a field H of only about 3A/cm, which is less than that calculated above for a paramagnetic by a factor of 10^6. There are obviously important factors in ferromagnetism which are additional to any so far considered. However, clues to these factors are provided when the foregoing theory, derived for gases on the assumption that there are no magnetic or electric interactions between the atoms or molecules, is applied to experimental measurements of the susceptibility of paramagnetic crystalline solids. Some of the conclusions reached are described in the next section.

4.2.3 *Susceptibility of Paramagnetic Materials*

When a is small it is easy to show that equation (4.9) becomes

$$M_T/M_0 = B(a) = (J + 1)a/3J$$

Putting $a = gJ\mu_B H/kT$ and $M_0 = ngJ\mu_B$ where n is now the number of atoms per unit volume, the volume susceptibility κ at temperature T for a constant volume of gas becomes

$$\kappa = M_T/H = ng^2J(J + 1)\mu_B^2/3kT$$

Now the magnetic moment per atom $\mu_J = gJ^*\mu_B = g\{J(J + 1)\}^{\frac{1}{2}}\mu_B = p\mu_B$ where in this case

$$p^2 = g^2J(J + 1) \tag{4.11}$$

and p is the number of Bohr magnetons per atom.

$$\therefore\ \kappa = np^2\mu_B^2/3kT = C'/T \tag{4.12}$$

If this formula is applied to a solid subject to change of volume with temperature it is better expressed in the form

$$\chi_A = Np^2\mu_B^2/3kT = C/T \tag{4.13}$$

where N is Avogadro's number and χ_A is the atomic mass susceptibility, that is, the magnetic moment divided by H for a mass of the substance corresponding to the atomic weight.

According to equations (4.12) and (4.13) the susceptibility is inversely proportional to the absolute temperature. This relation is found to occur for some solids and for oxygen gas and is known as CURIE's Law [3].

Many ionic crystals, however, follow the Curie–Weiss Law where

$$\chi_A = C/(T - \theta_W) \tag{4.14}$$

where C and θ_W are constants and where θ_W is sometimes negative. This, of course, reduces to Curie's Law when $\theta_W = 0$.

An explanation of the Curie–Weiss Law was proposed by WEISS [4], who suggested that in a crystal an aligning interaction between neighbouring ions, neglected in equations (4.12) and (4.13), might exist and be superimposed upon the applied field H. This interaction was assumed to be equivalent to a magnetic field of amount $N_W M_T$ where N_W was the Weiss 'molecular field' constant. The resultant field in the material when H was applied then amounted to $H + N_W M_T$.

In the absence of the Weiss field equation (4.12) may be written

$$M_T = (n\mu_J^2/3kT)H$$

If the molecular field exists then we may substitute $H + N_W M_T$ for H and then

$$M_T = (n\mu_J^2/3kT)(H + N_W M_T)$$

Whence on solving for M_T and dividing by H the volume susceptibility becomes

$$\kappa = M_T/H = n\mu_J^2/3k(T - n\mu_J^2 N_W/3k) = C'/(T - \theta_W)$$

or

$$\chi_A = Np^2\mu_B^2/3k(T - \theta_W) = C/(T - \theta_W) \tag{4.15}$$

This is the Curie–Weiss Law given in equation (4.14) where

$$C = Np^2\mu_B^2/3k \quad \text{and} \quad C' = np^2\mu_B^2/3k$$

and

$$\theta_W = np^2\mu_B^2 N_W/3k = (N_W\rho/A)C = C'N_W$$

where ρ is the density and A the atomic weight.

In equation (4.15) p is the number of Bohr magnetons per ion in the case of an ionic crystal, and from the observed relationship between χ_A and T experimental values of p may be determined. It is of particular interest to find if these results agree with the theoretical values of p given by equation (4.11).

4.2.4 *Susceptibility Measurements on Ionic Crystals*

In the case of the rare earth elements, starting with lanthanum, $Z = 57$, and ending with lutecium, $Z = 71$, and forming ionic crystals with the elements as trivalent ions, the magnetic moments of the ions are due to incompleted $4f$ sub-shells. (See Appendix 1.) The elements are listed in Table 4.1. The number of electrons in the atoms and the ions, respectively, are given in the first two columns of figures, and the number of electrons in the $4f$ sub-shells of the ions are in the next column. The normal state of the ion in each case is given, and from this J, g and p may be calculated.

As an example take Nd^{3+} ($Z = 60$). The spectroscopic state is $^4I_{\frac{9}{2}}$. Therefore $J = 9/2$, and for term I, $L = 6$. From the superscript we have

$2S + 1 = 4$, or $S = 3/2$. Whence $J^* = \{J(J+1)\}^{\frac{1}{2}} = 3\sqrt{11}/2$:

$$L^* = \{L(L+1)\}^{\frac{1}{2}} = \sqrt{42}\colon S^* = \{S(S+1)\}^{\frac{1}{2}} = \sqrt{15}/2$$

From equation (2.26)

$$g = 1 + \frac{(J^*)^2 + (S^*)^2 - (L^*)^2}{2(J^*)^2} = \frac{8}{11}$$

Therefore $$p = g\{J(J+1)\}^{\frac{1}{2}} = (8/11)(3/2)\sqrt{11} = 3{\cdot}62$$

The theoretical values of $p = g\{J(J+1)\}^{\frac{1}{2}}$ are given in the last column of

TABLE 4.1

Element	Number of Electrons in		Ion				
	Atom (Z)	Trivalent Ion	Electrons in 4f Shell	Normal State	J	g	$p=g\{J(J+1)\}^{\frac{1}{2}}$
La	57	54	0	$^1S_0$	0	—	0
Ce	58	55	1	$^2F_{\frac{5}{2}}$	$\frac{5}{2}$	$\frac{6}{7}$	2·54
Pr	59	56	2	$^3H_4$	4	$\frac{4}{5}$	3·58
Nd	60	57	3	$^4I_{\frac{9}{2}}$	$\frac{9}{2}$	$\frac{8}{11}$	3·62
Pm	61	58	4	$^5I_4$	4	$\frac{3}{5}$	2·68
Sm	62	59	5	$^6H_{\frac{5}{2}}$	$\frac{5}{2}$	$\frac{2}{7}$	0·84
Eu	63	60	6	$^7F_0$	0	—	0
Gd	64	61	7	$^8S_{\frac{7}{2}}$	$\frac{7}{2}$	2	7·94
Tb	65	62	8	$^7F_6$	6	$\frac{3}{2}$	9·72
Dy	66	63	9	$^6H_{\frac{15}{2}}$	$\frac{15}{2}$	$\frac{4}{3}$	10·65
Ho	67	64	10	$^5I_8$	8	$\frac{5}{4}$	10·60
Er	68	65	11	$^4I_{\frac{15}{2}}$	$\frac{15}{2}$	$\frac{6}{5}$	9·58
Tm	69	66	12	$^3H_6$	6	$\frac{7}{6}$	7·56
Yb	70	67	13	$^2F_{\frac{7}{2}}$	$\frac{7}{2}$	$\frac{8}{7}$	4·53
Lu	71	68	14	$^1S_0$	0	—	0

Table 4.1, and are plotted in Fig. 4.5 in relation to the number of electrons in the 4f sub-shell of the ion. The short vertical lines show approximately the range of the values obtained by experiment. The agreement between calculation and observation [5] is seen to be satisfactory except in the case of Sm^{3+}, $Z = 62$, and Eu^{3+}, $Z = 63$. VAN VLECK [6] has, however, shown from wave mechanics that another term should be added to that given in equation (4.15), a term which, in fact, becomes significant in these two particular cases and fully accounts for the discrepancy.

The experimental results just given show that the theory of paramagnetism, derived for gases with the assumption of no electric or magnetic interaction between the particles, is also satisfactory for solid crystals in which the magnetic moments are due to rare earth ions. However, similar measurements made on ionic crystals where the moments are due to ions of the iron group

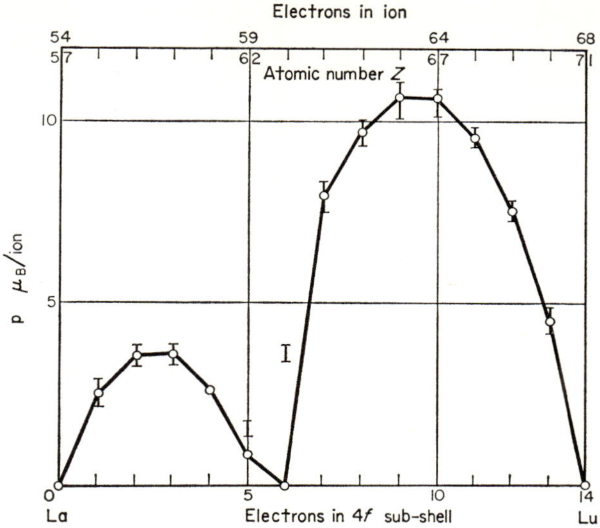

FIG. 4.5. Number of Bohr magnetons per ion for trivalent rare earth ions. ⊙ theoretical, $p = g\{J(J+1)\}^{\frac{1}{2}}$: I range of observed values.

TABLE 4.2

Ion	Number of Electrons in		Normal State	J	L	S	g	$g\{J(J+1)\}^{\frac{1}{2}}$	p Observed	$2\{S(S+1)\}^{\frac{1}{2}}$
	Ion	3d Sub-shell								
$K^+Ca^{2+}Sc^{3+}Ti^{4+}V^{5+}$	18	0	$^1S_0$	0	0	0	—	0	0	0
$Ti^{3+}V^{4+}$	19	1	$^2D_{\frac{3}{2}}$	$\frac{3}{2}$	2	$\frac{1}{2}$	$\frac{4}{5}$	1·55	1·70	1·73
V^{3+}	20	2	$^3F_2$	2	3	1	$\frac{2}{3}$	1·63	2·61	2·83
$V^{2+}Cr^{3+}$	21	3	$^4F_{\frac{3}{2}}$	$\frac{3}{2}$	3	$\frac{3}{2}$	$\frac{2}{5}$	0·77	3·85	3·88
$Cr^{2+}Mn^{3+}$	22	4	$^5D_0$	0	2	2	—	0	4·82	4·90
$Mn^{2+}Fe^{3+}$	23	5	$^6S_{\frac{5}{2}}$	$\frac{5}{2}$	0	$\frac{5}{2}$	2	5·91	5·82	5·91
Fe^{2+}	24	6	$^5D_4$	4	2	2	$\frac{3}{2}$	6·71	5·36	4·90
Co^{2+}	25	7	$^4F_{\frac{9}{2}}$	$\frac{9}{2}$	3	$\frac{3}{2}$	$\frac{4}{3}$	6·63	4·90	3·88
Ni^{2+}	26	8	$^3F_4$	4	3	1	$\frac{5}{4}$	5·59	3·12	2·83
Cu^{2+}	27	9	$^2D_{\frac{5}{2}}$	$\frac{5}{2}$	2	$\frac{1}{2}$	$\frac{6}{5}$	3·55	1·83	1·73
Cu^+Zn^{2+}	28	10	$^1S_0$	0	0	0	—	0	0	0

do not show the same agreement and bring to light an important effect in these crystals.

In Table 4.2 are listed a series of ions starting with monovalent potassium and ending with divalent zinc, covering a range of atomic numbers from $Z = 19$ to $Z = 30$. The series includes the elements iron, cobalt and nickel. The ions may be placed in order according to the total number of electrons in the ion and the number in the 3*d* sub-shell. These numbers may readily be checked by reference to Appendix 1. It is the incompleted 3*d* sub-shell which is responsible in these ions for the magnetic moment. The normal state of the

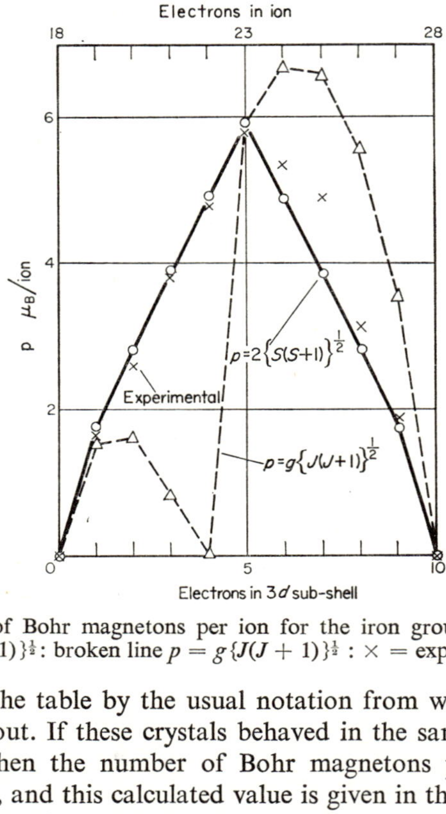

FIG. 4.6. Number of Bohr magnetons per ion for the iron group elements. Full line $p = 2\{S(S+1)\}^{\frac{1}{2}}$: broken line $p = g\{J(J+1)\}^{\frac{1}{2}}$: $\times$ = experimental values.

ions is given in the table by the usual notation from which J, L, S and g may be worked out. If these crystals behaved in the same way as the rare earth elements then the number of Bohr magnetons per ion should be $p = g\{J(J+1)\}^{\frac{1}{2}}$, and this calculated value is given in the table and plotted in Fig. 4.6 together with the observed values of p. It will be seen that there is almost complete disagreement between observation and theory in this case. However, if the orbital quantum number L is put equal to zero for each ion, so that then $J = S$ and $g = 2$, the value of p becomes $p = 2\{S(S+1)\}^{\frac{1}{2}}$. This new value of p is given in the table and in Fig. 4.6 and is in remarkably good agreement, on the whole, with the experimental values.

An explanation of the different behaviour of ions of the iron group from those of the rare earths was put forward by STONER [7]. In the iron group ions the magnetic moments are due to electrons in the incompleted $3d$ sub-shell, which, as reference to Appendix 1 will show, is the outermost sub-shell of the ions. There are in the ionic crystals strong electric fields between neighbouring ions. The interaction of these crystalline fields upon the orbital motions of the $3d$ electrons is very much greater than that of any applied magnetic field. There is no resultant bulk magnetization in any particular direction due to the electric fields and so the orbital moments are said to be 'quenched'. On the other hand the spin moments are not affected by the crystalline fields but react to the applied magnetic field in the Brillouin manner. In this case $J = S$ and there would be $2S + 1$ allowed positions for the axes of the ionic magnetic moments in the field H. However, for ions of the rare earth group the orbital motions are not quenched. Reference to Appendix 1 shows that for these trivalent ions in the range $Z = 57$ to $Z = 71$, the magnetic moments are due to the incompleted $4f$ sub-shell. This is an inner sub-shell of the ions since it is surrounded by complete $5s$ and $5p$ sub-shells. The $4f$ electrons are, for this reason, screened from the effects of the crystalline fields which therefore do not have the same discriminating action upon the orbital motions as in the case of the iron group ions.

4.2.5 *Observed Magnetic Saturation in Paramagnetic Materials*

It was pointed out in Section 4.2.2 that unattainably high magnetic field strengths would be needed to produce saturation in a paramagnetic material at room temperature. However, the Langevin and Brillouin magnetization curves are functions of the quantity H/T. Hence if the temperature is reduced the field strength to produce a given mean intensity of magnetization is also reduced. The whole of the magnetization curve up to saturation has, indeed, been observed for a number of paramagnetic materials when temperatures only a few degrees above absolute zero were employed and in maximum field strengths no greater than about 40,000 A/cm.

HENRY [8] reported results for gadolinium sulphate octahydrate, for ferric ammonium alum and for potassium chromium alum. The carriers of the magnetic moments in these cases are trivalent ions of gadolinium, iron and chromium, respectively. Gd is a rare earth element but for Gd^{3+}, $L = 0$ as may be deduced from the normal state of this ion shown in Table 4.1. Also $L = 0$ for Fe^{3+} as given in Table 4.2. In both cases therefore $J = S$ and $g = 2$, with $J = 7/2$ for Gd^{3+} and $J = 5/2$ for Fe^{3+}, and from equation (4.7), putting $a = (2S\mu_B/k)(H/T)$

$$\bar{\mu}/\mu_B = (2S + 1) \coth \{(2S + 1)\mu_B/k\}(H/T) - \coth (\mu_B/k)(H/T) \tag{4.16}$$

where $\bar{\mu}$ is the mean magnetic moment per ion for given values of H and T.

However, in the case of Cr^{3+}, the quantum numbers in the normal state are, from Table 4.2, $J = 3/2$, $L = 3$, $S = 3/2$ and then $g = 2/5$, but if the orbital motions are quenched we put $L = 0$, and then $J = S = 3/2$ and $g = 2$ once more, and equation (4.16) also applies to this ion.

Equation (4.16) is plotted by the full lines in Fig. 4.7 for the three cases of $J = S = 7/2$, $5/2$ and $3/2$. For the Cr^{3+} ion with quenching equation (4.16) becomes

$$\bar{\mu}/\mu_B = 4 \coth (0{\cdot}338H/10^5T) - \coth (0{\cdot}0845H/10^5T)$$

and is given by curve (*c*). Without quenching, and using equation (4.7), the expression is

$$\bar{\mu}/\mu_B = (4/5) \coth (0{\cdot}0676H/10^5T) - (1/5) \coth (0{\cdot}0169H/10^5T)$$

and this relation is plotted as the broken line (*d*). There is clearly a vast difference between curves (*c*) and (*d*) and indeed, since the initial susceptibility is proportional to g^2 (see Section 4.2.3), its value for curve (*c*) is 25 times that for curve (*d*).

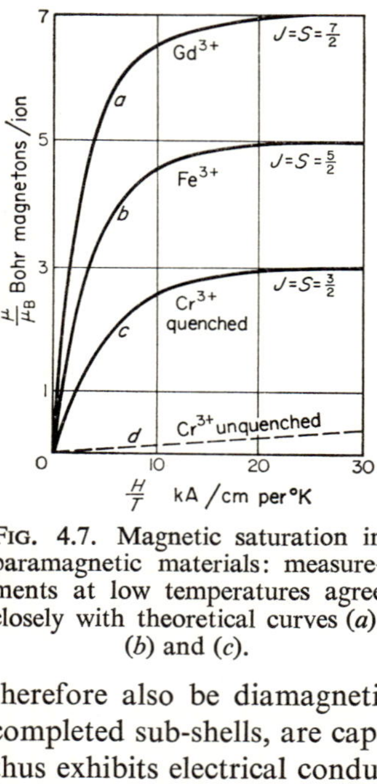

FIG. 4.7. Magnetic saturation in paramagnetic materials: measurements at low temperatures agree closely with theoretical curves (*a*), (*b*) and (*c*).

The experimental results obtained at temperatures of 1·30, 2·00, 3·00 and 4·21°K in field strengths up to about 40,000 A/cm fall very closely upon the three full lines (*a*), (*b*) and (*c*) of Fig. 4.7 in striking confirmation of Brillouin's theory and, for Cr^{3+}, demonstrating complete orbital quenching. It is, indeed, interesting to find that measurements could be made up to about 99·5% of the saturation value in these paramagnetic materials.

4.2.6 *Paramagnetism of Metals*

In metal crystals the inner electrons, in closed shells, remain in close association with their respective nuclei. This inner core of the atom, besides having a resultant positive charge, will therefore also be diamagnetic. However, the outer valency electrons, in incompleted sub-shells, are capable of movement through the crystal which thus exhibits electrical conductivity in an applied electric field.

In the quantum theory the conduction electrons occupy electronic quantum states, with discrete energy levels, in energy bands which may overlap. This has been discussed in Sections 3.2 and 3.3 For an unfilled energy band in a metal crystal it was shown in Section 3.3 that the density of states dn/dE was given by

$$dn/dE = 3n_0\sqrt{E}/2E_F^{\frac{3}{2}} \tag{4.17}$$

In this expression dn is the number of electrons in allowed energy states in an energy increment between E and $E + dE$. Two electrons, one of positive and one of negative spin, may occupy each energy level, so that dn electrons will occupy $dn/2$ energy levels. In the formula n_0 is the total number of free electrons in a unit volume of the crystal.

In deriving equation (4.17) it was assumed that the electrons could move freely in the crystal lattice and had only kinetic energy, that is to say that no account was taken of any potential energy of the electrons in the electric field of the atomic cores. There is in this case the parabolic relation given above between dn/dE and E, which is shown in Fig. 3.10 and repeated in Fig. 4.8(*a*). It should, however, be stated that, when the potential energy is allowed for, this relationship is changed and some other function must be substituted for equation (4.17). This has already been discussed and expressed diagrammatically in Fig. 3.15 at (*b*) and (*c*). We shall, however, make use of equation (4.17) in what follows.

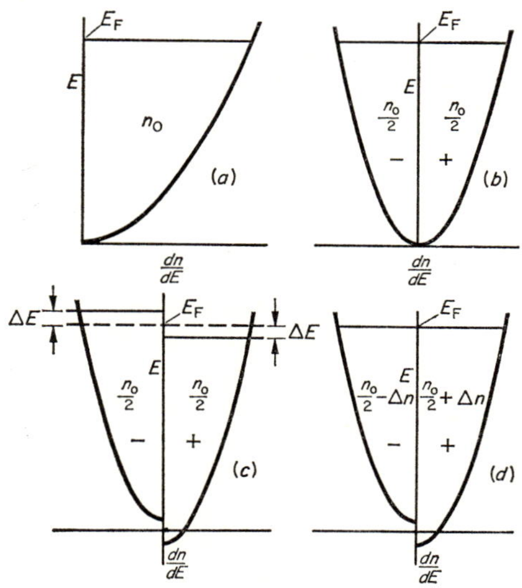

FIG. 4.8. Paramagnetism of an electron gas.

Electrons will occupy all energy levels up to the Fermi energy E_F where

$$E_F = (h^2/2m)(3n_0/8\pi)^{\frac{2}{3}} \tag{4.18}$$

E_F is the kinetic energy of electrons in the highest occupied level and, for reasons given in Section 3.3, it may be assumed that, at normal temperatures, there is a sharp cut-off at this level. The total number of electrons in occupied levels in a unit volume of the crystal is n_0 and this number is represented by the enclosed area below E_F in Fig. 4.8(*a*).

The electron gas in which the diamagnetic cores are immersed is paramagnetic. In a field H each electron has either positive or negative spin with spin moments in the field direction of $\pm 1\mu_B$ and magnetic potential energy $\mp H\mu_B$. Since positive spins have lower potential energy than negative spins in the field, the energy of the system is reduced if there is an excess of positive spins, but any redistribution of the relative number of either kind of spin is conditioned by the availability of allowed quantum states. However, assuming the distribution of states given by equation (4.17) it is a relatively simple matter to calculate the paramagnetic susceptibility.

Figure 4.8(*b*) shows the distribution of states for positive and negative spins separately, with $n_0/2$ electrons of positive spin on the right of the diagram and $n_0/2$ of negative spin on the left. In a field H an increment of potential energy $H\mu_B$ must be added to the electrons of negative spin, and subtracted from those of positive spin. The allowed energy levels are similarly moved as shown in Fig. 4.8(*c*) where the curve representing the distribution of states is raised by $H\mu_B$ on the left of the diagram and depressed by the same amount on the right, and where E now represents the sum of the kinetic energy and the potential energy.

This re-positioning of the allowed energy levels may be understood by referring back to the Zeeman effect, discussed in Section 2.5 and illustrated in Fig. 2.17. The *s* electrons in an atom have, as in the case of the free electrons in the crystal, no orbital motions and the *s* energy level is split into two allowed levels when a field H is applied. If both levels are occupied the electrons will be of opposite spin with an energy difference from the mean level of $\mp H\mu_B$.

Figure 4.8(*c*) is shown with $n_0/2$ electrons of negative spin and an equal number of positive spin. This, however, is not an equilibrium condition since there are allowed vacant levels for positive spin of lower energy than the upper occupied levels for negative spin. For any practical value of H the increment $\Delta E = H\mu_B$ will be very small compared with E_F. Hence equilibrium is reached, as shown in Fig. 4.8(*d*), when the electrons of negative spin in levels above E_F change over to positive spin and fill the unoccupied allowed levels for positive spin up to E_F.

The condition shown at (*c*) corresponds to zero resultant magnetic moment. At (*d*), however, there is an excess of positive spins and a resultant moment Δm in the field direction. If the number of electrons transferring from negative to positive spin is Δn, then bearing in mind that the magnetic moments reverse, we have $\Delta m = 2\,\Delta n\,\mu_B$. The density of states at the Fermi level is, from equation (4.17),

$$(dn/dE)_F = 3n_0/2E_F$$

And

$$2\Delta n = \Delta E(dn/dE)_F = H\mu_B 3n_0/2E_F$$

$$\therefore \Delta m = H\mu_B^2 3n_0/2E_F$$

Substituting for E_F from equation (4.18) gives for the volume susceptibility

$$\kappa = \Delta m/H = (\mu_B^2 3n_0/2)(2m/h^2)(8\pi/3n_0)^{\frac{2}{3}}$$

$$\kappa = (4m\mu_B^2/h^2)(3\pi^2 n_0)^{\frac{1}{3}} \tag{4.19}$$

Equation (4.19) gives the paramagnetic susceptibility, which is seen to be independent of temperature, of a free-electron gas in a crystal. PAULI [9] first drew attention to this kind of paramagnetism, but LANDAU [10] showed that, because the electrons would move helically in a magnetic field, there was also, when quantum considerations were taken into account, a diamagnetic effect, and a diamagnetic susceptibility equal to one-third of the value given in equation (4.19). The resultant paramagnetic susceptibility is thus two-thirds of the Pauli value.

The total susceptibility of a metal is the difference between the diamagnetic susceptibility of the cores and the resultant paramagnetic susceptibility of the electron gas. These quantities turn out to be comparable in magnitude so that metals may be diamagnetic or paramagnetic depending upon which is the larger. Agreement between the calculated and observed susceptibilities of metals is at present not very good.

PROBLEMS

4.1. Calculate the relative volume susceptibility of a gas under n.t.p. conditions if each molecule has a magnetic moment of 2 Bohr magnetons.

4.2 A gas under n.t.p. conditions consists, hypothetically, of atoms the normal state of which is given by $^2D_{\frac{3}{2}}$. Calculate the saturation value of the intensity of magnetization for the gas, and determine the applied field which would be necessary to produce an intensity equal to one half of the saturation value. (An estimate of the applied field may be made by making use of Fig. 4.4, and then a more accurate value obtained from trial values of H.)

SUGGESTED READING

STONER, E. C. *Magnetism and Matter*, Methuen, London (1934).

BATES, L. F. *Modern Magnetism*, 4th edn., Cambridge University Press (1961).

REFERENCES

[1] LANGEVIN, P. 'Magnétisme et théorie des électrons', *Annls Chim. Phys.* **5,** 70 (1905).

[2] BRILLOUIN, L. *Die Quantenstatistik*, Springer, Berlin (1931).

[3] CURIE, P. 'Propriétés magnétiques des corps à diverses températures', *Annls Chim. Phys.* **5,** 289 (1895).

[4] WEISS, P. 'L'hypothèse du champ moléculaire et la propriété ferromagnétique', *J. de Phys.* **6,** 661 (1907).

[5] GORTER, C. J. 'Paramagnetische Eigenschaften von Salzen', *Archs Mus. Teyler*, **7,** 183 (1932).

[6] VAN VLECK, J. H. *The Theory of Electric and Magnetic Susceptibilities*, The Clarendon Press, Oxford (1932).

[7] STONER, E. C. 'Ionic magnetic moments', *Phil. Mag.* **8,** 250 (1929).

[8] Henry, W. E. 'Spin paramagnetism of Cr^{3+}, Fe^{3+} and Gd^{3+} at liquid helium temperature and in strong magnetic fields', *Phys. Rev.* **88,** 559 (1952).

[9] Pauli, W. 'Über Gasentartung und Paramagnetismus', *Z. Phys.* **41,** 81 (1927).

[10] Landau, L. 'Diamagnetismus der Metalle', *Z. Phys.* **64,** 629 (1930).

Chapter 5

FERROMAGNETISM AND ANTIFERROMAGNETISM

5.1 The Problem of Ferromagnetism

Ferromagnetism is a phenomenon which occurs especially in the metals iron, cobalt and nickel. It occurs also in many alloys, including some whose separate constituent elements are not ferromagnetic. The rare earth elements are also ferromagnetic at lower temperatures, gadolinium and terbium below 16°C and −55°C, respectively, and the elements dyprosium, holmium, erbium and thulium at temperatures below 85°K, 20°K, 20°K and 22°K, respectively.

Paramagnetism and diamagnetism, discussed in Chapter 4, are comparatively very feeble effects. If we take for example the metal manganese, which has in fact an unusually high positive susceptibility, the observed mass susceptibility in CGS units, is $\chi' = 11{\cdot}8 \times 10^{-6}$ e.m.u./g with a volume susceptibility $\kappa' = 87 \times 10^{-6}$ e.m.u./cm^3. In rationalized MKS units the relative susceptibility is $\kappa_R = 4\pi\kappa' = 1{\cdot}10 \times 10^{-3}$ with an absolute value of $\kappa = \mu_0\kappa_R$. The value for the relative permeability of manganese thus becomes $\mu_R = 1 + \kappa_R = 1{\cdot}0011$, a value which is only slightly greater than unity. On the other hand a sample of iron may have an initial relative permeability of 10^2 or more, a value which may rise to 10^5 in carefully-prepared ferromagnetic specimens.

This great difference in the magnetic properties of ferromagnetic and paramagnetic materials has already been referred to in Section 4.2.2 where it was shown that magnetic fields applied to a paramagnetic substance would require to be about 10^6 times greater than in the case of iron to produce a comparable degree of magnetization. In a paramagnetic substance, such very high fields are necessary to overcome the disordering effects of the thermal motions of the atoms or molecules unless the temperature is very low. Since in a ferromagnetic material an orderly alignment of the magnetic axes of the elementary magnets is achieved with little external effort, it therefore appears that a very high internal field or its equivalent must in some way spontaneously come into existence. In 1907 Weiss postulated that a high internal field existed in any ferromagnetic material, and indeed this concept of a 'molecular field' has already been mentioned in Section 4.2.3 to account for the observed Curie–Weiss Law in paramagnetic materials.

The Weiss theory, which will be discussed in the next section, leads to certain fascinating conclusions which have since been well confirmed by experimental evidence. However, in spite of these successes it might be mentioned here that some aspects of the theory of ferromagnetism are far from satisfactory. Difficulties arise because the materials are metals. One approach is to treat the metal as though it were an ionic crystal. In this case the experimental values of the magnetic moment per ion, obtained from the saturation value at 0°K, do not agree with the theoretical ionic moments, even when orbital quenching is allowed for. An alternative is the energy-band or collective electron treatment. This method of attack is like that used in Section 4.2.6 for Pauli paramagnetism, from which is determined the temperature-independent susceptibility of the electron gas in a metal. The valency electrons for an entire crystal are in allowed levels in energy bands. The bands for positive and negative spin, respectively, are separated by an energy increment arising, in effect, from the presence of the Weiss molecular field. To be satisfactory the theory would require to account for the observed variation of magnetic saturation values with temperature. There are very great mathematical difficulties here which have not yet been overcome.

5.2 The Weiss Theory of Ferromagnetism

WEISS [1] proposed that in a crystal a 'molecular field' existed which, he assumed, was proportional in magnitude to the intensity of magnetization in the material. The Weiss field is thus given by $H_W = N_W M_T$ where N_W is the molecular field constant. In Section 4.2.3 it was shown that this concept leads to the Curie–Weiss Law which applies to many paramagnetic materials. However, if the constant N_W is large enough, the effect of the resulting high molecular field, the origin of which we disregard at this stage, is easily shown to produce an instability in a paramagnetic material which makes it ferromagnetic.

The paramagnetic magnetization curve of a material made up of ions, which each have a magnetic moment μ_J, is given by $M_T/M_0 = B(a)$ where $B(a)$ is the Brillouin function given in equation (4.8). M_T is the intensity of magnetization at temperature T, M_0 the saturation value at $T = 0$°K with $M_0 = ngJ\mu_B$ where n is the number of ions per unit volume, and

$$a = gJ\mu_B H/kT$$

The relation between M_T/M_0 and a has been plotted for several values of J in Fig. 4.4, and one such curve is repeated in Fig. 5.1(a). For a particular material, with the ions having given values of g and J, a series of curves relating M_T and H may be plotted, as shown in Fig. 5.1(b), corresponding to different absolute temperatures, zero, T_1, T_2, and so on. In this figure the straight line OA, with the equation $H = N_W M_T$, represents the Weiss field. It is obviously more difficult to magnetize the material as the temperature increases. The line OA through the origin will intersect each magnetization curve at another point, as shown in the figure, so long as the temperature is below a critical value $T = \theta_C$, but not when $T > \theta_C$.

An expression for θ_C may be derived by noting that the slope of the magnetization curve at the origin when $T = \theta_C$ is equal to $1/N_W$. As mentioned in Section 4.2.3, $B(a) = (J + 1)a/3J$ when a is small, and therefore at the origin $M_T/M_0 = (J + 1)a/3J$. Substituting for a and putting $T = \theta_C$ we have

$$M_T/H = (J + 1)(gJ\mu_B M_0)/3Jk = 1/N_W$$

$$\therefore \theta_C = (J + 1)(g\mu_B M_0 N_W)/3k \tag{5.1}$$

Or substituting for M_0

$$\theta_C = J(J + 1)nN_W g^2\mu_B^2/3k = C'N_W \tag{5.2}$$

where C' is the same as in Section 4.2.3 for the Curie–Weiss Law.

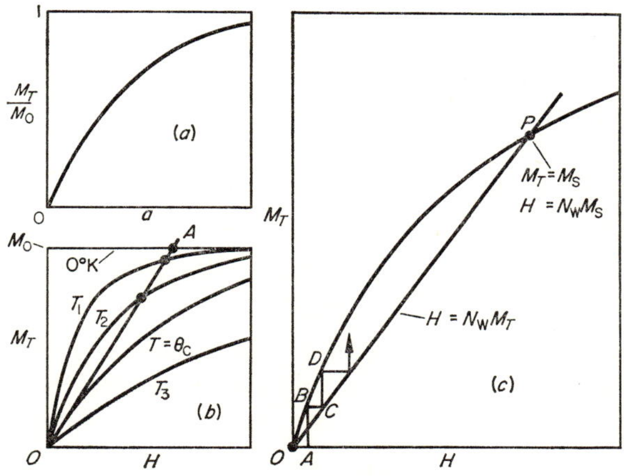

FIG. 5.1. Spontaneous magnetic saturation in a ferromagnetic material.

Figure 5.1(c) shows the magnetization curve for temperature T with the Weiss molecular field line intersecting the curve at the point P. In this case the material is not stable in the unmagnetized state at the origin. For if a small field corresponding to point A is accidentally applied, the magnetization will, according to the theory of paramagnetism, rise to a point B, but this intensity will spontaneously induce the Weiss field at C, and this in turn raises M_T to D. The magnetization will, in fact, finally increase to a value M_S at the stable point at P. It is therefore concluded that, if the molecular field exists, a material which would otherwise be paramagnetic becomes spontaneously magnetized to a saturation value M_S, which is a function of the temperature T, so long as T is less that θ_C.

The relation between M_S and T is readily obtained by noting that at the point P, $M_T = M_S$ and $H = N_W M_S$ and substituting these values in the equation $M_T/M_0 = B(a)$. Whence

$$\frac{M_S}{M_0} = \frac{2J+1}{2J}\coth\left\{\left(\frac{2J+1}{2J}\right)gJ\mu_B N_W M_S/kT\right\} - \frac{1}{2J}\coth\left\{\left(\frac{1}{2J}\right)gJ\mu_B N_W M_S/kT\right\}$$

By making use of equation (5.1) this becomes

$$\frac{M_S}{M_0} = \frac{2J+1}{2J}\coth\left\{\left(\frac{3}{2}\frac{2J+1}{2J}\right)\frac{Ms}{M_0}\frac{\theta_C}{T}\right\} - \frac{1}{2J}\coth\left\{\frac{3}{2(J+1)}\frac{M_S}{M_0}\frac{\theta_C}{T}\right\} \tag{5.3}$$

When $J = 1/2$, equation (5.3) reduces to

$$\frac{M_S}{M_0} = 2\coth\left(2\frac{M_S}{M_0}\frac{\theta_C}{T}\right) - \coth\left(\frac{M_S}{M_0}\frac{\theta_C}{T}\right)$$

which becomes

$$\frac{M_S}{M_0} = \tanh\left(\frac{M_S}{M_0}\frac{\theta_C}{T}\right) \tag{5.4}$$

If, on the other hand, we put $J = \infty$, then noting that

$$\coth x = 1/x + x/3 - x^3/45 + \ldots,$$

equation (5.3) becomes

$$\frac{M_S}{M_0} = \coth\left(3\frac{M_S}{M_0}\frac{\theta_C}{T}\right) - \frac{1}{3}\frac{M_0}{M_S}\frac{T}{\theta_C} \tag{5.5}$$

The hypothetical quantum case of $J = \infty$ is one which would allow all positions of the magnetic axes in the field and corresponds with the classical Langevin solution in which there are, in fact, no quantum restrictions.

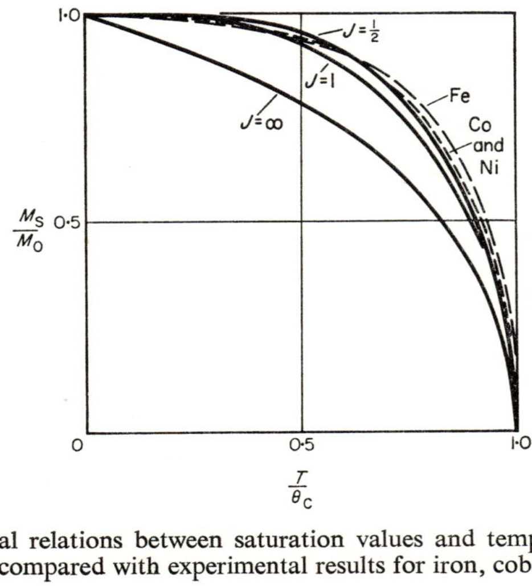

FIG. 5.2. Theoretical relations between saturation values and temperature for different values of J, compared with experimental results for iron, cobalt and nickel.

Equation (5.5) might have been obtained by using the Langevin relation $M_T/M_0 = L(a)$ instead of the Brillouin function.

The relations between M_S/M_0 and T/θ_C, calculated from equation (5.5) for $J = \infty$, from equation (5.4) for $J = 1/2$, and from equation (5.3) for $J = 1$, are shown in Fig. 5.2. The curves show that M_S, which equals M_0 at 0°K, falls to zero when $T = \theta_C$. θ_C is known as the magnetic change point or Curie temperature.

Measurements made on iron are shown by the broken line, and on cobalt and nickel by the dotted line. It will be noticed first that the classical curve, represented by $J = \infty$, departs considerably from the experimental values. The theoretical curve which best agrees with those determined experimentally is that for $J = 1/2$, although the agreement is not perfect. Reference to the normal states in Appendix 1 and Table 4.2 shows, however, that none of the free atoms or ions of iron, cobalt or nickel has a value $J = 1/2$, not even with orbital quenching or $L = 0$. This is an important apparent discrepancy which will be further discussed in the following sections.

Observed values of the saturation values, Curie temperatures and the mean magnetic moment per atom for iron, cobalt and nickel are given in Table 5.1. In the table n_e is the effective number of Bohr magnetons per atom at 0°K, that is $n_e = M_0/n\mu_B$ where n is the number of atoms per unit volume at 0°K. n_e for each metal is seen from experiment to be non-integer and is not equal either to gJ or $2S$ which are values expected in the ionic crystals discussed in Section 4.2.4.

TABLE 5.1

Metal	Saturation Magnetization Wb/m²		n_e μ_B/atom	Curie Temperature	
	M_0 0°K	M_S 20°C	0°K	°K	°C
Iron	2·20	2·15	2·221	1043	770
Cobalt	1·82	1·76	1·716	1400	1127
Nickel	0·641	0·609	0·606	631	358

If now we take $J = 1/2$ for these ferromagnetic materials, then since the number of allowed positions of the axes of the magnetic carriers is $2J + 1$, only two positions, that is parallel and antiparallel to the field direction, are possible. The values which should be assigned to L and S are not immediately apparent, but this information comes from the experiments described in the next section.

In the foregoing account of the Weiss theory the conclusion is reached that any ferromagnetic material is spontaneously magnetized to a saturation value M_S corresponding to the particular temperature. This value is taken to be the same as that which may be observed for the material in bulk when a high magnetic field H is applied. If, however, the material is in a weak field, or in a 'demagnetized' condition, it is not immediately apparent that this spontaneous saturation exists. Weiss suggested that any ferromagnetic material was composed entirely of 'domains', or small subdivisions, each of which was magnetically saturated in a particular direction. The demagnetized condition then corresponded with a random distribution of the vectors representing the domain magnetizations. The effect of an externally-applied field was to rearrange these vectors so that there was a resultant bulk magnetization in the material in the field direction, this process of alignment requiring only a small field strength when compared with that required in a paramagnetic material. The domain theory is now well established by experiment and, in fact, the boundaries between the domains are observable under a microscope.

5.3 Gyromagnetic Experiments

As described in Chapter 2 the magnetic moment and the angular momentum or mechanical moment of an atom, or ion, arise from the orbital and spin motions of electrons in incompleted shells in the electronic structure. The carriers of the moments in the atom are regarded as having gyroscopic properties, as depicted for a free atom of iron in Fig. 2.20. The values of the moments may be written down in terms of the quantum numbers which specify the state of the atom. Thus the magnetic moment $\mu_J = gJ^*\mu_B$ and the angular momentum $p_J = J^*h/2\pi$. In an applied field H the resolved values in the field direction are $\mu = gM\mu_B$ and $p = Mh/2\pi$ where M has the allowed values $0, \pm 1 \ldots \pm J$ or $\pm 1/2, \pm 3/2 \ldots \pm J$ depending on whether J is whole number or half-integer, respectively. g is the Landé factor and

$$g = 1 + \frac{(J^*)^2 + (S^*)^2 - (L^*)^2}{2(J^*)^2} \tag{5.6}$$

The magnetic moment is, in the general case, contributed to by both the orbital and spin motions, represented by the quantum numbers L and S, respectively. The value of g is an indication of the relative contributions to the magnetic moment of the orbital and spin motions. Since J^* is the vector sum of L^* and S^* then, in particular, if $S^* = 0, J^* = L^*$ and if $L^* = 0, J^* = S^*$. Therefore if there were no component in the magnetic moment due to spin, $S^* = 0$ and from equation (5.6) $g = 1$. On the other hand if the moment is entirely due to spin, that is $L^* = 0$, $g = 2$.

The ratio of magnetic moment to angular momentum for any atom or ion is $\mu_J/p_J = \mu/p = g(2\pi/h)\mu_B$, and since $\mu_B = \mu_0(e/2m)(h/2\pi)$,

$$\therefore \mu/p = g\mu_0 e/2m \tag{5.7}$$

where $\mu_0 e/2m = 1{\cdot}1053 \times 10^5$. Equation (5.7) gives the ratio of the resolved components of magnetic moment and angular momentum with respect to the direction of an applied field for that part of an atom or ion which is the carrier of the magnetic and mechanical moments.

Suppose now that in a magnetic field H the axis of an ion jumps from one allowed angular position to another, that is the magnetic quantum number M changes from one value to another, there will be corresponding increments of magnetic moment and angular momentum, resolved in the field direction, of $\Delta M'$ and ΔP and, from what has been said, $\Delta M'/\Delta P = \mu/p$. If on application of the field H to the material the total induced magnetic moment is M' then $M' = \Sigma \Delta M'$, and the total induced angular momentum due to the realignment of the axes of all the gyroscopic carriers is $P = \Sigma \Delta P$. Clearly

$$M'/P = \Sigma \Delta M'/\Sigma \Delta P = \mu/p = g\mu_0 e/2m \tag{5.8}$$

If now the specimen is conveniently in the form of a vertical rod, hanging on a fine suspension, and field H is suddenly applied in the vertical direction, the induced angular momentum of the magnetic carriers produces an equal and opposite impulse upon the rod itself causing it to rotate. The values of M' and P occurring when field H is applied may be measured, and thus a g value for the magnetic carriers, whatever they may be, is obtainable by use of equation (5.8).

The phenomenon of rotation produced by magnetization is known as the RICHARDSON [2], or as the EINSTEIN–DE HAAS Effect [3]. Alternatively if a rod is suddenly rotated the axes of the elementary gyroscopes will tend to line up with the axis of rotation thus producing a resultant magnetization in the axial direction. This is called the BARNETT Effect [4]. Experiments based on these effects have been carried out using several different measurement techniques, and in spite of the very small values of angular momentum occurring, g has been determined with considerable accuracy for many ferromagnetic and paramagnetic materials. Table 5.2 gives some of the observed values obtained for $(\mu_0 e/m)(P/M')$, that is for $2/g$, and the corresponding value of g, in the case of iron, cobalt and nickel.

The results given in the table indicate that for iron, cobalt and nickel g is on the average a little less than 2, from which it is inferred that the carriers of the magnetic moment in these ferromagnetic materials are spinning electrons, although there is a contribution from orbital motions which is not entirely negligible.

TABLE 5.2

Material	$2/g$	g	Observer
Iron	1·00 ± 0·01	2·0	CHATTOCK and BATES [5]
	1·032 ± 0·003	1·94	Barnett
Cobalt	1·030 ± 0·064	1·94	Sucksmith
	1·078 ± 0·007	1·84	Barnett
Nickel	1·002 ± 0·004	2·00	SUCKSMITH and BATES [6]
	1·046 ± 0·004	1·91	Barnett

5.4 Discussion of the Saturation-Temperature and Gyromagnetic Results

In discussing the Weiss theory of ferromagnetism in Section 5.2 the metals have been regarded as made up of ions so that use could be made of the Brillouin function. If we continue with this view and neglect the small departure from complete orbital quenching, the results in Table 5.2 give a value $L = 0$ for the ions, and so $J = S$. The relation between M_S and T, however, indicates that the appropriate value of J is $1/2$ and so we put $S = 1/2$. The resolved values of the magnetic moment for the ions for each metal would therefore be $gS\mu_B = 2(1/2)\mu_B = 1\mu_B$ with only two positions, parallel and antiparallel, for the magnetic axes. This, at least, is consistent with the carriers being individual and independent spinning electrons, but a moment of $1\mu_B$ per ion is not the expected value, either for free atoms, or for ions, of iron, cobalt and nickel.

In the case of iron, for example, $S = 2$ for the free atom, and $S = 5/2$ and 2, respectively, for the ions Fe^{3+} and Fe^{2+}. With orbital quenching, the values of J then become 2, 5/2 and 2, and not the required value of 1/2. The magnetic moments $gS\mu_B$ are $4\mu_B$ for the atom and 5 and $4\mu_B$ for the trivalent and divalent ions, respectively. These values of S and of the magnetic moments arise from the Russell–Saunders coupling between spins which was discussed in Section 2.7. The moment of the iron atom, assuming orbital quenching, is due to four uncompensated spins in the incompleted $3d$ sub-shell, which are coupled in the same sense to give four units of magnetic moment to the atom, and making $S = 4(1/2) = 2$. Since, from the experimental results, we must take $J = S = 1/2$, it is concluded that in the ferromagnetic metals the Russell–Saunders coupling between spins is broken down. Theoretically this may happen if a very strong magnetic field is applied to the atoms, as described in Section 2.5 and illustrated in Fig. 2.15(*b*), producing the phenomenon called the Paschen–Back Effect. The postulated Weiss molecular field is such a strong field.

We may therefore take the view that, in a ferromagnetic material, there is almost complete orbital quenching, that the Russell–Saunders coupling between spins in the atoms is broken down in the presence of the Weiss field, and that the units of magnetic moment are spinning electrons in positions which are either parallel or antiparallel with respect to the resultant magnetic field in the material.

However, as already mentioned, the experimental values given in Table 5.1 for the magnetic moment in Bohr magnetons per atom are not whole numbers. In view of this and of the considerations just mentioned it appears that the spinning magnetic carriers are not closely associated with particular atoms, as they would be in an ionic crystal, but are shared throughout the crystal. Beyond this point therefore the ionic model of a ferromagnetic substance is not helpful. We return to the metallic state, envisaged for the paramagnetic metals, with diamagnetic atomic cores immersed in a swarm of valency electrons, but with the added concept of superimposed forces which give rise to the Weiss molecular field.

5.5 The Collective Electron Theory

Intensive efforts have been made to account for the observed spontaneous saturation in ferromagnetic metals where, as already stated, the mean magnetic moment per atom is not a whole number of Bohr magnetons. In the collective electron theory the valency electrons belong to the crystal as a whole and occupy energy levels in two overlapping energy bands. These bands may in turn be thought of as divided into two half-bands, one occupied by electrons of positive and the other of negative spin, respectively. However, the mathematical difficulties are great and solutions involve the use of approximations. In fact, doubt is increasingly being thrown on the validity of the results achieved from this particular approach, but the ideas are of great interest and some account of them will be given.

Consider first the case of solid copper. The free atom of copper, with $Z = 29$, has an electronic structure in which all the sub-shells up to and including $3d$ are completely full and there is one electron in the $4s$ sub-shell. In the solid crystal the energy levels corresponding to the outermost sub-shells $3d$ and $4s$ spread into broad overlapping bands of discrete levels as described in Chapter 3 and shown in the case of nickel, for example, in Fig. 3.8. Instead of a few electrons occupying sharp energy levels in a single free atom, we now have a very great number of electrons, corresponding to the whole crystal, occupying successive levels in these broad energy bands. The electron density in relation to the energy at any level may be depicted by curves in the manner shown, for example, in Fig. 3.15(*c*). The curves show the allowed density of states at any energy level, and the area enclosed by the curves equals the total number of states in the band. The whole of the energy band may or may not be fully occupied, depending on the number of valency electrons available in the particular crystal.

The results of some calculations are given in Fig. 5.3. In the case of solid copper ten electrons per atom are in $3d$ states and one electron per atom in $4s$ states, as in the free atom. The $3d$ and $4s$ bands overlap as shown by the block diagram in Fig. 5.3(*a*). According to HOWARTH [7] the relative positions of the bands are as shown and the $3d$ band with ten electrons per atom has an energy depth of 3·46 eV. The $4s$ band with one electron per atom is half full, since two electrons per atom would be required to fill it completely, and the depth of the band from the highest occupied level, or Fermi level, is calculated to be 7·1 eV. The highest occupied level in $4s$ is 3·7 eV above the top of the $3d$ band. The block diagrams do not show the density of allowed states in the bands, but Fig. 5.3(*b*) gives the results of calculations for solid copper by KRUTTER [8] and SLATER [9]. The ordinates in this diagram represent energy, as in the previous diagram, and the abscissae the density of allowed states, where dn is the number of states per unit volume in an increment of energy *dE*. The energy bands, in these calculations again, overlap so that the 3*d* band is completely filled with ten electrons per atom and the 4*s* band is half filled with one electron per atom up to the level denoted by the figure 11 in the diagram. If it is assumed that the density-of-states curves remain the same for nickel, cobalt and iron then for these metals the bands would be filled up to the levels marked 10, 9 and 8, respectively, as shown in the diagram, where these figures denote the total number of electrons per atom in the 3*d* and 4*s* sub-shells jointly. In these cases neither the 4*s* nor the 3*d* bands would be

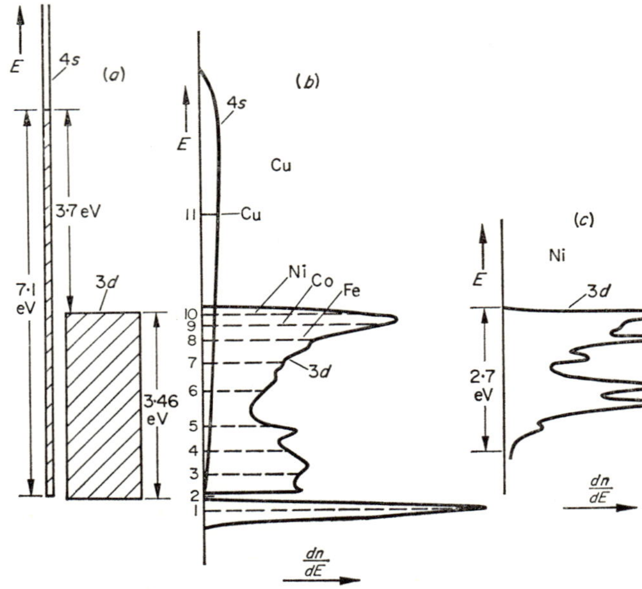

FIG. 5.3. Energy bands in copper and nickel crystals.

completely full. Figure 5.3(*c*), however, shows the shape of the 3*d* band for nickel as calculated by KOSTER [10], with a band depth of about 2·7 eV. The curves for nickel and copper, according to these calculations are, indeed, not identical, and some inconsistencies in the energy values are also apparent in comparing the three diagrams. These differences reflect the approximate nature of the results achieved.

The observed diamagnetism of metallic copper may be expected from the full 3*d* bands resulting in diamagnetic positive ion cores, the numerical value of the negative susceptibility being reduced somewhat by a weak Pauli paramagnetism from the electrons in the unfilled 4*s* band.

In Section 4.2.6 the paramagnetism arising from the valency electrons in a metal was accounted for in terms of unfilled energy bands which, in the presence of an applied field, were divided into two bands containing, respectively, electrons of positive and negative spin. The effect of the field was to raise the negative band by an energy increment $H\mu_B$ and to depress the positive band by the same amount, as in the Zeeman effect. The resulting excess of positive spins then gave to the electron gas a positive susceptibility. The same approach may now be made to the ferromagnetic metals, a useful physical picture being obtained if we suppose that the applied field in this case is the very high spontaneously-existing Weiss molecular field H_W. Again the half-bands are displaced, upwards for negative spin and downwards for positive spin, by an amount $H_W\mu_B$. In this case the total number of unpaired positive spins is relatively high, since H_W is a strong field, leading to the high average

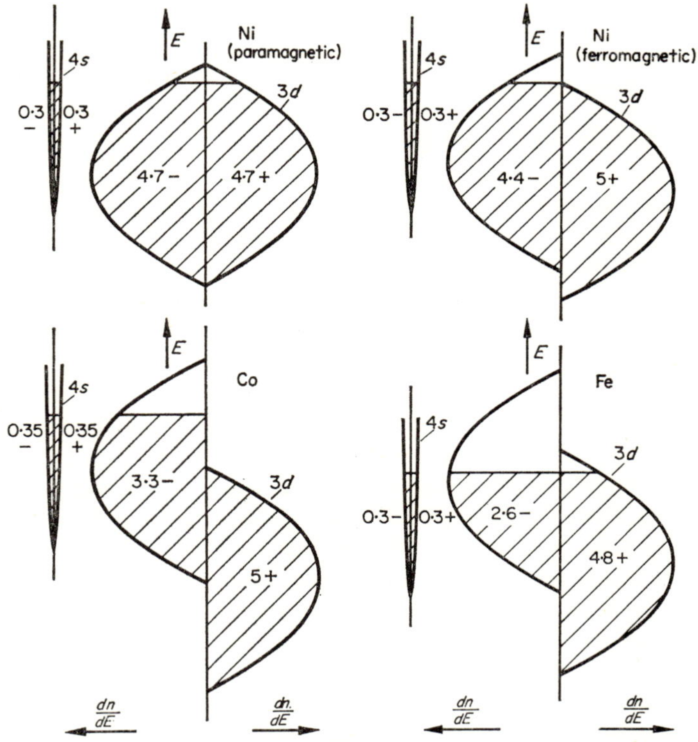

FIG. 5.4. Electron distribution in energy bands in paramagnetic nickel and in ferromagnetic nickel, cobalt and iron.

magnetic moment per atom observed in the ferromagnetic metals. These atomic moments are aligned in the direction of the hypothetical H_W giving spontaneous magnetic saturation in one direction.

In Fig. 5.4 any attempt to delineate the precise shape of the energy-band curves, as was done in Fig. 5.3, is abandoned. It is supposed also that the Weiss field acts only on the 3*d* electrons and not on the 4*s* electrons. The justification for this assumption is that the 3*d* electrons are more closely associated with the atomic cores, and more under the influence of each other, than the 4*s* electrons from the outermost sub-shell. This will be discussed in

Section 5.6 where the origin of the Weiss field is ascribed to a mutual interaction between the $3d$ electrons. Displacement of the half-bands therefore occurs only for the $3d$ electrons while the numbers of positive and negative spins remain equal in the $4s$ band. The shaded areas in Fig. 5.4 indicate the extent to which the energy bands are occupied and the numbers refer to the number of electrons of positive and negative spin per atom. In the case of ferromagnetic nickel and cobalt the positive $3d$ band is completely filled, but calculations seem to show that for iron neither the positive nor negative band is completely full. The difference between the number of electrons per atom of positive and negative spin, respectively, gives the number of Bohr magnetons per atom and these differences are seen to correspond approximately with the values of n_e in Table 5.1.

5.6 Origin of the Weiss Molecular Field

In Section 3.1.1 the formation of a hydrogen molecule, consisting of a pair of positive nuclei and a pair of electrons, was ascribed to forces of exchange, or in effect a strong coupling between the two spinning electrons of the two atoms. This particular problem is soluble in wave mechanics and it is found that the system consisting of two hydrogen atoms has minimum energy when the distance between the nuclei has an optimum value and the electrons are of opposite spin. The forces of interaction producing antiparallel spins in this case are said to be negative forces. They are very much greater in magnitude than the forces between dipoles and a simple physical interpretation of their character does not seem to be possible.

HEISENBERG [11] first suggested that the hypothetical Weiss molecular field in ferromagnetism was similarly due to the forces of interaction leading to coupling between unpaired spinning electrons in adjacent atoms of the material, but that the interaction was now positive, that is the forces acted to produce parallel spins in neighbouring atoms thus causing spontaneous magnetic saturation.

For a crystal made up of atoms each with one unpaired electron, and each having z neighbours, STONER [12] derived a relation as follows between the Weiss field constant N_W and the exchange energy J_e where J_e is the energy for one pair of electrons.

$$N_W = zJ_e/2M_0\mu_B$$

If H_{W0} is the Weiss field corresponding to the saturation intensity M_0 at 0°K then

$$H_{W0} = zJ_e/2\mu_B \tag{5.9}$$

A relation between J_e and the Curie temperature θ_C is obtainable by using equation (5.1), from which

$$H_{W0} = M_0 N_W = 3k\theta_C/(J+1)g\mu_B \tag{5.10}$$

Therefore, from equations (5.9) and 5.10)

$$J_e = 6k\theta_C/(J+1)zg \tag{5.11}$$

Values of the Weiss field and exchange energy may be derived from equations (5.10) and (5.11), although in view of the assumptions made and the difficulties pointed out in Section 5.4 with the ionic model the numerical results will show only orders of magnitude. However, supposing we take for a ferromagnetic substance $J = 1/2$ and $g = 2$, then $H_{W0} = k\theta_C/\mu_B$ and $J_e = 2k\theta_C/z$. For iron putting $\theta_C = 1043°K$ and $z = 8$ with Boltzmann's constant $k = 1{\cdot}38 \times 10^{-23}$ J/°K and $\mu_B = 1{\cdot}165 \times 10^{-29}$ Wb.m, $H_{W0} = 12 \times 10^6$ A/cm and $J_e = 0{\cdot}36 \times 10^{-20}$ J.

The Weiss field and the exchange energy are of electrostatic origin and their values are much greater than would be possible if they were due to the purely magnetic interaction between magnetic dipoles. For if we consider a small spherical cavity inside a material in which the saturation intensity is M_0, then the magnetic field in the cavity is $H = M_0 N_D$ where N_D is the demagnetization factor for a sphere and, as given in Chapter 1, $N_D = 1/3\mu_0$. For iron $M_0 = 2{\cdot}20$ Wb/m², and then $H = 2{\cdot}20 \times 10^7/(3 \times 4\pi)$ A/m $=$ 5800 A/cm. The maximum energy of a dipole of magnetic moment $1\mu_B$ placed in the cavity would be $H\mu_B = 5{\cdot}8 \times 10^5 \times 1{\cdot}165 \times 10^{-29} = 0{\cdot}68 \times 10^{-23}$ J. These values are thus lower than those due to the interaction between spins by a factor greater than 10^3.

There seems little doubt that the Weiss molecular field in ferromagnetic materials is due to the exchange forces between spins, but the question now arises as to why iron, cobalt and nickel, in particular, are ferromagnetic while other elements such, for example, as manganese are not. A reason for this has been suggested by BETHE [13].

The calculation of the interaction energy J_e for the hydrogen molecule involves the integration of an expression containing six terms. There are two positive terms which arise from the forces of repulsion between the two positive nuclei, and between the two electrons, respectively. The remaining four terms are negative and come from the attraction between electron 1 and the two nuclei (two terms), and electron 2 and the nuclei (two terms). The charge density around the nuclei may be considered as a negative cloud of charge as described in Section 3.1.1 and illustrated in Fig. 3.1. The calculations show that when the stable molecule is formed the electrons are of opposite spin and the interaction energy J_e is negative with a numerical value of about 6×10^{-19} J. On the other hand the required condition to give ferromagnetism is that J_e should be positive. For this to be the case the positive repulsion terms in the exchange integral must exceed the negative attraction terms. It appears that this may be possible if the distance between the nuclei is large compared with the effective mean diameter of the electron density distribution in the sub-shell responsible for the magnetic moment of the atoms, that is in the 3*d* sub-shell in the ferromagnetic elements iron, cobalt and nickel. A positive value of J_e leads to parallel spins and spontaneous saturation. The value of J_e estimated above for iron from the Curie temperature is numerically smaller than that for the hydrogen molecule by a factor of about 10^2 and this also may be expected from the calculations which involve a difference between positive and negative terms. Now SLATER [14] has shown that the ratio of interatomic distance of nearest neighbours to the effective diameter of the incompleted $3d$ sub-shell is greater for the ferromagnetic elements than for any other. Values of the ratio are given in Table 5.3 where, however, the ratio for gadolinium refers to the $4f$ sub-shell.

TABLE 5.3

Element	Cr	Mn	Fe	Co	Ni	Gd
Ratio of interatomic distance to sub-shell diameter	1·45	1·47	1·63	1·82	1·97	3·1
Curie temperature θ_C, °K	—	—	1043	1400	631	289
Néel temperature θ_N, °K	475	100	—	—	—	—

According to these ideas [14] then it appears that a ratio of interatomic distance to the diameter of the unfilled sub-shell should be greater than about 1·5 for ferromagnetism to occur, for it will be noticed from the table that the ratio for manganese is a little lower than this. Chromium also has a lower ratio and both manganese and chromium are, in fact, now known to have a property, when the temperature is low enough, known as antiferromagnetism.

5.7 Antiferromagnetism

Ferromagnetism occurs, according to the foregoing discussion, when the exchange integral is positive. The forces of interaction between the spins in neighbouring atoms are then positive, and produce parallel alignment of the magnetic moments and spontaneous magnetic saturation. The proposed necessary condition for the spins to be parallel is that the ratio of atom spacing to the effective diameter of the $3d$ (or $4f$) sub-shells shall be greater than about 1·50. When the ratio is less than this, negative interaction is to be expected causing the magnetic moments of neighbouring atoms to be antiparallel. This is the condition just referred to known as antiferromagnetism, a phenomenon which occurs in the metal chromium at temperatures below 475°K, and in manganese below 100°K. The ratios for these elements are less than 1·50 as already pointed out. A great many chemical compounds are also antiferromagnetic. The effect was, indeed, first observed in the oxide MnO [15].

A body-centred cubic crystal such as chromium, for example, may be considered to consist of two interlocking simple cubic systems or sub-lattices A and B, as shown in Fig. 5.5(a). With negative interaction between nearest neighbours, and neglecting the effects of temperature or an applied field, the atomic magnetic axes in system A would be parallel with each other and opposite in direction from those in B as illustrated. Systems A and B are then magnetically saturated equally but in opposite directions. The material has a domain structure but a resultant intensity of magnetization of zero. The dif-

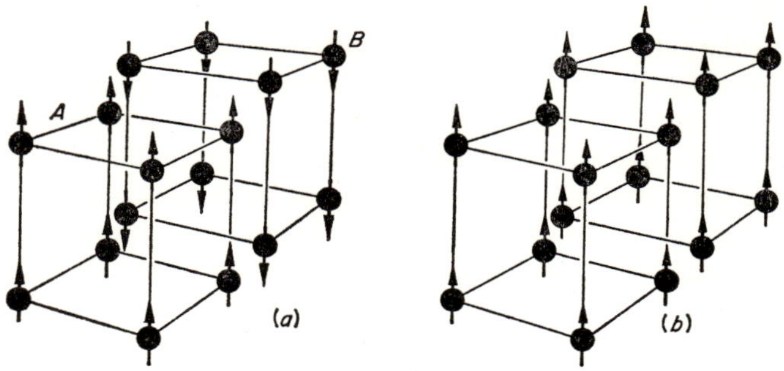

FIG. 5.5. Comparing antiferromagnetic chromium (*a*) with ferromagnetic iron (*b*).

ference between antiferromagnetic chromium and ferromagnetic iron, in the arrangement of the atomic magnets, may be seen by comparing Figs. 5.5(*a*) and 5.5(*b*).

The existence of antiferromagnetism in a material may be observed experimentally in two ways: (a) from measurements of susceptibility in relation to temperature and (b) from the spectra obtained from the material when using neutron beams rather than X-rays.

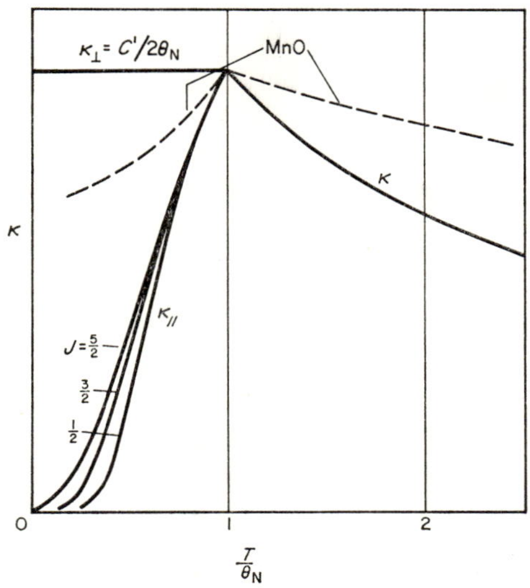

FIG. 5.6. Theoretical relations between susceptibility and temperature for an antiferromagnetic, and observed relation for MnO.

The broken line in Fig. 5.6 shows the way in which the susceptibility of MnO varies with temperature. This is a typical relation for an antiferromagnetic material. There is a sharp discontinuity in the curve at a critical temperature θ_N known as the Néel point where the susceptibility reaches a maximum. The Néel temperature for MnO is 122°K but θ_N varies widely for different materials from a little above absolute zero to about 800°K. We may explain the variation of susceptibility with temperature by considering a magnetic field applied to an antiferromagnetic domain. At low temperatures the negative exchange forces firmly hold the antiparallel positions of the ions and the susceptibility is low. As the temperature increases thermal agitation disturbs the ordering action of the exchange forces and the susceptibility rises. At the Néel point thermal agitation completely overcomes the exchange forces so that, at still higher temperatures, the material behaves like a paramagnetic substance with the susceptibility decreasing with increasing temperature.

The detection of antiferromagnetism by the use of neutron beams is similar to the detection of a superlattice in alloy systems by the use of X-rays. In crystallography, the crystal structures and interatomic distances may be determined with a monochromatic X-ray beam of suitable wavelength impinging on a powdered crystal specimen. The X-rays, diffracted in rings at different angles by different sets of crystal planes, may be recorded as a series of lines on a strip of photographic film. If the crystal is a solid solution of two elements a superlattice may exist. For example, the alloy Cu_3Au, suitably heat-treated, is face-centred cubic with gold on the cube corners and copper at the centres of the cube faces. An ordered arrangement such as this, equivalent to two superimposed lattices, is observable by the appearance of extra lines on the X-ray photograph. In the case of an antiferromagnetic material the superimposed lattices will consist of atoms of the same kind but with reversed magnetic moments as in Fig. 5.5(*a*). This magnetic ordering may be detected when a neutron beam impinges on the crystal due to the interaction between the neutron spin and the magnetic ions, in this case also by the appearance of extra lines in the diffracted pattern. These lines are present below the Néel point but disappear at higher temperatures.

5.8 Susceptibility of Antiferromagnetic Materials

We first consider a domain, at temperatures below the Néel point, in an antiferromagnetic crystal in which there are two equal superimposed lattices *A* and *B* with antiparallel magnetization. At a low temperature the ionic magnets may be arranged as in Fig.5.5(*a*). In this case, with no externally-applied field, each lattice has a saturation intensity of magnetization M_S, the resultant intensity for the whole domain which contains both lattices, of course, being zero. M_S in the *A* system is due to the Weiss molecular field $N_W M_S$ in the *A* direction set up by the reversely-pointing magnetization M_S in the *B* lattice, since there is a negative *A–B* interaction. Similarly the *B* system is saturated by the Weiss field produced by lattice *A*.

We thus have, in effect, two opposing ferromagnetic lattices, and each may be treated as in Section 5.2 for a ferromagnetic material. Thus we may take

Fig. 5.1 to apply, for example, to lattice *A*, where, however, the intensity of magnetization relates only to the *A* atoms, that is to half the total number in the crystal. The *A* atoms will follow a Brillouin function and the Weiss field, due to the *B* atoms, will produce a saturation value M_S appropriate to the temperature *T* of the material. The same applies to the *B* system. As *T* increases the Brillouin magnetization curve will be shifted to the right in the manner shown in Fig. 5.1(*b*) until, when the critical temperature $T = \theta_N$ is reached, where the magnetization curve becomes tangential at the origin to the Weiss field line, the material ceases to be antiferromagnetic and reverts to a paramagnetic condition. θ_N is the Néel temperature which has just been mentioned and which corresponds with the Curie temperature given in equation (5.2), except that now, since only half the total atoms are contained in each lattice, we substitute $n/2$ for *n*. If orbital quenching is assumed then *S* might be put in place of *J* with $g = 2$. Whence

$$\theta_N = J(J+1)(n/2)N_W g^2 \mu_B^2/3k = C'N_W/2 \tag{5.12}$$

where $C' = nJ(J+1)g^2\mu_B^2/3k$. C' is the same constant as that obtained in Section 4.2.3 for a paramagnetic following the Curie Law, or the Curie–Weiss Law, and given again in equation (5.2) for a ferromagnetic material. It will be noted that in the case of the Curie–Weiss Law for a paramagnetic substance $\kappa = C'/(T - \theta_W)$ where $\theta_W = C'N_W$, and in the case of a ferromagnetic the Curie temperature $\theta_C = C'N_W$. Thus θ_N, θ_W and θ_C all involve the Weiss molecular field constant N_W appropriate to the particular material.

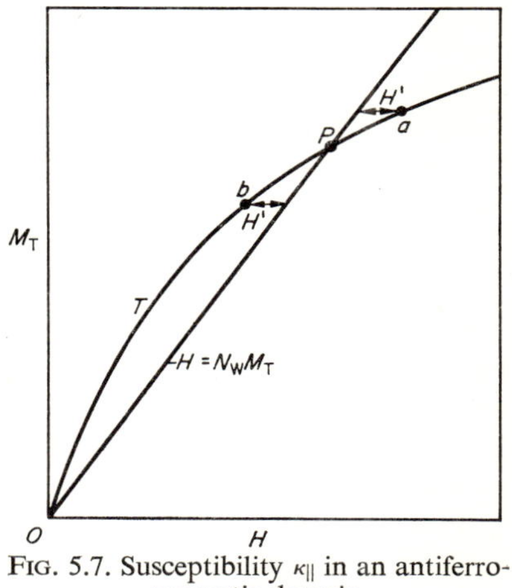

FIG. 5.7. Susceptibility $\kappa_{\parallel}$ in an antiferromagnetic domain.

To determine the susceptibility of an antiferromagnetic domain at temperatures below the Néel point, consider first the effect of an applied field acting in the direction of the saturation magnetization in the *A* lattice, and thus in the opposite direction to the magnetization in the *B* lattice. In Fig. 5.7 the curve *OP* represents the Brillouin function for the *A* lattice and the straight line *OP* the Weiss field acting on the *A* atoms due to the negative interaction of the *B* atoms. As in the case of a ferromagnetic substance, the lattice will be spontaneously saturated at the point *P* with a saturation intensity M_S. The figure equally represents conditions in the *B* lattice for the reverse direction. If now a magnetic field H' is applied in the *A* direction then the magnetization in the *A* system will increase to the point *a* on the curve as shown. Since H' opposes the Weiss field on the *B* lattice the magnetization in it will fall to point *b*. The susceptibility is thus given by $\kappa_{\parallel} = (M_a - M_b)/H'$, and clearly depends on the slope of the Brillouin curve at P. The slope of the curve depends on the value adopted for J and the point of intersection P also depends on the temperature T. The slope is zero at 0°K and increases as the temperature rises until the Néel temperature θ_N is reached. When $T = \theta_N$ the Weiss field line is tangential to the Brillouin curve at the origin and, for either lattice, the equation of the magnetization curve near the origin becomes $H = N_W M_T$ so that the susceptibility corresponding to θ_N is $\kappa = M_T/H = 1/N_W = C'/2\theta_N$. The relation between $\kappa_{\|}$ and T has been calculated for different values of J by VAN VLECK [16] and the form of the curves is shown by the full lines in Fig. 5.6.

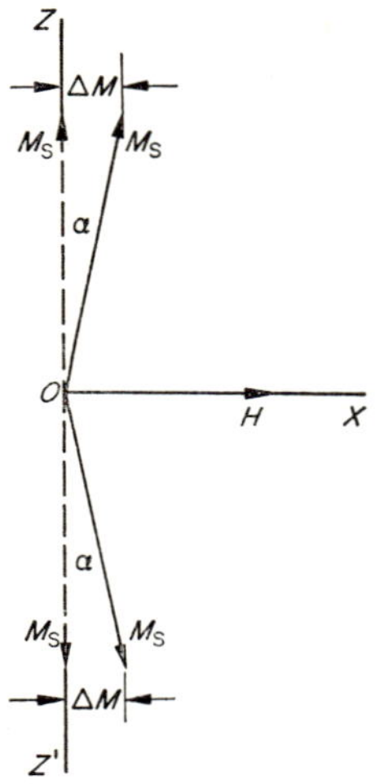

FIG. 5.8. Susceptibility $\kappa_{\perp}$ in an antiferromagnetic domain.

The spontaneous saturation in a ferromagnetic domain is found from experiment to prefer particular directions in a crystal. For example in iron M_S will set itself along a cube-edge or [100] direction of the crystal and a component of magnetic field normal to this direction is required to deflect the M_S vector away from the preferred direction. This is discussed in Chapter 6. Similar behaviour occurs in an antiferromagnetic domain, except that we now have to consider two sub-lattices and a negative interaction between them. The saturation magnetization in Fig. 5.5(a) for the two lattices is, for example, also directed along a cube-edge direction. In calculating $\kappa_{\|}$ it was supposed that the external field was applied along the preferred direction, and it may be expected that the susceptibility at right-angles to this will differ from $\kappa_{\|}$. The susceptibility $\kappa_{\perp}$ in the perpendicular direction may readily be calculated. If in Fig. 5.8 for an antiferromagnetic below the Néel point, the A and B lattices in a domain are saturated in the directions OZ and OZ', respectively, and a field H is applied along OX, then the effect of the applied field is to produce a small rotation of M_S away from the Z axis. For a rotation α the increment of magnetization in the X direction for each of the A and B lattices is $\Delta M = M_S \alpha$ and the resultant susceptibility is $\kappa_{\perp} = 2\,\Delta M/H$. Now for either the OZ or OZ' directions $H_Z = M_S N_W$. It will be noted that, in this case, the increment ΔM is in the same direction OX in each lattice, and therefore each is acted upon by the reverse interaction force $-\Delta M\, N_W$ set up by the other. Therefore for the OX direction the resultant field acting on either lattice is $H - \Delta M\, N_W = H_X$. If it is assumed that the forces of interaction are large compared with any other crystalline forces then for either the A or B lattice $H_X/H_Z = \Delta M/M_S$. Whence $\Delta M/M_S = (H - \Delta M\, N_W)/M_S N_W$.

$$\therefore \kappa_{\perp} = 2\,\Delta M/H = 1/N_W = C'/2\theta_N$$

Thus $\kappa_\perp$, shown in Fig. 5.6, is theoretically independent of temperature and equal to the value of $\kappa_\parallel$ at the Néel point. In the practical case of a multi-domain crystal or a polycrystalline specimen the observed values of κ are found to lie between the theoretical values for $\kappa_\parallel$ and $\kappa_\perp$ as may be expected. This is exemplified for the case of MnO in Fig. 5.6.

At temperatures above the Néel point the Weiss field is no longer able to maintain a spontaneous saturation magnetization in the sub-lattices. Each now behaves like a paramagnetic. This is similar to the behaviour of a ferromagnetic at temperatures above the Curie point except that now we suppose that each sub-lattice exerts a negative interaction upon the other.

For a paramagnetic material with no Weiss field the susceptibility in low fields is given, from equation (4.12), as

$$\kappa = M_T/H = ng^2J(J+1)\mu_B^2/3kT \tag{5.13}$$

However, if M_A and M_B are the intensities of magnetization in the A and B lattices, respectively, in the direction of an applied field H, the total field acting in the two lattices will be $H - N_W M_B$ and $H - N_W M_A$, respectively, and the number of atoms per unit volume in each lattice will be $n/2$. From equation (5.13) we may therefore write for lattice A,

$$M_A = \{(n/2)g^2J(J+1)\mu_B^2/3kT\}(H - N_W M_B)$$

or

$$M_A = (C'/2T)(H - N_W M_B) \tag{5.14}$$

and for lattice B

$$M_B = (C'/2T)(H - N_W M_A) \tag{5.15}$$

The resultant intensity is $M = M_A + M_B$. Whence adding equation (5.14) and (5.15) and rearranging the terms, we have

$$\kappa = M/H = C'/(T + C'N_W/2)$$

or

$$\kappa = C'/(T + \theta_N) \tag{5.16}$$

This expression for the susceptibility of an antiferromagnetic above the Néel point may be compared with that for a paramagnetic following the Curie–Weiss Law, namely $\kappa = C'/(T - \theta_W)$ where $\theta_W = C'N_W$, while now $\theta_N = C'N_W/2$. Equation (5.16) is shown by the full line in Fig. 5.6 and the corresponding observed values of κ for MnO are shown by the broken line.

Experimentally θ_N may be observed by noting the temperature when the susceptibility has its maximum value. It should be possible to deduce θ_N also using equation (5.16) and the observed relation between κ and T above the Néel point. There is not good agreement between the two methods. If θ is the value obtained in the second case, θ is always found to be greater than θ_N, the ratio θ/θ_N being generally greater than about 1·4 and, in the case of MnO, $\theta/\theta_N = 5{\cdot}0$. If exchange interactions between atoms which are not nearest neighbours are also assumed to exist this discrepancy can be explained.

5.9 Superexchange Forces

The arrangement of the ionic magnets in MnO, as determined by neutron diffraction experiments, is shown in Fig. 5.9. The structure may be regarded as a face-centred cubic system of oxygen ions, shown by the large circles, with a superimposed face-centred cubic lattice of manganese ions. Each manganese atom gives up a pair of electrons, and each oxygen ion receives two electrons. Thus the O^{2-} ions have the $2p$ sub-shell completed with six electrons and so have no magnetic moment. The Mn^{2+} ions, however, are five electrons short of a full $3d$ sub-shell and so, with orbital quenching, each manganese ion has a moment of $5\mu_B$. The A system is made up of manganese ions in alternate (111) planes with parallel magnetic axes in these planes. The B layers of ions in (111) planes, with reversely-pointing magnets, are sandwiched between the A planes.

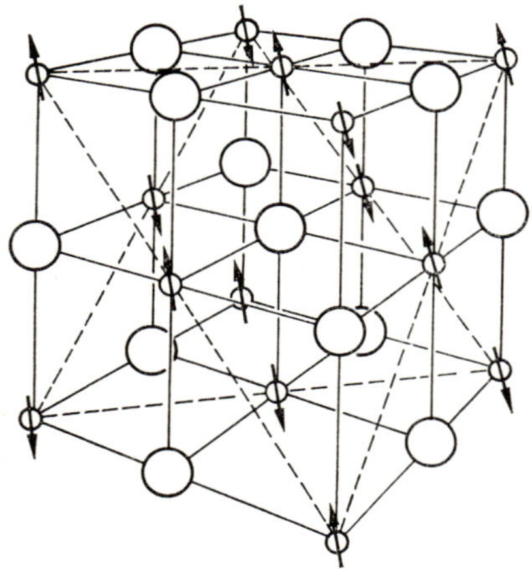

FIG. 5.9. Structure of MnO: larger spheres represent oxygen ions.

The diameter of the oxygen ions is about 2·64 Å and that of the Mn^{2+} ions about 1·82 Å, and it is apparent that the distance between the metal ions in the crystal is too great for the direct exchange interaction mentioned in Section 5.6 to be effective. In any case negative interaction would require the ratio of interatomic distance to the diameter of the $3d$ sub-shell of the manganese ions to be less than 1·5, whereas in MnO the ratio is substantially greater than this.

Reference to Fig. 5.9 shows that, in the cube-edge directions, the manganese ion magnetic moments are alternately antiparallel with an oxygen ion between them. It is considered that, in effect, the electronic structure of the oxygen ions diametrically transmits the negative interaction between the electron spins of the manganese ions on either side. This effect is referred to as superexchange and occurs in many oxides including the synthetic oxides known as ferrites which are discussed in Chapter 9.

PROBLEMS

5.1. A magnetic field is suddenly applied along the vertical axis of a freely-suspended iron rod. The rod experiences an impulse of angular momentum of 6×10^{-11} kg m^2 s^{-1}. If the volume of the rod is 8 cm^3 and $g = 2$, calculate the intensity of magnetization induced.

5.2. Calculate the angular velocity about the axis which must be imparted to a ferromagnetic rod to induce in it a magnetic field intensity of 40 A/m. Assume $g = 2$.

(Barnett Effect. *Note*: the torque on an atomic magnet is $H\mu_J \sin\theta$ and the torque is also $\omega p_J \sin\theta$.)

5.3. Estimate the magnitude of the Weiss molecular field at room temperature in iron, cobalt and nickel by means of the data in Table 5.1. Assume $g = 2$.

SUGGESTED READING

STONER, E. C. *Magnetism and Matter*, Methuen, London (1934).

KITTEL, C. *Solid State Physics*, 2nd edn., Wiley, New York (1957).

REFERENCES

[1] WEISS, P. 'L'hypothèse du champ moléculaire et la propriété ferromagnétique', *J. Phys. théor. appl.* **6,** 661 (1907).

[2] RICHARDSON, O. W. 'A mechanical effect accompanying magnetization', *Phys. Rev.* **26,** 248 (1908).

[3] EINSTEIN, A. and DE HASS, W. J. 'Experimental proof of Ampère's molecular currents', *Verh. dt. phys. Ges.* **17,** 152 (1915).

[4] BARNETT, S. J. 'Magnetization by rotation', *Phys. Rev.* **6,** 239 (1915).

[5] CHATTOCK, A. P. and BATES, L. F. 'On the Richardson gyromagnetic effect', *Trans. R. Soc.* A, **223,** 257 (1923).

[6] SUCKSMITH, W. and BATES, L. F. 'On a null method of measuring the gyromagnetic ratio', *Proc. R. Soc.* A, **104,** 499 (1923).

[7] HOWARTH, D. J. 'Electronic eigenvalues of copper, *Proc. R. Soc.* A, **220,** 513 (1953).

[8] KRUTTER, H. 'Energy bands in copper', *Phys. Rev.* **48,** 664 (1935).

[9] SLATER, J. C. 'The ferromagnetism of nickel', *Phys. Rev.* **49,** 537 (1936).

[10] KOSTER, G. F. 'Density of states curve for nickel', *Phys. Rev.* **98,** 901 (1955).

[11] HEISENBERG, W. 'On the theory of ferromagnetism', *Z. Phys.* **49,** 619 (1928).

[12] STONER, E. C. *Magnetism and Matter*, p. 358, Methuen, London (1934).

[13] BETHE, H. see STONER, Reference 12, p. 428.

[14] SLATER, J. C. 'Atomic shielding constants', *Phys. Rev.* **36,** 57 (1930).

[15] BIZETTE, H., SQUIRE, C. F. and TSAI, B. 'Transition point of magnetic susceptibility of MnO', *C. r. Acad. Sci., Paris*, **207,** 449 (1938).

[16] VAN VLECK, J. H. 'Theory of antiferromagnetism', *J. Chem. Phys.* **9,** 85 (1941).

Chapter 6
MAGNETIC ANISOTROPY

6.1 Directional Effects in Solids

The physical properties of crystals are in some cases strongly directional which, considering the geometry of crystal structures, is not surprising. Thus the elasticity and strength may depend markedly on crystallographic directions. Furthermore, crystals of the ferromagnetic metals and the ferrites are, in most cases, magnetically anisotropic to a remarkable degree, and have easy and difficult directions of magnetization. Apart, however, from the crystalline anisotropy a ferromagnetic body in bulk may also have, so far as an externally-applied field is concerned, easy and difficult directions depending on its shape. A small elastic deformation of a material in a particular direction may introduce further directional effects. We shall first consider the magnetic anisotropy of ferromagnetic metals arising from the crystal structure.

6.2 Magnetic Anisotropy Energy in Crystals of Cubic Structure

In a crystal of cubic structure we suppose that any physical property of the crystal has cubic symmetry. Thus the property in any cube-edge direction, for example, would be the same as that in the direction of any other cube edge. There would thus be six such directions corresponding, if the cube edges are taken as coordinate axes, to OX, OY and OZ and each of these directions reversed. Putting this in terms of the Miller Indices the equivalent directions would be [100], [010], [001], $[\bar{1}00]$, $[0\bar{1}0]$ and $[00\bar{1}]$. There would be eight directions equivalent to [111] and twelve equivalent to [110]. For any arbitrary direction [*abc*] it is easy to work out that there would be a total of 48 directions in which the particular physical property was the same.

Because of the Weiss field a ferromagnetic domain in a crystal is spontaneously magnetized to a saturation value appropriate to the temperature. The experimental evidence shows that the direction of the spontaneous magnetization, in the absence of any external restraints, will coincide with a particular crystal axis. For example, in the case of body-centred cubic iron the preferred direction for the vector representing M_S, is a cube edge. As just stated, there are six such equivalent directions. A crystal of iron is composed of a great many saturated domains. When the crystal is in a demagnetized condition, that is the resultant magnetization in bulk in any direction is zero, the M_S vectors in the constituent domains may be distributed among the six equivalent directions, although not necessarily equally in all these directions. In the case of face-centred cubic nickel the preferred directions for M_S are the eight directions corresponding to [111], a result which, in fact, complicates the geometrical pattern of the domains.

A state of minimum energy exists in a domain when the magnetization vector M_S lies along one of its preferred directions. If, by application of a magnetic field, M_S is turned away from this position, work must be done against the magnetocrystalline forces which tend to hold M_S in the easy direction, and energy is stored in the system. If for a cubic crystal the cube edges are taken as coordinate axes and the direction of M_S is specified by direction cosines l, m, n then the magnetocrystalline energy, or energy of magnetic anisotropy, of the domain is a function of l, m, n. Thus $W_A = f(l, m, n)$ where W_A is the magnetic anisotropy energy per unit volume.

The energy W must have cubic symmetry. The direction cosines in the equation must appear in even powers to give equivalence in positive and negative directions and there must also then be symmetry between l^2, m^2 and n^2. These conditions are fulfilled by the terms $l^2 + m^2 + n^2$, $l^2m^2 + m^2n^2 + n^2l^2$ and $l^2m^2n^2$. Further terms of higher even powers of l, m and n may be written down, but agreement with experiment is obtainable without proceeding further. Since $l^2 + m^2 + n^2 = 1$, we may thus write down the magnetic anisotropy energy as

$$W_A = K_0 + K_1(l^2m^2 + m^2n^2 + n^2l^2) + K_2(lmn)^2 \quad (6.1)$$

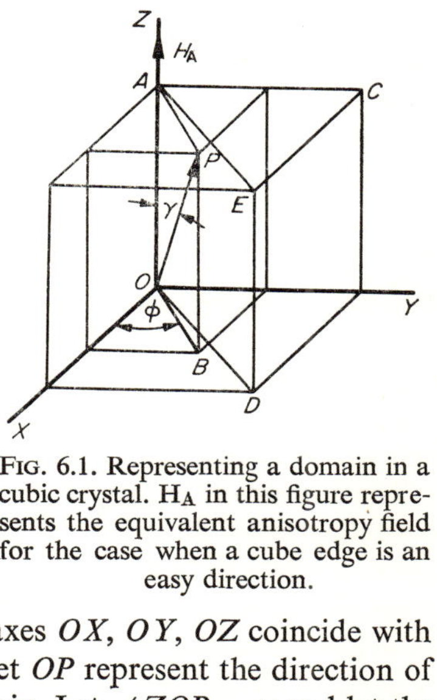

FIG. 6.1. Representing a domain in a cubic crystal. H_A in this figure represents the equivalent anisotropy field for the case when a cube edge is an easy direction.

The component of energy K_0 is independent of direction and so the directional properties of the domain are dependent upon the anisotropy constants K_1 and K_2 which, in different materials, may have either positive or negative values.

In Fig. 6.1 let the rectangular coordinate axes OX, OY, OZ coincide with cube edges of a domain in a cubic crystal. Let OP represent the direction of the saturation magnetization M_S of the domain. Let $\angle ZOP = \gamma$ and let the plane containing OZ and OP be inclined to the XZ plane at an angle ϕ as shown in the figure. The direction cosines for M_S may be determined from the figure and are $l = \sin\gamma\cos\phi$, $m = \sin\gamma\sin\phi$ and $n = \cos\gamma$. Substituting these values in equation (6.1), neglecting the constant term K_0, and simplifying, gives the anisotropy energy per unit volume as follows:

$$W_A = (K_1/64)\{(3 - 4\cos 2\gamma + \cos 4\gamma)(1 - \cos 4\phi) + 8(1 - \cos 4\gamma)\} + (K_2/256)\{(2 - \cos 2\gamma - 2\cos 4\gamma + \cos 6\gamma)(1 - \cos 4\phi)\} \quad (6.2)$$

The energy, when M_S lies along any of the three principal directions in the crystal, [100], [110] or [111], may be obtained by substituting the appropriate values of l, m and n in equation (6.1). For the three directions we have

$$[100],\quad l = 1,\quad m = 0,\quad n = 0,\quad W_A = 0$$
$$[110],\quad l = 0,\quad m = n = 1/2^{\frac{1}{2}},\quad W_A = K_1/4$$
$$[111],\quad l = m = n = 1/3^{\frac{1}{2}},\quad W_A = K_1/3 + K_2/27$$

For [100] to be the lowest energy direction of these three we must have $K_1 > 0$ and $K_2 > -9K_1$. [110] is of lowest energy if $K_1 < 0$ and $K_2 > -9K_1/4$, while [111] is the direction of lowest energy if $K_2 < -9K_1/4$ when K_1 is

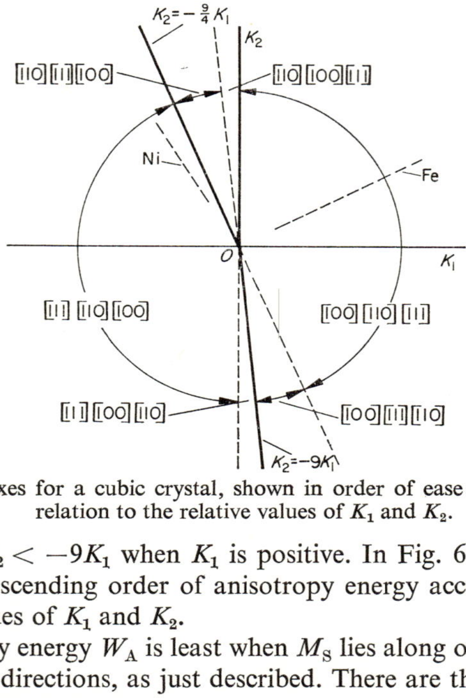

FIG. 6.2. Crystal axes for a cubic crystal, shown in order of ease of magnetization, in relation to the relative values of K_1 and K_2.

negative, and $K_2 < -9K_1$ when K_1 is positive. In Fig. 6.2 the crystal axes are marked in ascending order of anisotropy energy according to the sign and relative values of K_1 and K_2.

The anisotropy energy W_A is least when M_S lies along one of the principal crystallographic directions, as just described. There are therefore crystalline forces tending to hold M_S in the preferred direction and, for small displacements of M_S from this direction, these crystalline forces may be imagined as equivalent to a magnetic field H_A along the preferred direction, thus acting to restore M_S to the equilibrium position. The value of this equivalent field H_A for small displacements of M_S is readily obtained from equation (6.2).

In Fig. 6.1 let the anisotropy field H_A, for the case when the cube edge is the easy direction, act along OZ and let M_S move in the plane $OAPB$. The anisotropy restoring torque per unit volume of the domain acting upon M_S is $dW_A/d\gamma$. This must be equated to the restoring torque due to H_A. Therefore $H_A M_S \sin\gamma = dW_A/d\gamma$. Whence, using equation (6.2) with ϕ constant,

$$H_A M_S \sin\gamma = (K_1/64)\{(8\sin 2\gamma - 4\sin 4\gamma)(1 - \cos 4\phi) + 32\sin 4\gamma\} + (K_2/256)\{(2\sin 2\gamma + 8\sin 4\gamma - 6\sin 6\gamma)(1 - \cos 4\phi)\}$$

We are concerned only with the value of H_A when γ is small, and then

$$H_A M_S \gamma = 2K_1 \gamma$$

H_A is therefore, for a [100] direction, independent of K_2 and of ϕ and

$$H_A = 2K_1/M_S$$

If [110] is the easy direction the equivalent anisotropy field H_A for this direction may be found similarly. Now, however, the value of H_A depends upon the plane in which M_S is rotated. In Fig. 6.1 let M_S be in equilibrium in the direction OC and suppose rotation of M_S is in the YZ plane. If M_S is displaced by an angle θ in this plane from the direction OC then we may put $\gamma = \pi/4 + \theta$ with $\phi = \pi/2$. Equation (6.2) then becomes

$$W_A = (K_1/8)(1 + \cos 4\theta)$$

As before $$H_A M_S \sin\theta = dW_A/d\theta = -(K_1/2)\sin 4\theta$$

Thus for rotation of M_S in a (100) plane and when θ is small

$$H_A = -2K_1/M_S$$

K_1 must be negative for [110] to be an easy direction, and so H_A remains a positive quantity.

If, however, M_S is to be rotated in a (110) plane, we may suppose in Fig. 6.1 that M_S lies at rest along OD and is to be rotated in the plane $OAED$. We may now put $\phi = \pi/4$ and $\gamma = \pi/2 + \theta$ and then

$$W_A = (K_1/32)(7 + 4\cos 2\theta - 3\cos 4\theta) + (K_2/128)(2 + \cos 2\theta - 2\cos 4\theta - \cos 6\theta)$$

If now H_A is the equivalent anisotropy field for [110] and rotation in the plane mentioned then, once more, $H_A M_S \sin\theta = dW_A/d\theta$ and

$$dW_A/d\theta = (K_1/8)(-2\sin 2\theta + 3\sin 4\theta) + (K_2/64)(-\sin 2\theta + 4\sin 4\theta + 3\sin 6\theta)$$

From which, when θ is small

$$H_A = (K_1 + K_2/2)/M_S$$

When [111] is the direction of lowest energy for M_S the equivalent field H_A for this direction and rotation of M_S in a (110) plane, may again be obtained from equation (6.2). Suppose that M_S lies along the direction OE and is to rotate in the plane $OAED$. Putting $\phi = \pi/4$ and $\gamma = \delta + \theta$ where $\delta = \angle AOE = \cos^{-1}(1/3^{\frac{1}{2}})$ equation (6.2) becomes

$$W_A = (K_1/32)\{7 - 4\cos(2\delta + 2\theta) - 3\cos(4\delta + 4\theta)\} + (K_2/128)\{2 - \cos(2\delta + 2\theta) - 2\cos(4\delta + 4\theta) + \cos(6\delta + 6\theta)\}$$

Then $H_A M_S \sin\theta = dW_A/d\theta$. Carrying out the differentiation, substituting for δ and making θ small gives

$$H_A = -4(K_1 + K_2/3)/3M_S$$

The foregoing relations for a domain in a cubic crystal are summarized in Table 6.1.

TABLE 6.1

CUBIC CRYSTAL STRUCTURES

Crystal Direction	[100]	[110]	[111]
Anisotropy energy W_A (the constant term K_0 omitted)	0	$K_1/4$	$K_1/3 + K_2/27$
Conditions for least energy	$K_1 > 0$ $K_2 > -9K_1$	$K_1 < 0$ $K_2 > -9K_1/4$	$K_1 < 0$ $K_2 < -9K_1/4$ $K_1 > 0$ $K_2 < -9K_1$
Equivalent anisotropy field H_A	$2K_1/M_S$	(100) plane $-2K_1/M_S$ (110) plane $(K_1 + K_2/2)/M_S$	$-4(K_1+K_2/3)/3M_S$

6.3 Magnetic Anisotropy Energy in Crystals of Hexagonal Structure

The directional magnetic properties of hexagonal crystals are of interest in connection with cobalt and some ferrites. Figure 6.3 represents part of a domain in a ferromagnetic hexagonal crystal in which the axes OW, OX, OY, 120° apart are in the basal plane and OZ, perpendicular to this plane, coincides with the c-axis of the crystal. In the figure the saturation vector M_S is in the position OP where $\angle ZOP = \theta$, and the vertical plane containing OP is rotated through an angle ϕ from the OW axis. Crystal symmetry requires that the magnetic properties should have a twofold symmetry for rotation of M_S in the plane ZOP, but a sixfold symmetry for a given value of θ and rotation about the c-axis. In practice it is found that the variation of magnetic properties with direction in the latter case is small, and the magnetic anisotropy energy may be written to a close approximation as

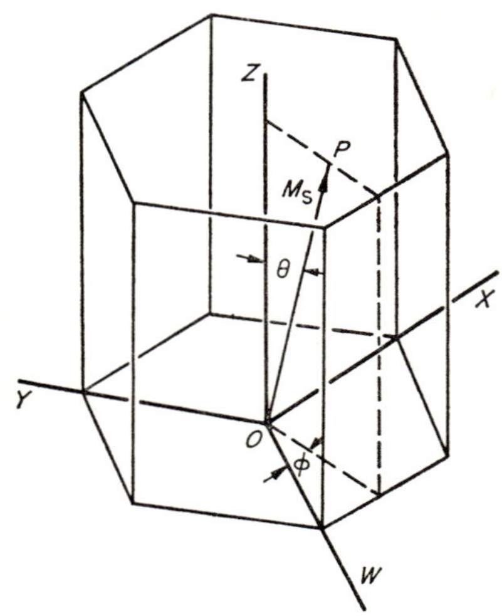

FIG. 6.3. Illustrating a domain in a hexagonal crystal.

$$W_A = K_0 + K_1 \sin^2 \theta + K_2 \sin^4 \theta \qquad (6.3)$$

It is convenient to use terms in $\sin \theta$, rather than the direction cosines, and

K_1 and K_2 are not now related to the constants used for a cubic crystal. To preserve the twofold symmetry already mentioned $\sin\theta$ must appear in the equation in even powers.

In the absence of an externally-applied field M_S will take up a stable or easy position at some value of θ which makes W_A a minimum. This value of θ may be 0°, 90°, or indeed an intermediate value, depending on the relative magnitudes and signs of K_1 and K_2.

Neglecting the constant term K_0, which does not affect the directional properties, equation (6.3) may be rewritten as

$$W_A = (K_1/2 + 3K_2/8) - (K_1/2 + K_2/2)\cos 2\theta + (K_2/8)\cos 4\theta$$

$$\therefore dW_A/d\theta = (K_1 + K_2)\sin 2\theta - (K_2/2)\sin 4\theta \qquad (6.4)$$

and

$$d^2W_A/d\theta^2 = 2(K_1 + K_2)\cos 2\theta - 2K_2\cos 4\theta \qquad (6.5)$$

W_A is a minimum when $dW_A/d\theta = 0$ and $d^2W_A/d\theta^2 > 0$. $dW_A/d\theta$ is zero when $\theta = 0°, 90°$ or $\sin^{-1}(-K_1/2K_2)^{\frac{1}{2}}$. The values of W_A and of $d^2W_A/d\theta^2$ for these values of θ are given in Table 6.2. From these quantities it is found that $\theta = 0°$ is the direction of lowest energy if $K_1 + K_2 > 0$ or $K_2 > -K_1$ and if $K_1 > 0$. The relative values of K_1 and K_2 which make $\theta = 0°$ the easy direction are shown diagrammatically in Fig. 6.4.

If $\theta = \sin^{-1}(-K_1/2K_2)^{\frac{1}{2}}$ is to be the easy direction, then W_A in this direction must be less than W_A when $\theta = 0°$, and therefore $-K_1^2/4K_2 < 0$. It follows that K_2 must be positive and, for a minimum of W_A, $d^2W_A/d\theta^2$ must be positive. Thus $-(2K_1/K_2)(K_1 + 2K_2) > 0$. With K_2 positive this condition is met if K_1 is negative and $(K_1 + 2K_2)$ is positive. In that case $K_1 + 2K_2 > 0$ or $K_2 > -K_1/2$ and, with K_2 positive, K_1 must be negative. The required conditions for W_A to be a minimum are therefore $K_1 < 0$ and $K_2 > -K_1/2$.

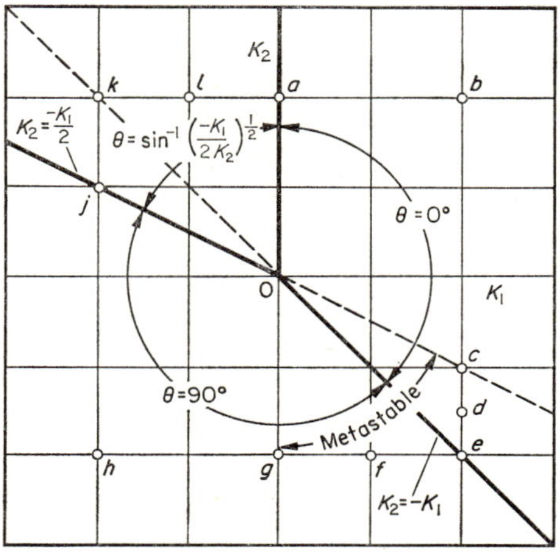

FIG. 6.4. Showing values of θ for lowest anisotropy energy in a domain of a hexagonal crystal in relation to the relative values of K_1 and K_2.

The conditions required, so that $\theta = 90°$ gives the lowest value of W_A, follow from the foregoing and are given in the table. Figure 6.4 shows the results diagrammatically.

TABLE 6.2

HEXAGONAL CRYSTAL STRUCTURE

Crystal Direction	$\theta = 0°$	$\theta = \sin^{-1}\sqrt{(-K_1/2K_2)}$	$\theta = 90°$
Anisotropy energy W_A	0	$-K_1^2/4K_2$	$K_1 + K_2$
$d^2W_A/d\theta^2$	$2K_1$	$-(2K_1/K_2)(K_1 + 2K_2)$	$-2(K_1 + 2K_2)$
Conditions for least energy	$K_1 > 0$ $K_2 > -K_1$	$K_1 < 0$ $K_2 > -K_1/2$	$\begin{cases} K_1 < 0 \\ K_2 < -K_1/2 \end{cases}$ $\begin{cases} K_1 > 0 \\ K_2 < -K_1 \end{cases}$
Equivalent anisotropy field H_A	$2K_1/M_S$	$-2K_1(K_1+2K_2)/K_2M_S$	$-2(K_1+2K_2)/M_S$

The relations between W_A and θ for a number of different relative values of K_1 and K_2 are given by the curves in Fig. 6.5. Since the shape of these curves depends only on the ratio of K_1 to K_2, arbitrary units have been chosen for these quantities. The curves are lettered *a* . . . *l* to correspond with the arbitrary points similarly marked on the diagram in Fig. 6.4. It will be seen that the curves from *c* to *g* in Fig. 6.5 have two minima. The domain may be said to be metastable in these cases, since it would have two stable angular positions for the vector M_S, namely when $\theta = 0°$ and $\theta = 90°$. These energy minima are, however, unequal except at the change-over point *e* where K_1 is positive and $K_1 = -K_2$.

The lower part of Fig. 6.5 gives energy curves in which W_A is a minimum when $\theta = \sin^{-1}(-K_1/2K_2)^{\frac{1}{2}}$. In this case the stable positions of the vector M_S lie on the surface of a cone which has a semi-vertical angle of this amount. For example, curve *l* corresponds to $K_2 = -2K_1$ and the cone swept out by M_S has then a semi-vertical angle $\theta = \sin^{-1}(1/2) = 30°$.

When the preferred direction of M_S in the domain is along the *c*-axis, that is with $\theta = 0°$, the crystalline forces acting on the vector may again be considered, for small deviations of M_S from this minimum energy position, as equivalent to an axial magnetic field H_A. As in the case of the cubic crystals we may put for small values of θ, $H_A M_S \sin\theta = dW_A/d\theta$. Thus from equation (6.4)

$$H_A M_S \sin\theta = (K_1 + K_2)\sin 2\theta - (K_2/2)\sin 4\theta$$

When θ is small this gives

$$H_A = 2K_1/M_S$$

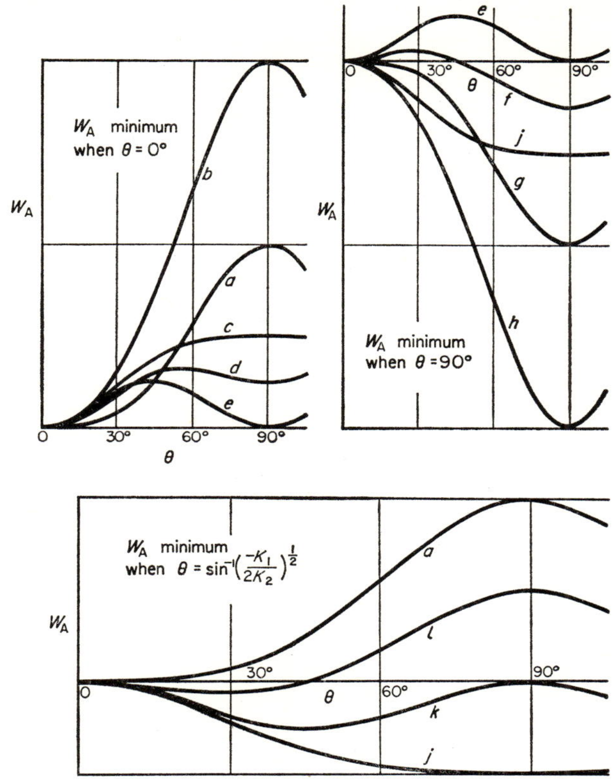

FIG. 6.5. Anisotropy energy in a hexagonal domain in relation to θ for different relative values of K_1 and K_2. Letters on the curves correspond with those on the diagram in Fig. 6.4.

When $\theta = 90°$ is the preferred direction for M_S, then the equivalent anisotropy field H_A acts at 90° to the c-axis. In this case M_S will be in equilibrium in the basal plane but may rotate freely in it. To find H_A put $\theta = 90° + \theta'$ and then

$$H_A M_S \sin \theta' = dW_A/d\theta'$$

$$\therefore H_A M_S \sin \theta' = (K_1 + K_2) \sin 2(90° + \theta') - (K_2/2) \sin 4(90° + \theta')$$

When θ' is small this gives

$$H_A = -2(K_1 + 2K_2)/M_S$$

The equivalent anisotropy field H_A for the third case, when the preferred direction is given by $\theta = \sin^{-1}(-K_1/2K_2)^{\frac{1}{2}}$ is obtainable similarly. In this case put $\theta = \gamma + \theta'$ where $\sin \gamma = (-K_1/2K_2)^{\frac{1}{2}}$.
We then have

$$H_A M_S \sin \theta' = dW_A/d\theta'$$

$$\therefore H_A M_S \sin \theta' = (K_1 + K_2) \sin 2(\gamma + \theta') - (K_2/2) \sin 4(\gamma + \theta')$$

When θ' is small this reduces to

$$H_A M_S = 2(K_1 + K_2) \cos 2\gamma - 2K_2 \cos 4\gamma$$

and on substituting for γ we have

$$H_A = -2K_1(K_1 + 2K_2)/K_2 M_S$$

The field H_A now acts to keep the vector M_S on the surface of the cone of semi-vertical angle $\sin^{-1}(-K_1/2K_2)^{\frac{1}{2}}$, but M_S may freely sweep around on this surface.

The values of H_A in the three cases are given in Table 6.2. Bearing in mind the signs and relative magnitudes of K_1 and K_2 it will be found, as may be expected, that H_A is always a positive quantity.

6.4 Magnetization Curves of Single Crystals of Cubic Structure

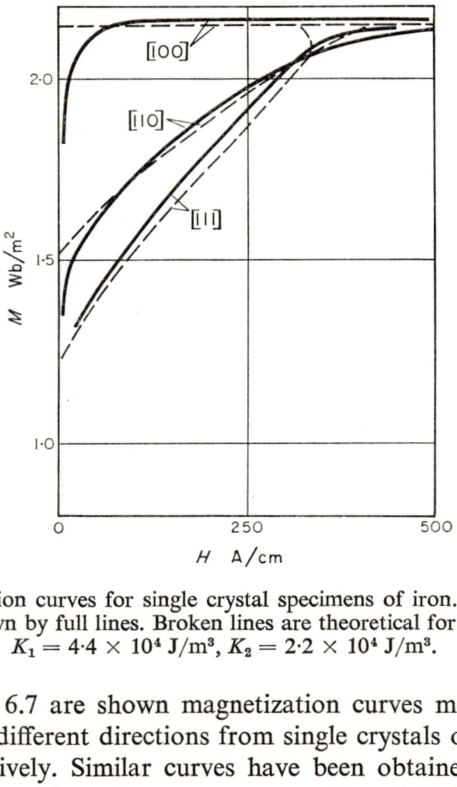

FIG. 6.6. Magnetization curves for single crystal specimens of iron. Honda and Kaya's observed results shown by full lines. Broken lines are theoretical for $M_S = 2{\cdot}14$ Wb/m², $K_1 = 4{\cdot}4 \times 10^4$ J/m³, $K_2 = 2{\cdot}2 \times 10^4$ J/m³.

In Figs. 6.6. and 6.7 are shown magnetization curves measured on small specimens cut in different directions from single crystals of iron [1] and of nickel [2], respectively. Similar curves have been obtained also for many ferromagnetic alloys. The anisotropy constants K_1 and K_2 for these materials may be determined in the manner mentioned in this and the following sections. In the case of iron K_1 and K_2 are both positive. The broken line, marked Fe in Fig. 6.2, shows that iron appears on the chart in the region where the domain anisotropy energy W_A for different crystal directions is in the sequence

[100], [110], [111]. The magnetization curves for iron follow the same order, with [100] the easiest direction of magnetization, and [111] the most difficult.

The curves for nickel are in precisely the opposite order, [111] being easiest and [100] most difficult. For nickel at room temperature K_1 is negative and K_2 positive, and the approximate position of nickel in Fig. 6.2 is shown by the broken line marked Ni. This is in the region of the chart where the sequence for W_A is [111], [110], [100], which is the same as that of the magnetization curves.

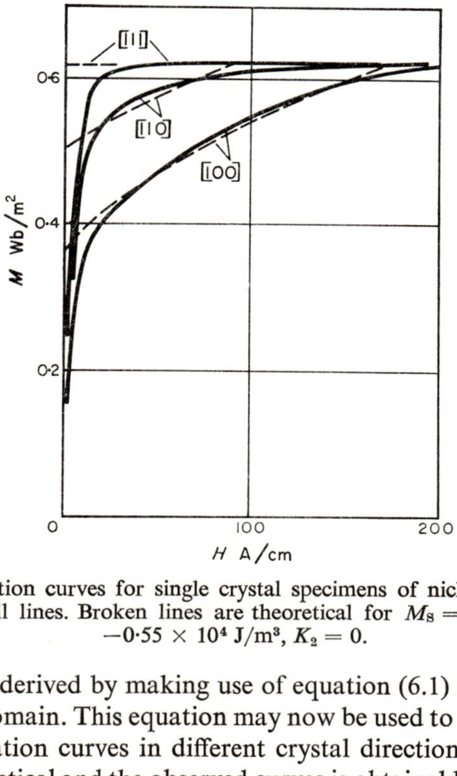

FIG. 6.7. Magnetization curves for single crystal specimens of nickel. Kaya's observed results shown by full lines. Broken lines are theoretical for $M_S = 0{\cdot}62$ Wb/m², $K_1 = -0{\cdot}55 \times 10^4$ J/m³, $K_2 = 0$.

Figure 6.2 was derived by making use of equation (6.1) for the anisotropy energy W_A of a domain. This equation may now be used to derive expressions for the magnetization curves in different crystal directions. Fair agreement between the theoretical and the observed curves is obtainable if suitable values are selected for the constants K_1 and K_2, fitting the curves being, indeed, one method of experimentally determining the values of the constants. It should, however, be noted that theory and experiment in the cases here considered relate only to the upper part of the magnetization curves and give no detailed information on the form of the curves for lower intensities of magnetization and relatively weak fields.

Figure 6.8(a) shows, diagrammatically, a domain in which a field H is applied in a direction having direction cosines l_1, m_1, n_1, and the vector M_S has the direction cosines $l = \sin\gamma \cos\phi$, $m = \sin\gamma \sin\phi$, $n = \cos\gamma$. The magnetic anisotropy energy per unit volume W_A is given by equation (6.2).

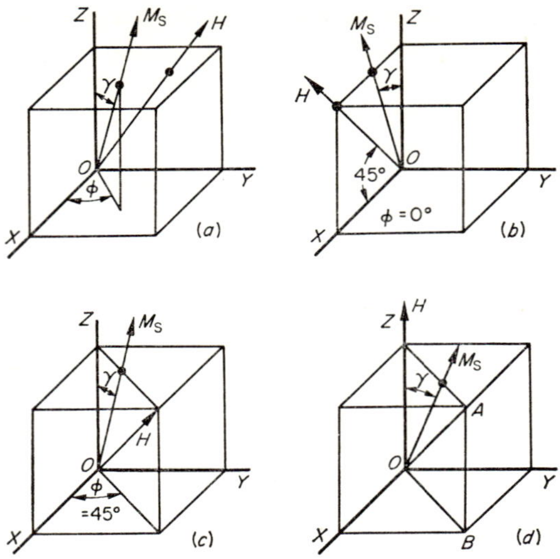

FIG. 6.8. Diagrams representing domains.

If θ is the angle between H and M_S there is also magnetic potential energy per unit volume $W_P = -HM_S \cos \theta$. Whence

$$W_P = -HM_S(l_1 l + m_1 m + n_1 n)$$

or

$$W_P = -HM_S(l_1 \sin \gamma \cos \phi + m_1 \sin \gamma \sin \phi + n_1 \cos \gamma) \quad (6.6)$$

The total energy is given by

$$W = W_P + W_A \quad (6.7)$$

For a given applied field H, M_S will take up an equilibrium position which makes the total energy W a minimum. The intensity of magnetization M in the direction of H is $M = M_S \cos \theta$. Thus if θ can be found then a relation between H and M is obtainable, that is, the magnetization curve of the domain can, in principle, be determined. In the quite general case, and for any positive or negative values of K_1 and K_2, this calculation turns out to be difficult, since M_S may not always rotate in one plane towards H as H increases. The difficulty, however, does not arise in many particular cases which include those we shall consider for iron, and also for nickel, at least in the [110] direction.

Consider a crystal like iron consisting of domains in which the preferred direction for the saturation vector M_S is a cube edge. Initially the M_S vectors will be distributed in the multi-domain crystal among the six possible cube-edge directions. If, however, a field H, assumed small in the first instance, is applied to the crystal, the magnetic potential energy of the individual domains will depend on the angle between H and M_S. The whole system will have minimum potential energy if all the domain M_S vectors take up the cube-edge direction nearest to the direction of H. Thus if the field were applied along a cube edge the energy would be a minimum when M_S in all the domains was aligned with the field. Thus theoretically, in this case, the application of a small H might be expected to cause the magnetization M to rise from zero to M_S, and to remain at this value for any subsequent increase in H.

In Fig. 6.8(*b*) let the field H be applied to the crystal in a [101] direction as shown. The magnetic anisotropy energy of the crystal will be a minimum when M_S in the domains is aligned along the directions OX and OZ, when H is small. As H is increased, the vectors M_S will rotate towards the direction of the field. We need consider only one such domain in which we suppose M_S is initially along OZ, and in the presence of the field H, is rotated through the angle γ with respect to OZ as shown in the figure. We require to find first whether M_S will rotate in the plane containing H and OZ, that is in the (010) plane.

The direction of H is given by $l_1 = n_1 = 1/2^{\frac{1}{2}}$, $m_1 = 0$, whence, from equation (6.6)

$$W_P = -(HM_S/2^{\frac{1}{2}})(\sin\gamma\cos\phi + \cos\gamma)$$

The total energy $W = W_P + W_A$, where W_A is given by equation (6.2). M_S will be in stable equilibrium in the (010) plane, for a given applied field H and angle γ if, when $\phi = 0$, $dW/d\phi = 0$ and $d^2W/d\phi^2 > 0$ and if also $dW/d\gamma = 0$. When these criteria are applied it is found that, as H increases, M_S will stay in the (010) plane for all values of K_1 and K_2 within the region marked [100], [110], [111] in the chart of Fig. 6.2, that is when K_1 is positive and $K_2 > -9K_1/4$.

For rotation of M_S in the (010) plane, we put $\phi = 0$ and then

$$W = -(HM_S/2^{\frac{1}{2}})(\sin\gamma + \cos\gamma) + (K_1/8)(1 - \cos 4\gamma)$$

Equilibrium is reached when $dW/d\gamma = 0$. Whence

$$-(HM_S/2^{\frac{1}{2}})(\cos\gamma - \sin\gamma) + (K_1/2)\sin 4\gamma = 0 \tag{6.8}$$

If M is the intensity of magnetization resolved in the direction of H then

$$M = M_S\cos(\pi/4 - \gamma) \tag{6.9}$$

Eliminating γ between equations (6.8) and (6.9) gives

$$H = (2K_1/M_S)(M/M_S)\{2(M/M_S)^2 - 1\} \tag{6.10}$$

Equation (6.10) represents the magnetization curve for a crystal in a [110] direction when K_1 is positive and $K_2 > -9K_1/4$. The curve starts from the point where $H = 0$, and $M = M_S/2^{\frac{1}{2}}$, and reaches saturation, that is $M = M_S$, when $H = 2K_1/M_S$.

In Fig. 6.8(*c*) the field H is applied in a [111] direction. We suppose that a small field H is enough to cause the M_S vectors in the domains to take up the nearest cube-edge directions to the direction of H. The vectors would thus be initially aligned along the axes OX, OY, OZ. We need consider only one such domain. When H is small let M_S in this domain be along OZ and rotate towards H as H increases. Then $l_1 = m_1 = n_1 = 1/3^{\frac{1}{2}}$ and, for M_S in any position,

$$W_P = -(HM_S/3^{\frac{1}{2}})(\sin\gamma\cos\phi + \sin\gamma\sin\phi + \cos\gamma) \tag{6.11}$$

The total energy is $W = W_P + W_A$ where W_A is again given by equation (6.2). In the figure M_S is shown to lie in the diagonal plane containing OZ and H. If the vector M_S is to be in equilibrium and to remain in this plane as H increases, the conditions to be satisfied are that, for a given value of H and γ, and when $\phi = \pi/4$, $dW/d\phi = 0$, $d^2W/d\phi^2 > 0$, and $dW/d\gamma = 0$. Calculations made for a number of values of γ between 0° and $\cos^{-1}(1/3^{\frac{1}{2}})$ show that these criteria are met when K_1 is positive and K_2 lies between the limits given approximately by $10K_1 > K_2 > -2K_1$. The vector M_S will therefore rotate in the plane containing OZ and H for nearly the whole of the sector of the chart in Fig. 6.2 which is marked [100], [110], [111], and which contains the values of K_1 and K_2 appropriate to iron.

Bearing in mind these limitations on the values of K_1 and K_2, we may therefore put $\phi = \pi/4$ and then, from equations (6.11) and (6.2),

$$W = -(HM_S/3^{\frac{1}{2}})(\sqrt{2}\sin\gamma + \cos\gamma) + (K_1/32)(7 - 4\cos 2\gamma - 3\cos 4\gamma) \\ + (K_2/128)(2 - \cos 2\gamma - 2\cos 4\gamma + \cos 6\gamma)$$

The vector M_S will be in equilibrium when

$$dW/d\gamma = 0 \tag{6.12}$$

Also the intensity of magnetization M in the direction of H is given by

$$M = M_S \cos(\delta - \gamma) \tag{6.13}$$

where $\delta = \cos^{-1}(1/3^{\frac{1}{2}})$.

Carrying out the differentiation and eliminationg γ between equations (6.12) and (6.13) gives the following relation between M and H, where for convenience we put $a = M/M_S$:

$$HM_S = K_1(\sqrt{2}/3)(1 - a^2)^{\frac{1}{2}}(4a^2 - 1) + K_1(a/3)(7a^2 - 3) \\ - K_2(\sqrt{2}/18)(1 - a^2)^{\frac{1}{2}}(10a^4 - 9a^2 + 1) + K_2(a/18)(23a^4 - 16a^2 + 1) \tag{6.14}$$

The magnetization curve for a [111] direction of the crystal is given by equation (6.14). The curve starts from the point $H = 0$, $M = M_S/3^{\frac{1}{2}}$ and ends when $M = M_S$, or $a = 1$, where $HM_S = (4/3)(K_1 + K_2/3)$.

Equations (6.10) and (6.14) are plotted in Fig. 6.9. The [110] curve is independent of K_2 but the effect of different values of K_2 on the [111] curve is shown in the figure. The way in which the [111] curves approach saturation will be noted.

A value of K_1 for iron may be found by fitting the theoretical curve for the [110] direction to the experimental curve in Fig. 6.6. The broken line, obtained from equation (6.10), is drawn for $K_1 = 4{\cdot}4 \times 10^4$ J/m³ and $M_S = 2{\cdot}14$ Wb/m² and is seen to fit the observed curve fairly well. In the case of the [111] direction the broken line in Fig. 6.6, obtained from equation (6.14), is drawn for the same values of K_1 and M_S, and for $K_2 = 2{\cdot}2 \times 10^4$ J/m³. The theoretical and observed curves cannot be fitted closely enough in this case to get a very accurate value for K_2.

In the case of a crystal like nickel in which [111] is the lowest energy direc-

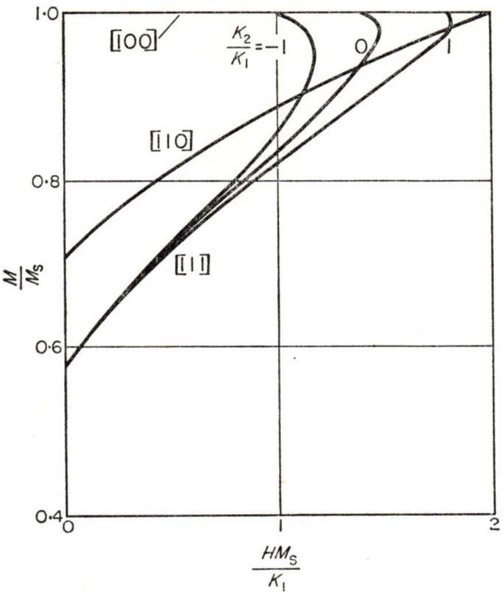

FIG. 6.9. Theoretical magnetization curves for single crystals when [100] is the easy and [111] the most difficult direction, showing also the effect of K_2 on the [111] curve.

tion for M_S in a domain, we are concerned with deriving the equations of the magnetization curves in the [100] and [110] directions. Let Fig. 6.8(*d*) represent a crystal with H applied in the direction OZ, that is in an [001] direction. A small field, we suppose, is enough to cause all the domain vectors to distribute themselves in angular positions about the [001] axis corresponding to the low energy [111] direction. There would be four such positions in this case, of which OA in Fig. 6.8(*d*) is one. We need consider only the case shown in the figure where, as H increases, M_S for a domain is rotated from the initial position OA towards the direction of H. Applying the same criteria as before it is found that M_S will remain in the diagonal plane $OZAB$ provided that $K_1 < 0$ and K_2 is within the limits given by $0 < K_2 < -1{\cdot}2K_1$, approximately. With these limitations on the anisotropy constants in mind we have $W_P = -HM_S \cos\gamma$ while W_A is again given by equation (6.2) with $\phi = \pi/4$ and the total energy $W = W_P + W_A$. As before we put $dW/d\gamma = 0$. Also $M = M_S \cos\gamma$. When γ is eliminated from these two equations the equation of the magnetization curve for the [001] direction is found to be

$$HM_S = a(3a^2 - 1)\{-K_1 + (K_2/2)(a^2 - 1)\} \qquad (6.15)$$

where $a = M/M_S$.

When H is applied along OB in Fig. 6.8(*d*), that is in a [110] direction, we put $W_P = -HM_S \sin\gamma$ and $M/M_S = \sin\gamma$. In this case M_S will remain in one plane for values of K_1 and K_2 in the whole of the sector marked [111], [110], [100] in Fig. 6.2. Following the same procedure as before the equation of the magnetization curve for the [110] direction is found to be

$$HM_S = a(3a^2 - 2)\{-K_1 - K_2(a^2/2)\} \qquad (6.16)$$

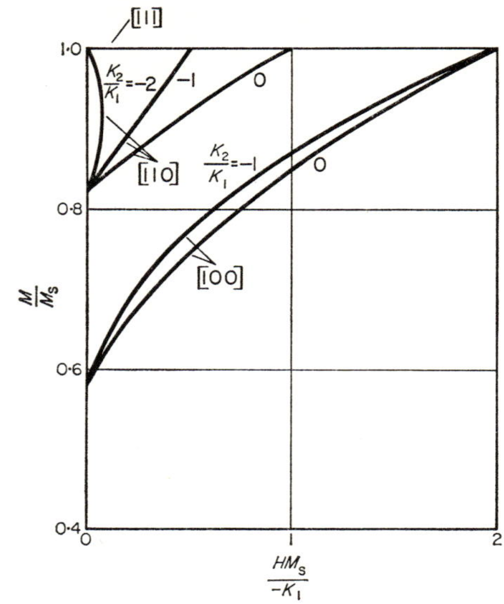

FIG. 6.10. Theoretical magnetization curves for single crystals when [111] is the easy and [100] the most difficult direction, showing also the effect of K_2 on the curves.

Equation (6.15) is plotted in Fig. 6.10 for the cases where $K_2/K_1 = 0$ and -1, and equation (6.16) when $K_2/K_1 = 0, -1$ and -2. It will be remembered that K_1 is negative in these equations.

Equations (6.15) and (6.16) are also plotted, for the [100] and [110] directions, respectively, in Fig. 6.7, for comparison with Kaya's observed curves for nickel. There is only moderately good agreement between the observed and theoretical curves.

6.5 Magnetization Curves of Single Crystals of Hexagonal Structure

Figure 6.11 shows the magnetization curves observed by KAYA for single crystal specimens of cobalt [3]. It will be seen that the easy direction for cobalt, which is of hexagonal crystal structure, is along the c-axis. Magnetization is very much more difficult for directions in the basal plane. The curves observed by Kaya for the directions B and C in the figure were very close together and are shown by a single curve. The very strong magnetic anisotropy of cobalt may be seen by comparing the scales of H in Fig. 6.11 with those for iron and nickel in Figs. 6.6 and 6.7.

It will be seen in Fig. 6.3 and Section 6.3. that, in hexagonal crystals, the directions of lowest anisotropy energy in a domain are when $\theta = 0°$, $90°$ or $\sin^{-1}(-K_1/2K_2)^{\frac{1}{2}}$ depending on the values of the anisotropy constants K_1 and K_2. We will first consider the case when the preferred direction of M_S in a domain is along the c-axis, corresponding in fact to the case of cobalt. If a small field H is applied to a crystal in the axial direction the magnetic potential energy will be a minimum when M_S in all the domains coincides in direc-

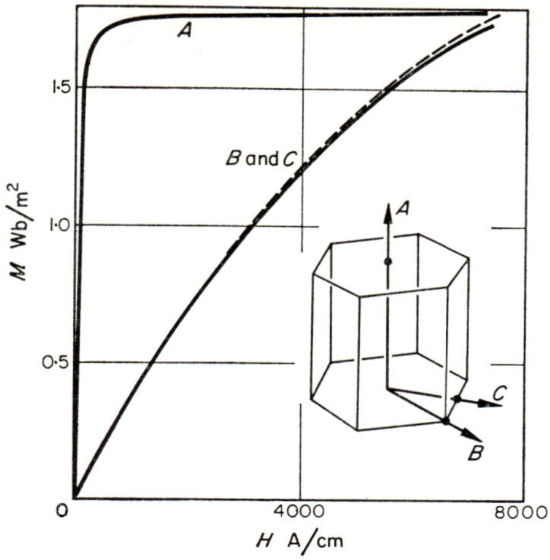

FIG. 6.11. Magnetization curves for cobalt. Full lines are Kaya's observed curves. Broken lines are calculated for $M_S = 1{\cdot}79$ Wb/m², $K_1 = 41 \times 10^4$ J/m³, $K_2 = 13 \times 10^4$ J/m³.

tion with H. Theoretically, then, we suppose that the magnetization rises from zero to saturation on the application of a small field.

When the field is applied in the basal plane M_S in each domain will initially be directed axially and, as H is increased, M_S in each domain will rotate away from the axis in the direction of the field. In Fig. 6.3 if a field H causes M_S to rotate through the angle θ from the axis then the magnetic potential energy per unit volume is $W_P = -HM_S \sin\theta$ and, from equation (6.3), the magnetic anisotropy energy is

$$W_A = K_0 + K_1 \sin^2\theta + K_2 \sin^4\theta$$

or $W_A = K_0 + K_1/2 + 3K_2/8 - (1/2)(K_1 + K_2)\cos 2\theta + (K_2/8)\cos 4\theta$

and the total energy $W = W_P + W_A$. M_S reaches a position of equilibrium when $dW/d\theta = 0$. Whence

$$-HM_S\cos\theta + (K_1 + K_2)\sin 2\theta - (K_2/2)\sin 4\theta = 0 \qquad (6.17)$$

The intensity of magnetization in the direction of H is

$$M = M_S \sin\theta \qquad (6.18)$$

Eliminating θ from equations (6.17) and (6.18) gives

$$HM_S = 2(M/M_S)\{K_1 + 2K_2(M/M_S)^2\} \qquad (6.19)$$

This is the equation of the magnetization curve for directions in the basal plane. The theoretical curve fits the observed curve very closely in Fig. 6.11, as shown by the broken line, for the values

$$M_S = 1{\cdot}79 \text{ Wb/m}^2,\ K_1 = 41 \times 10^4 \text{ J/m}^3, \text{ and } K_2 = 13 \times 10^4 \text{ J/m}^3.$$

When the easy direction for the domains is in the basal plane and the field H is applied axially, the magnetization curve for the axial direction may be obtained by putting $W_P = -HM_S \cos\theta$. When $dW/d\theta = 0$ we have

$$HM_S \sin\theta + (K_1 + K_2)\sin 2\theta - (K_2/2)\sin 4\theta = 0$$

Also

$$M = M_S \cos\theta$$

Eliminating θ between these two equations gives

$$HM_S = -2(M/M_S)[K_1 + 2K_2\{1 - (M/M_S)^2\}] \qquad (6.20)$$

Equation (6.20) represents the magnetization curve for the axial direction when the basal plane is easy. The corresponding range of values of K_1 and K_2 are given in Table 6.2 and Fig. 6.4.

When the easy direction for M_S in a domain corresponds with $\theta = \sin^{-1}(-K_1/2K_2)^{\frac{1}{2}}$ we may derive expressions for the magnetization curves

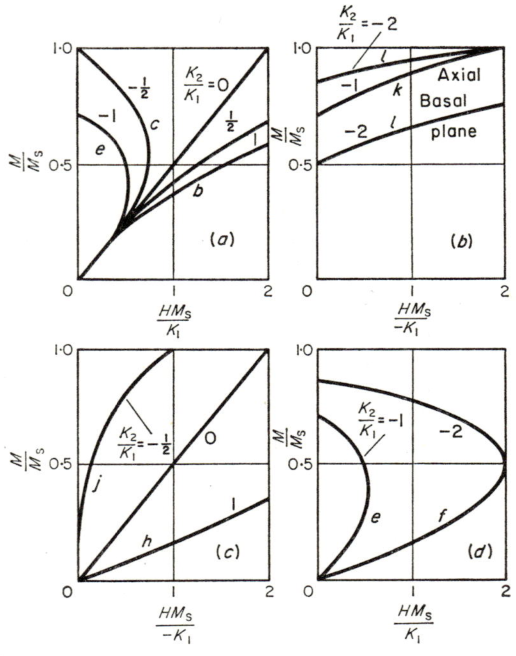

FIG. 6.12. Theoretical magnetization curves for crystals of hexagonal structure for different values of K_2/K_1. (*a*) in basal plane: $\theta = 0^0$ easy: (*b*) along the *c*-axis and in basal plane: $\theta = \sin^{-1}\{-K_1/(2K_2)\}^{\frac{1}{2}}$ easy: (*c*) along the *c*-axis: $\theta = 90°$ easy: $K_1 < 0$: (*d*) along the *c*-axis: $\theta = 90°$ easy: $K_1 > 0$.

in the axial direction and in the basal plane, respectively. It is easily seen that equation (6.20) also represents the magnetization curve in the axial direction, and equation (6.19) the magnetization curve in the basal plane. The range of values of K_1 and K_2 in this case are again given in Table 6.2 and shown diagrammatically in Fig. 6.4.

The theoretical magnetization curves in the direction of the *c*-axis or in the basal plane are plotted in Fig. 6.12 for a number of different ratios of K_2 to

K_1. The lettering on the curves corresponds with that given in Fig. 6.4 and Fig. 6.5.

6.6 Determination of Anisotropy Constants from Magnetization Curves

In Fig. 6.13(*a*) let OA represent a crystal axis along which a field H is applied and let OB be the easy direction for M_S in a domain of the crystal, so that $\angle AOB = \delta$. As H increases M_S rotates through an angle θ away from the easy direction in the plane AOB. We have already calculated the magnetization curve in the direction of an applied field for a number of particular cases. Let the magnetization curve for the direction OA be as shown at (*b*) in the figure. The torque acting per unit volume is given by $HM_S \sin(\delta - \theta)$

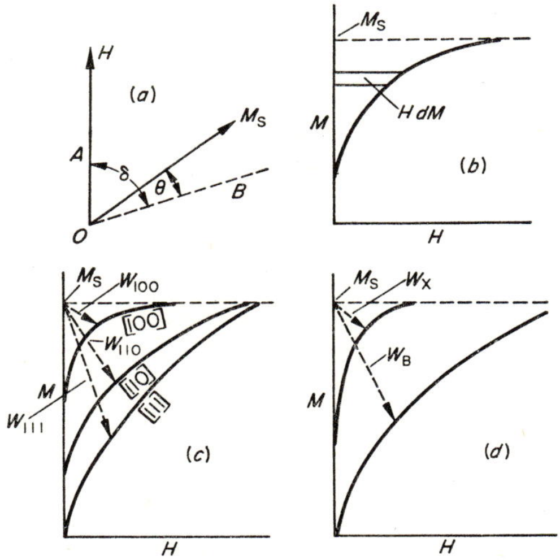

FIG. 6.13. Stored anisotropy energy in crystals.

and the intensity of magnetization is given by $M = M_S \cos(\delta - \theta)$. When θ increases by $d\theta$ the increment of magnetization is $dM = M_S \sin(\delta - \theta)d\theta$ and the work done to rotate M through the angle $d\theta$ is

$$dW = HM_S \sin(\delta - \theta)d\theta.$$

Therefore the element of energy stored in the crystal is $dW = H\,dM$ and this is shown at (*b*). The total amount of work done, which is equal to the total energy stored in the crystal per unit volume, when the crystal is magnetized from zero to saturation, is given by $W = \int_0^{M_S} H\,dM$ and this is clearly represented by the area between the magnetization curve in Fig. 6.13(*b*) and the ordinate axis. This energy is stored as a consequence of the crystalline forces which are acting to restore M_S to its easy direction in the crystal. Theoretically this energy would be recovered in full by the source which provides the field if H were returned to zero. It is here supposed that the magnetization curve is perfectly reversible, so that no losses are incurred in magnetizing to saturation and returning to the demagnetized state. The subject of losses will be dealt with later.

Magnetization curves for a crystal of cubic structure are represented in Fig. 6.13(*c*) for the [100], [110] and [111] directions, respectively. The total stored energy of magnetic anisotropy, when magnetized to saturation in each case, is then given by the areas in the figure denoted W_{100}, W_{110} and W_{111}. The energy in each case may, however, also be derived in terms of the anisotropy constants K_0, K_1 and K_2 by using equation (6.1). W_{100} is obtained by substituting $l = 1$, $m = 0$, $n = 0$ in the equation, W_{110} by putting $l = m = 1/2^{\frac{1}{2}}, n = 0$, and W_{111} from $l = m = n = 1/3^{\frac{1}{2}}$. We therefore have $W_{100} = K_0$, $W_{110} = K_0 + K_1/4$ and $W_{111} = K_0 + K_1/3 + K_2/27$. Whence $K_0 = W_{100}$, $K_1 = 4(W_{110} - W_{100})$ and $K_2 = 27(W_{111} - W_{100}) - 36(W_{110} - W_{100})$.

The anisotropy constants may therefore be determined in this manner if the magnetization curves of a crystal are known for the several directions, and provided that the criteria mentioned when calculating the magnetization curves are satisfied. In so far as the magnetization curves for iron and nickel in Figs. 6.6 and 6.7 are concerned, it will be observed, for example, that W_{100} is relatively small for iron but large for nickel, so that K_1 turns out to be positive for iron and negative for nickel as we have already found.

In the case of crystals of hexagonal structure W_X and W_B in Fig. 6.13(*d*) represent the stored energy for magnetization to saturation in the axial direction and in the basal plane, respectively. Using equation (6.3) we put $\theta = 0°$, and $\theta = 90°$, in the two cases, respectively, so that $W_X = K_0$ and $W_B = K_0 + K_1 + K_2$. Then $K_1 + K_2 = W_B - W_X$. Thus K_0 can be found but K_1 and K_2 cannot be separated simply by taking areas on the diagram.

6.7 Determination of Anisotropy Constants from Torque Curves

The best way of determining the constants K_1 and K_2 is from single crystal specimens in the form of oblate ellipsoids of revolution cut in particular crystal planes. The ideal shape of specimen is ellipsoidal although frequently thin discs of uniform thickness are used.

Consider first an ellipsoid of revolution placed in a steady magnetic field H_E which acts in the direction of a diameter, as shown in Fig. 6.14(*a*). If the material is magnetically isotropic and of constant permeability there will be induced in the specimen a uniform intensity of magnetization M. The uniform demagnetizing field produced by M will be $H_D = N_D M$ where N_D is the demagnetizing factor of the specimen in the direction of a diameter. N_D is dependent upon the ratio of minor to major axis of the ellipsoidal cross-section, as discussed in Chapter 1, and will be small if the ratio is small. The resultant field producing M in the material of the specimen is $H = H_E - N_D M$ and, in this isotropic case, H, H_E and M will be parallel with one another while the demagnetizing field $N_D M$ is antiparallel. To produce saturation in the specimen the externally-applied field H_E must exceed the saturating field derived

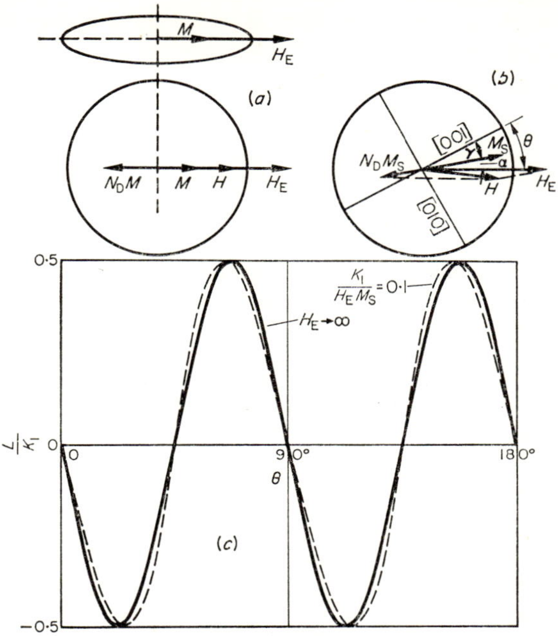

FIG. 6.14. Torque curve for an oblate ellipsoidal specimen in a (100) plane.

from the magnetization curve of the material itself by the amount of the demagnetizing field $N_D M$.

Considering as an example a sample of iron in the form of a sphere, for which $N_D = 1/3\mu_0$, the component of applied field to overcome the demagnetizing field at saturation, with $M_S = 2{\cdot}16$ Wb/m^2 for iron, would be $(1/3\mu_0)2{\cdot}16 = 2{\cdot}16 \times 10^7/(3 \times 4\pi)$ A/m $= 5730$ A/cm. On the other hand if the iron specimen were in the form of a thin ellipsoid of revolution with a ratio of minor to major axis of, say, 0·05 the value of N_D in the direction of a diameter is $0{\cdot}036/\mu_0$, approximately, and $N_D M_S = 620$ A/cm. The torque curves to be discussed are measured under conditions of magnetic saturation and it is therefore an advantage to have thin specimens, since then the required applied fields are not inconveniently large.

Figure 6.14(*b*) represents a thin ellipsoidal specimen cut in a (100) plane for a single crystal of cubic structure. If a strong field H_E is applied in the direction of a diameter at an angle θ to the cube-edge direction [001] as shown, then the crystal may be magnetized to saturation with the saturation vector M_S at an angle γ to [001]. A small angle $\alpha = \theta - \gamma$ will exist between the direction of H_E and M_S. The resultant field H in the specimen is then the vector difference of H_E and the demagnetizing field $N_D M_S$ as shown. The anisotropy energy when M_S is at an angle γ to a cube-edge direction in a (100) plane is obtainable from equation (6.2) and is given by $W_A = K_0 + (K_1/8)(1 - \cos 4\gamma)$. The torque per unit volume required to hold the specimen in the position shown in Fig. 6.14(*b*) is given by $dW_A/d\gamma$ and this is also equal to $H_E M_S \sin\alpha$.

$$\therefore\ H_E M_S \sin\alpha = (K_1/2)\sin 4\gamma$$

or
$$\alpha = \sin^{-1}\{(K_1/2H_E M_S)\sin 4\gamma\} \tag{6.21}$$

The restoring torque on the specimen per unit volume is $L = -dW_A/d\gamma$.

$$\therefore\ L = -(K_1/2)\sin 4\gamma$$

or
$$L = -(K_1/2)\sin 4(\theta - \alpha)$$

If H_E is made large enough, α, given by equation (6.21), becomes negligible and the relation between torque and the angle of rotation of the specimen is simply

$$L = -(K_1/2)\sin 4\theta \tag{6.22}$$

This equation is plotted by the full line in Fig. 6.14(*c*). The constant K_1 may thus be obtained from the experimental curve from the amplitude of the sine wave.

The broken line in Fig. 6.14(*c*) is drawn, as an example, for $K_1/H_E M_S = 0{\cdot}1$, and in this case there is a small distortion of the sine wave. For a sample of iron with $K_1 = 4{\cdot}4 \times 10^4$ J/m³ and $M_S = 2{\cdot}14$ Wb/m², H_E in this case would be 2050 A/cm. Increasing H_E would, of course, reduce the distortion.

K_1 only is obtainable from specimens cut in a (100) plane, but both K_1 and K_2 may be found if the ellipsoidal specimen is cut from a (110) plane. The specimen in Fig. 6.15(*a*) and (*b*) contains the directions [001], [111] and [110]. If θ is the angle between [001] and the direction of a strong field H_E, then as before M_S and H_E will be nearly coincident in direction. The aniso-

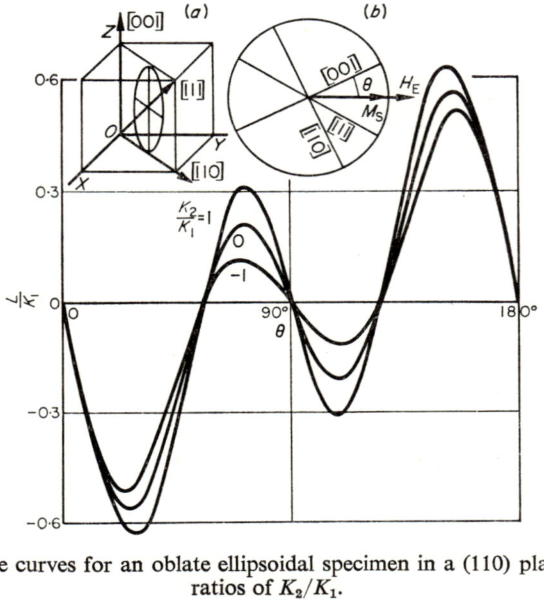

FIG. 6.15. Torque curves for an oblate ellipsoidal specimen in a (110) plane for different ratios of K_2/K_1.

tropy energy W_A is obtainable from equation (6.2) by putting $\phi = \pi/4$ and $\gamma = \theta$ and is

$$W_A = K_0 + (K_1/32)(7 - 4\cos 2\theta - 3\cos 4\theta) + (K_2/128)(2 - \cos 2\theta - 2\cos 4\theta + \cos 6\theta)$$

The restoring torque on the specimen is $L = -dW/d\theta$ whence

$$L = -(K_1/4 + K_2/64)\sin 2\theta - (3K_1/8 + K_2/16)\sin 4\theta + (3K_2/64)\sin 6\theta \tag{6.23}$$

Equation (6.23) is plotted in Fig. 6.15 for different ratios of K_2/K_1. The values of K_1 and K_2 may therefore be obtained by fitting a theoretical curve to the observed torque curve of a specimen.

The use of a (110) specimen is not a very accurate method of finding K_2 particularly if K_2 is relatively small. A better way is to measure the torque curve of a specimen cut from a (111) plane since K_1 does not then appear in the equation of the curve.

In Fig. 6.16 *PQO* represents a (111) type of plane in which the thin ellipsoidal specimen is cut. During the torque test the saturation vector M_S will be rotated in this plane. Let M_S be at an angle θ to *OA* which is perpendicular to *PQ*. From the geometry of the diagram the direction cosines of M_S are found to be $l = (1/6^{\frac{1}{2}})(\cos\theta - \sqrt{3}\sin\theta)$: $m = (1/6^{\frac{1}{2}})(\cos\theta + \sqrt{3}\sin\theta)$: $n = (\sqrt{2}/\sqrt{3})\cos\theta$. The magnetic anisotropy energy for M_S at any angle θ is obtained from equation (6.1) and is

$$W_A = K_0 + K_1(l^2m^2 + m^2n^2 + n^2l^2) + K_2(lmn)^2$$

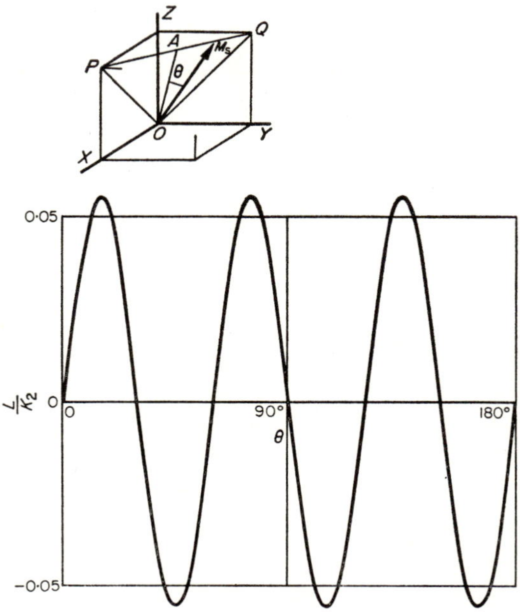

FIG. 6.16. Torque curve for an oblate ellipsoidal specimen in a (111) plane.

We find that $l^2m^2 + m^2n^2 + n^2l^2 = 1/4$, and $(lmn)^2 = (1/108)(1 + \cos 6\theta)$

$$\therefore\ W_A = K_0 + K_1/4 + (K_2/108)(1 + \cos 6\theta)$$

The torque is given by $L = -dW_A/d\theta$ as before. Whence

$$L = (K_2/18) \sin 6\theta \tag{6.24}$$

Equation (6.24) is plotted in Fig. 6.16 and from the amplitude of the sine wave K_2 may be directly obtained.

In the case of crystals of hexagonal structure we consider a thin ellipsoid of revolution cut so that the c-axis of the crystal coincides with a diameter. If,

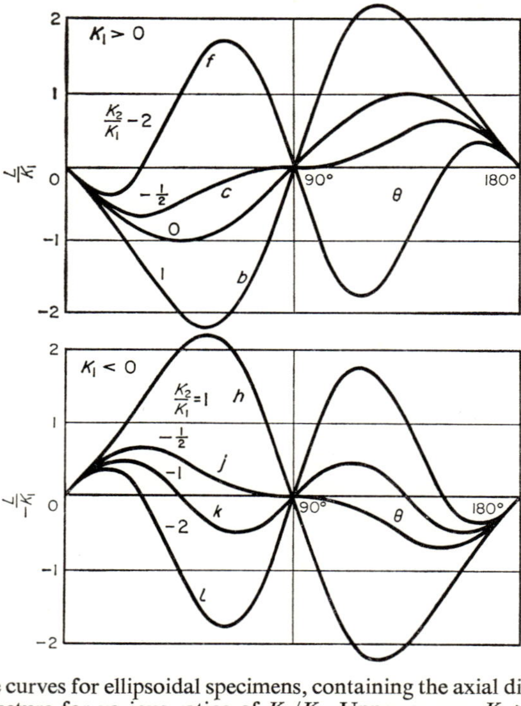

FIG. 6.17. Torque curves for ellipsoidal specimens, containing the axial direction, of crystals of hexagonal structure for various ratios of K_2/K_1. Upper curves, $K_1 > 0$: lower curves, $K_1 < 0$.

in a strong saturating magnetic field, the vector M_S lies in the plane of the circle of revolution at an angle θ to the c-axis then the anisotropy energy W_A is given by equation (6.3). Again the restoring torque on the specimen is $L = -dW_A/d\theta$ and from equation (6.4) this is

$$L = -(K_1 + K_2) \sin 2\theta + (K_2/2) \sin 4\theta \tag{6.25}$$

The shape of the torque curve depends on the sign of K_1 and the ratio K_2/K_1. Torque curves are plotted in Fig. 6.17 for $K_1 > 0$ in the upper part of the figure, for $K_1 < 0$ in the lower part and for a number of values of the ratio K_2/K_1. The lettering on the curves corresponds with that in Figs. 6.4 and 6.5.

For any specimen K_1 and K_2 can be found by fitting the observed torque curve with a theoretical curve.

The values of K_1 and K_2 for the metals, quoted by different observers, vary fairly widely, due perhaps to the effect of impurities and perhaps also, in some cases, to failure to get complete saturation in the specimens. The values also change with temperature. Some room-temperature values are given in Table 6.3.

TABLE 6.3

ANISOTROPY CONSTANTS OF METALS

Metal	$K_1/10^4 J/m^3$	$K_2/10^4 J/m^3$
	Cubic Structure	
Fe	4·4	2·1
96% Fe, 4% Si	3·2	1·0
Ni	−0·34	0·53
50% Ni, 50% Fe	0·33	−1·8
50% Co, 50% Fe	−0·68	−3·9
	Hexagonal Structure	
Co	40	20

6.8 Magnetostriction of Crystals

It is found experimentally that when ferromagnetic materials are magnetized they undergo small changes in dimensions. The general term magnetostriction is applied to this phenomenon but there are a number of effects, which include the JOULE effect [4, 5, 6] and volume magnetostriction [7].

A small change in volume in a strong magnetic field is found to occur which, in crystals of cubic structure, is isotropic, that is the accompanying linear strain is the same in all directions. In hexagonal crystals, however, the volume change may be accompanied also by a change of shape.

Additional to the high-field volume change is the Joule effect involving changes in linear dimensions in relatively low fields, which in most cases may be 10 to 100 times greater than those due to the volume magnetostriction. In this case an extension or contraction occurs in the direction in which the specimen is magnetized. Then considering the volume change as of second-order importance a longitudinal strain λ will be accompanied by a lateral strain $-\lambda/2$. The Joule linear magnetostriction is not only a function of the intensity of magnetization in the specimen but also varies, particularly in the case of iron, with direction in the crystal.

Volume magnetostriction is thought to be associated with the forces of interaction or the exchange energy between electron spins in the crystal, forces discussed in Section 5.6 which are responsible for the Weiss molecular field and the spontaneous saturation of the domains. The directional Joule effect, and the directional magnetic properties of crystals already discussed in this chapter, are considered to have a common cause. When the electron spins are rotated by a magnetic field away from their easy direction in the crystal an internal torque is exerted on the non-spherical atoms through spin–orbit coupling giving small dimensional changes in the lattice and causing energy to be stored. There should therefore be a close relation between the magnetostriction parameters and the anisotropy constants K_1 and K_2.

6.8.1 *Volume Magnetostriction*

Figure 6.18(*a*) shows how the volume of iron changes in a strong field. Here ω denotes the change in volume per unit volume. In these high fields the changes in length are the same in the direction of the applied field and laterally so that

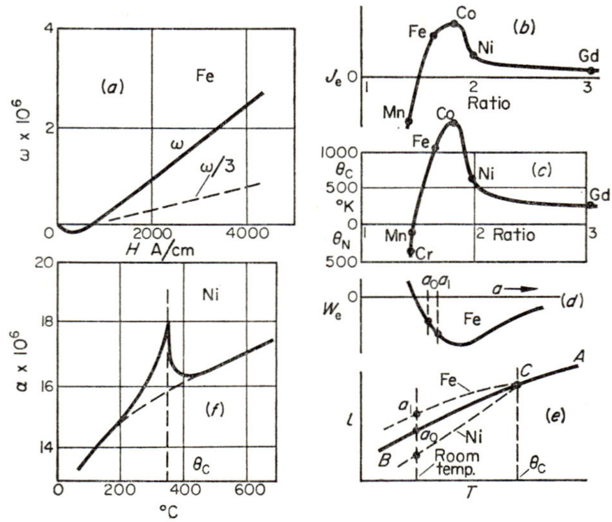

FIG. 6.18. Illustrating the origin of volume magnetostriction in iron and nickel.

the linear strain in any direction is $\omega/3$. For fields above about 1000 A/cm ω is positive and increases at the rate of about $0{\cdot}8 \times 10^{-6}$ per 1000 A/cm. The volume-change measurements are difficult and conflicting results have been obtained for nickel, but AZUMI and GOLDMAN [8] have found that nickel contracts, with the volumetric strain ω changing with field strength at the rate of about $-0{\cdot}07 \times 10^{-6}$ per 1000 A/cm.

A satisfactory explanation of the volume magnetostriction in these cases can be provided by considering how the elastic energy, and the exchange energy associated with the electron spins, vary as the lattice dimensions of the crystals change.

The saturation magnetization in a domain is the result of the forces of interaction or exchange between unpaired spinning electrons, causing parallel alignment of the spins in ferromagnetism and an antiparallel condition in antiferromagnetism. This has been discussed in Section 5.6 where the energy of exchange between a pair of electrons was given as J_e. This exchange interaction is responsible in effect for the Weiss molecular field and a relation between J_e and the Weiss field is given in equation (5.9). At the Curie temperature the thermal agitation is sufficient to break down the aligning action of the exchange forces and equation (5.11) gives a relation between J_e and the Curie temperature θ_C. When J_e is positive the potential energy associated with the effective Weiss field H_W and the saturation magnetization M_S is negative, since H_W and M_S are in the same direction.

According to BETHE and SLATER [9] the magnitude and sign of J_e depend upon the ratio of interatomic distance to the diameter of the sub-shell of the atom which is responsible for the atomic magnetic moment. This relationship is shown in Fig. 6.18(*b*) and the corresponding curve, plotted from the data in Table 5.3, for the Curie temperature θ_C and the Néel temperature θ_N, is shown at (*c*). The exchange energy per unit volume W_e for iron in relation to the interatomic spacing a will then have the form shown in Fig. 6.18(*d*). The actual known value of a for an iron crystal comes on this curve in the region where dW_e/da is negative, so that a small increase in a would lead to a reduction of the energy.

The full line AC in Fig. 6.18(*e*) is intended to show how the length l of a specimen of iron falls as the temperature decreases to the Curie point at C. If we imagine that the iron did not become ferromagnetic on further cooling then the length would follow the full line ACB. At room temperature the interatomic distance would be a_0. However, on cooling below the Curie temperature the actual condition is that the domains form and the exchange energy W_e is introduced. A small expansion of the crystal leads to a fall in W_e but causes elastic energy to be stored. The dimensions of the crystal will adjust themselves to make the sum of these components of energy a minimum with the result that the length of the ferromagnetic crystal will follow the broken line marked Fe in the figure. At room temperature the interatomic distance will then be $a_1 > a_0$. This increase in dimensions due to the ferromagnetism is estimated to be of the order of one part in 10^3.

Avoiding the complications arising from the three-dimensional nature of the foregoing problem we may consider it qualitatively as follows. Let J_e' be the exchange energy for one pair of electron spins and suppose there are, at room temperature, N' uncompensated spins per unit volume, then the exchange energy per unit volume is $-N'J_e$. If the interatomic distance is a, with a_0 the value in the assumed absence of ferromagnetism, then the linear elastic strain is $(a - a_0)/a_0$. If, for present purposes, we assume an extension in one axial direction only, and neglect the changes in the other two directions, then the elastic energy stored per unit volume is $(1/2)E_Y\{(a - a_0)/a_0\}^2$ where E_Y is Young's Modulus. The total energy per unit volume is then

$$W = -N'J_e + (1/2)E_Y\{(a - a_0)/a_0\}^2$$

Equilibrium is reached when $dW/da = 0$ or

$$N' \, dJ_e/da = (E_Y/a_0^2)(a - a_0)$$

so that the linear strain induced is

$$(a_1 - a_0)/a_0 = (N'a_0/E_Y)dJ_e/da \tag{6.26}$$

When a strong external field is applied, as in the case of Fig. 6.18(*a*), the number of uncompensated spins is increased by a small amount from N' to N'' and in that case the linear extension given by equation (6.26) is slightly increased. The observed positive volume magnetrostriction of iron may thus be accounted for.

It will be seen from Fig. 6.18(*b*) that, if the same procedure is followed for nickel, then dJ_e/da will be negative. The change in linear dimensions with temperature for nickel would then be given by the broken line marked Ni in Fig. 6.18(*e*), and equation (6.26) would indicate that ferromagnetism induces a contraction of the crystal. The application of a strong external field would then lead to a negative volume magnetostriction.

The coefficient of thermal expansion α for the metal is proportional to the slope of the characteristic shown in Fig. 6.18(*e*), and therefore a discontinuity in the relation between the coefficient of expansion and temperature should occur for a ferromagnetic in the neighbourhood of the Curie temperature. The coefficient observed for nickel is given in Fig. 6.18(*f*) and shows the effect very strongly. For iron the peak in the $\alpha - T$ curve would be inverted.

6.8.2 *Longitudinal Magnetostriction*

Magnetization curves of crystals of cubic structure were discussed in Section 6.4. In a demagnetized iron crystal it was supposed that the spontaneous saturation M_S vectors in the constituent domains were distributed amongst the six possible cube-edge directions in such a way that the bulk magnetization was zero. On the application of a relatively small magnetic field in any direction of the crystal the domain vectors then tended to line up in the cube-edge direction nearest to the direction of the field. In this case the unfavourably pointing M_S vectors will change direction by 90° or 180°. We have so far assumed that a negligibly small magnetic field was sufficient to complete this process and to raise the magnetization to the knee of the magnetization curve, and this would imply infinite permeability. In practice a finite and varying permeability is observed, and the phenomena occurring in this region of 90° and 180° changes are discussed in more detail in Chapter 7. On this part of the magnetization curve below the knee the effect of an increasing applied field is, in fact, progressively to increase the total fraction of the domains in which M_S is favourably directed with respect to the field, at the expense of the domains in which M_S is at 90° or at 180°. Above the knee a much stronger applied field produces an increase in magnetization by rotating the M_S vectors away from the easy cube-edge direction, and this rotational process we have already discussed in detail. The Joule effect in which a change of length occurs in the direction of the field, and laterally, accompanies this redistribution of the M_S vectors, and is in two parts. First the length changes as M_S swings from one cube-edge direction to another, and then a more complicated effect, causing a further change in length, occurs as M_S is rotated away from the easy direction.

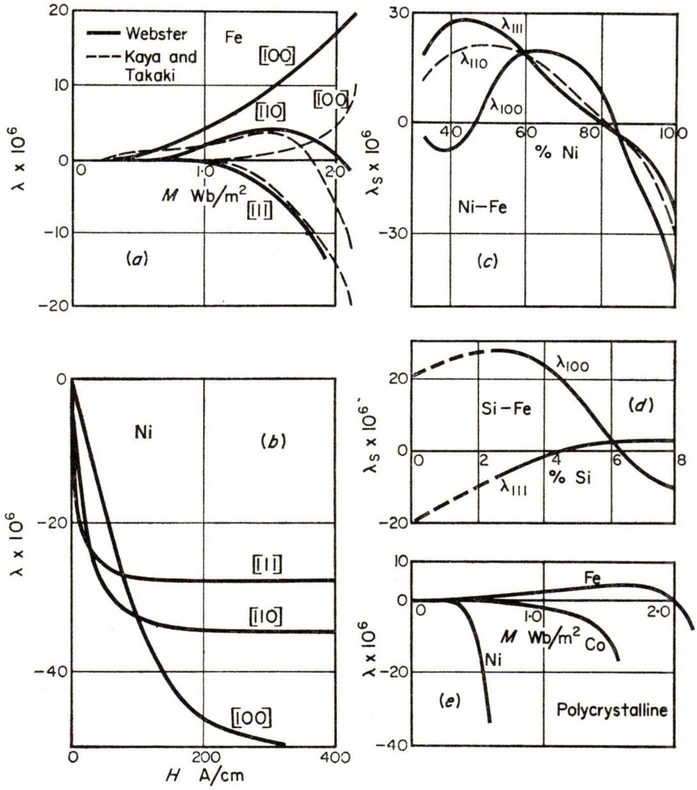

FIG. 6.19. Longitudinal magnetostriction in crystals and in polycrystalline materials.

Figure 6.19(*a*) shows the linear strain, or magnetostriction $\lambda = dl/l$ occurring during magnetization for crystals of iron in different crystal directions. The full lines give results due to WEBSTER [10] and the broken lines are due to KAYA and TAKAKI [11]. Figure 6.19(*b*) shows corresponding results for nickel plotted, however, in this case in relation to H. Observed values of λ_S, the magnetostriction at saturation, in different directions are shown in Fig. 6.19(*c*) for a range of nickel-iron crystals, and in Fig. 6.19(*d*) for crystals of silicon-iron. The curves in Fig. 6.19(*e*) show how λ varies with the intensity of magnetization in polycrystalline samples of iron, cobalt and nickel.

The experimental evidence seems to show that the spontaneous saturation M_S in a domain is accompanied by a small spontaneous strain in the crystal lattice. In a free domain of iron the lattice is extended in the direction of M_S, that is in a cube-edge direction, the linear strain $\lambda = \lambda_{100}$ being about 20×10^{-6}. At the same time the lattice is contracted laterally by an amount $\lambda_{100}/2$. The lattice is thus not quite cubic so that if M_S swings through 90° from one cube edge to another, then assuming no external restraints, the new direction of M_S experiences a change of strain of $+3\lambda_{100}/2$ and the old direction a change of $-3\lambda_{100}/2$. Thus rearrangements of the domain vectors in a crystal during magnetization will involve the changes in dimensions of the whole crystal which are observed as the linear magnetostriction, and which may be seen in the examples in Fig. 6.19. Volume changes we may disregard in the present discussion since these are appreciable only if very high field strengths are applied, as may be seen by reference to Fig. 6.18(*a*).

It is also inferred from measurements of magnetostriction in crystals that if the saturation vector M_S is rotated away from the easy axis there is a change in the magnetostrictive strain in the domain which, particularly for iron, is somewhat complex. In iron if M_S is rotated into the [111] direction there is now a contraction along that axis, and λ_{111} is about -20×10^{-6}. In nickel λ_{100} and λ_{111} are both negative. Enough is not known about the basic origins of magnetostriction to enable values of λ_{100} or λ_{111} to be determined theoretically for any material, and these constants can only be deduced from measurements on samples.

The magnetostrictive strains in any direction of a self-saturated domain depend upon the angular position of the vector M_S. Bearing in mind that the magnetostriction, like the magnetic properties, in a crystal of cubic structure, must have cubic symmetry, it is possible to write down an equation relating the strain in any direction with the direction of M_S. Agreement with experiment is adequate using only the two constants λ_{100} and λ_{111}, and the equation is as follows:

$$\lambda_S = (3/2)\lambda_{100}(l^2 l_1^2 + m^2 m_1^2 + n^2 n_1^2 - 1/3) + 3\lambda_{111}(l m l_1 m_1 + m n m_1 n_1 + n l n_1 l_1) \quad (6.27)$$

Taking the cube edges as coordinate axes, λ_S is the spontaneous magnetostrictive strain in a domain in the direction whose direction cosines are l_1, m_1, n_1, when the direction cosines of the vector M_S are l, m, n.

Equation (6.27) applies also to a multi-domain single crystal which has been brought to saturation from an ideally-demagnetized initial condition. The experimental value of λ_S depends upon the observed difference between a length at saturation and a length in the initial condition. The ideal initial condition is when all the easy directions for the domains are equally occupied. The resultant strain in the crystal as a whole is then zero. It should be mentioned, however, that the crystal could be in a demagnetized state, in the sense that the bulk magnetization in any direction was zero, with M_S in the domains distributed in some other way—in an infinite number of possible ways in fact. For example, if hypothetically in the demagnetized state of the sample the domains were equally distributed along [100] and in the reverse direction [$\bar{1}$00], respectively, then the observed value of λ_S measured in the [100] direction when also magnetized to saturation in that direction would be zero, while on the other hand, the observed value measured along [010] when saturated in the [010] direction would be $3\lambda_{100}/2$. Unfortunately in practice the initial distribution of M_S in the domains in a demagnetized sample is never known for sure, a fact which must introduce uncertainty into the observed values of the constants.

However, using constants λ_{100} and λ_{111}, the strain in any direction of a crystal may be calculated from equation (6.27), when the crystal is saturated in any other direction. It is assumed, of course, that no other constraints or forces are acting on the crystal.

If the crystal is saturated in a [100] direction, $l = 1$, $m = n = 0$, and $\lambda_S = (3/2)\lambda_{100}(l_1^2 - 1/3)$. The strain in the [100] direction, $l_1 = 1$, is then $\lambda_S = \lambda_{100}$, while at right-angles to this direction, $l_1 = 0$, the strain is $\lambda_S = -(1/2)\lambda_{100}$.

For saturation in a [111] direction $l = m = n = 1/3^{\frac{1}{2}}$. The first term of equation (6.27) then becomes zero for any values of l_1, m_1, n_1, and $\lambda_S = \lambda_{111}(l_1m_1 + m_1n_1 + n_1l_1)$. The strain along [111] with $l_1 = m_1 = n_1 = 1/3^{\frac{1}{2}}$ is then $\lambda_S = \lambda_{111}$. Values of λ_{100} and λ_{111} are thus obtainable from measurements on crystals.

With saturation in a [110] direction $l = m = 1/2^{\frac{1}{2}}$, $n = 0$, and

$$\lambda_S = (3/2)\lambda_{100}\{(1/2)(l_1^2 + m_1^2) - 1/3\} + (3/2)\lambda_{111}(l_1m_1 + m_1n_1)$$

The strain along [110], $l_1 = m_1 = 1/\sqrt{2}$, $n_1 = 0$, is then

$$\lambda_S = \lambda_{110} = (1/4)(\lambda_{100} + 3\lambda_{111})$$

There is thus a relatively simple theoretical relation between the saturation values of magnetostriction in the [100], [110] and [111] directions, respectively. This relation is quite well satisfied by the experimental data given in Fig. 6.19(*a*), (*b*) and (*c*) except in the case of Webster's results.

The saturation value of the magnetostriction in a polycrystalline material with a random distribution of the crystal axes is obtainable from equation (6.27). If in any crystal the direction cosines of M_S are l, m, n and we consider λ_S in the direction of M_S, then $l_1 = l$, $m_1 = m$, $n_1 = n$ and equation (6.27) becomes

$$\lambda_S = (3/2)\lambda_{100}(l^4 + m^4 + n^4 - 1/3) + 3\lambda_{111}(l^2m^2 + m^2n^2 + n^2l^2) \quad (6.28)$$

Since $(l^2 + m^2 + n^2)^2 = 1 = l^4 + m^4 + n^4 + 2(l^2m^2 + m^2n^2 + n^2l^2)$ equation (6.28) may be rewritten more conveniently as

$$\lambda_S = \lambda_{100} + 3(\lambda_{111} - \lambda_{100})(l^2m^2 + m^2n^2 + n^2l^2) \quad (6.29)$$

The saturation magnetostriction λ_m in the polycrystalline material will be the mean value of λ_S for all values of l, m, n. Referring to Fig. 6.8(*a*) we have $l = \sin\gamma\cos\phi$, $m = \sin\gamma\sin\phi$ and $n = \cos\gamma$. For any direction l, m, n

consider an element of solid angle $d\gamma$ ($\sin\gamma . d\phi$), then λ_m is obtainable from

$$\lambda_m = (1/4\pi)\int_0^{\gamma=\pi}\int_0^{\phi=\pi} \lambda_S . \sin\gamma . d\gamma . d\phi$$

Substituting for λ_S in terms of γ and ϕ, and carrying out the double integration between the limits given, leads to

$$\begin{aligned}\lambda_m &= \lambda_{100} + (1/4\pi)3(\lambda_{111} - \lambda_{100})2\pi(2/5)\\ &= \lambda_{100} + (3/5)(\lambda_{111} - \lambda_{100})\\ \therefore \lambda_m &= (2/5)\lambda_{100} + (3/5)\lambda_{111} \qquad (6.30)\end{aligned}$$

For polycrystalline iron putting $\lambda_{100} = 20 \times 10^{-6}$ and $\lambda_{111} = -20 \times 10^{-6}$, $\lambda_m = -4 \times 10^{-6}$, a value which, compared with experimental values, is at least of the right order of magnitude.

A polycrystalline material with a random distribution of the constituent crystals will be magnetically isotropic, with a saturation magnetostriction λ_m in any direction when magnetized to saturation in the same direction. It is of some interest to know in this case the value of the magnetostriction in any direction relative to the direction of magnetic saturation. An isotropic polycrystalline material may for this purpose be considered to be equivalent to an isotropic single crystal. In that case equation (6.27) may be applied by putting $\lambda_{100} = \lambda_{111} = \lambda_m$. We then have

$$\lambda_S = (3/2)\lambda\ (ll_1 + mm_1 + nn_1 - 1/3 + 2lml_1m_1 + 2mnm_1n_1 + 2nln_1l_1) \qquad (6.31)$$

where, with respect to a selected system of coordinate axes in the polycrystalline material, the material is magnetized to saturation in a direction given by l, m, n, while the magnetostriction λ_S is observed in a direction l_1, m_1, n_1. If θ is the angle between the direction of M_S and the direction in which λ_S is measured, then $\cos\theta = ll_1 + mm_1 + nn_1$ and equation (6.31) becomes

$$\lambda_S = (3/2)\,\lambda_m(\cos^2\theta - 1/3) \qquad (6.32)$$

Thus for the polycrystalline material we see that $\lambda_S = \lambda_m$ when $\theta = 0$, $\lambda_S = -\lambda_m/2$ when $\theta = 90°$, and $\lambda_S = 0$ when $\theta = \cos^{-1}(1/\sqrt{3})$. λ_m is obtainable, as we have shown, in terms of the constants of the individual crystals from equation (6.30).

Having considered the saturation values of magnetostriction in a number of cases we may now turn to the relation between the magnetostriction λ and the intensity of magnetization M in a crystal below saturation. The observed value of λ at any point of the magnetization curve must depend upon the initial distribution of the domains in the demagnetized state and on the way in which the domains become rearranged during magnetization. If we suppose that, in the case of iron, the crystal is initially in the ideal demagnetized condition the bulk magnetostrictive strain then is zero. If a field is applied along a [100] direction an increase in length in this direction will occur due to 90° swings of the domain vectors but there will be no change due to 180° reversals. A limiting case would be when the 180° reversals, involving one-sixth of the domains, occurred first. In this case the magnetization would rise to $M_S/3$ with $\lambda = 0$. As M was further increased, due to 90° turns, then λ would increase linearly with M up to the saturation value. This theoretical relation, based, however, on the assumption just stated, is shown in Fig. 6.20 and, taking $M_S = 2{\cdot}14$ Wb/m² and $\lambda_{100} = 20 \times 10^{-6}$, compares favourably with Webster's observed curve, but not at all well with the result due to Kaya and Takaki shown in Fig. 6.19(*a*).

When the field is applied along a [110] direction M reaches the value $M_S/2^{\frac{1}{2}}$ at the knee of the magnetization curve when the domain vectors have all taken up cube-edge positions at 45° to the direction of H. Below the knee 90°

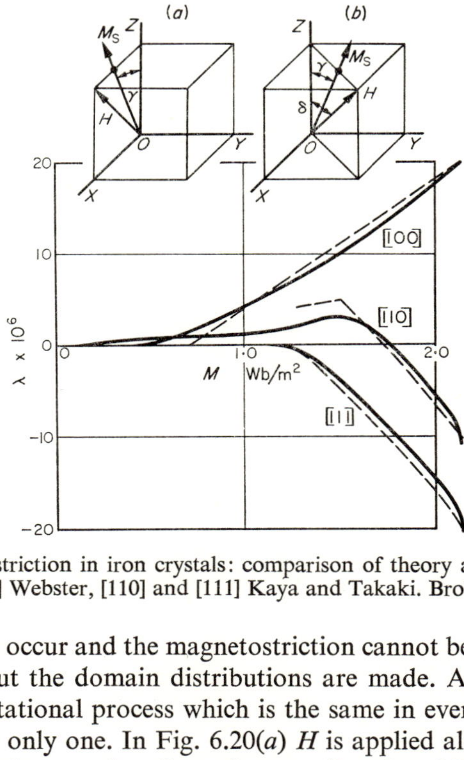

FIG. 6.20. Magnetostriction in iron crystals: comparison of theory and experiment. Full lines observed, [100] Webster, [110] and [111] Kaya and Takaki. Broken lines theoretical.

and 180° changes occur and the magnetostriction cannot be calculated unless assumptions about the domain distributions are made. Above the knee M increases by a rotational process which is the same in every domain so that we need consider only one. In Fig. 6.20(*a*) H is applied along [101] and M_S has rotated through an angle γ from the easy direction OZ. When $\gamma = 0$ the strain in the direction of H is obtainable from equation (6.27) by putting $l = m = 0$, $n = 1$ for the direction of M_S, and $l_1 = n_1 = 1/2^{\frac{1}{2}}$, $m_1 = 0$ for the direction of H. Then

$$\lambda_S = (3/2)\lambda_{100}(1/2 - 1/3) = (1/4)\lambda_{100}$$

If the crystal is initially in the ideal demagnetized condition with $\lambda = 0$, then the magnetostriction in the [101] direction for the crystal at the knee, when $M = M_S/2^{\frac{1}{2}}$, is $\lambda = (1/4)\lambda_{100}$. When M_S has rotated through the angle γ we put for the direction of M_S, $l = \sin\gamma$, $m = 0$, $n = \cos\gamma$. Substituting these values in equation (6.27) gives

$$\lambda = (1/4)\lambda_{100} + (3/4)\lambda_{111} \sin 2\gamma \qquad (6.33)$$

The intensity of magnetization in the crystal in the direction of H is given by $M = M_S \cos(\pi/4 - \gamma)$, and this may be rewritten as $\sin 2\gamma = 2(M/M_S)^2 - 1$. Substituting for $\sin 2\gamma$ in equation (6.33) gives

$$\lambda = (1/4)\lambda_{100} + (3/4)\lambda_{111}\{2(M/M_S)^2 - 1\} \qquad (6.34)$$

Equation (6.34) gives a relation between the magnetostriction and the intensity of magnetization in a [110] direction of the crystal.

When H is applied along a [111] axis, as in Fig. 6.20(*b*), the knee of the magnetization curve, with $M = M_S/3^{\frac{1}{2}}$, is reached when the domain vectors are all in the cube-edge directions OX, OY, OZ. Again for a further increase in M we need consider only one domain. In the figure suppose that M_S for a domain has rotated through an angle γ from the easy axis OZ in the diagonal plane containing OZ and H. The strain along [111] when $\gamma = 0$ is obtained by using equation (6.27) and putting $l = m = 0$, $n = 1$ for the direction of M_S and $l_1 = m_1 = n_1 = 1/3^{\frac{1}{2}}$ for the [111] direction. Substituting these values in equation (6.27) gives $\lambda = 0$. For any angle γ we have $l = m = (1/2^{\frac{1}{2}}) \sin\gamma$, $n = \cos\gamma$ for the direction of M_S. Putting these values in equation (6.27) gives

$$\lambda = (\lambda_{111}/2)(\sin^2\gamma + 2\sqrt{2} \sin\gamma \cos\gamma) \qquad (6.35)$$

Now $M = M_S \cos(\delta - \gamma)$ where $\cos\delta = 1/3^{\frac{1}{2}}$ and this may be rewritten in the form $\cos\gamma + \sqrt{2}\sin\gamma = \sqrt{3}(M/M_S)$. Squaring and rearranging gives

$$\sin^2\gamma + 2\sqrt{2}\sin\gamma\cos\gamma = 3(M/M)_S{}^2 - 1$$

Substituting in equation (6.35) gives

$$\lambda = (1/2)\lambda_{111}\{3(M/M_S)^2 - 1\} \qquad (6.36)$$

This equation gives a relation between the magnetostriction in a [111] direction and the intensity of magnetization in the crystal.

Equations (6.34) and (6.36) are plotted in Fig. 6.20 for the case of iron, assuming the values $\lambda_{100} = 20 \times 10^{-6}$, $\lambda_{111} = -20 \times 10^{-6}$ and $M_S = 2{\cdot}14$ Wb/m². There is quite good agreement between these theoretical curves and the observations by Kaya and Takaki for iron.

6.9 Magnetostriction and Stress

Let Fig. 6.21(*a*) represent a domain in a crystal of cubic structure in which the direction of the saturation vector M_S is represented by the direction cosines l, m, n. The magnetic anisotropy energy W_A of the domain is given by equation (6.1), and the magnetostriction λ_S in the domain in a direction corresponding to l_1, m_1, n_1 is, in this case, given by equation (6.27). We suppose now that there is a uniform tensile stress σ in the domain in the direction l_1, m_1, n_1, as shown in the figure. If we imagine the magnetostriction

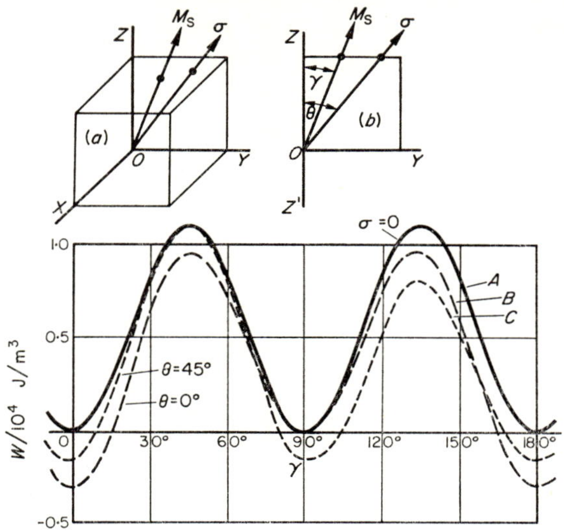

FIG. 6.21. Showing effect of tensile stress on the energy minima in a domain.

to come into existence in the presence of the stress then strain energy, or magnetoelastic energy, will be stored in the domain. For example, if the domain extends magnetostrictively against a compressive stress then energy will be stored elastically in the crystal. On the other hand if the stress is tensile and the magnetostriction is positive the stored energy will have a negative sign. The magnetoelastic energy per unit volume is, in fact, thus given simply by $-\lambda_S\sigma$.

The total energy W, due to magnetic anisotropy and to stress, corresponding to the condition represented in Fig. 6.21(*a*), is therefore $W = W_A - \lambda_S\sigma$ and substituting from equations (6.1) and (6.27) we have

$$\begin{aligned} W = K_0 &+ K_1(l^2m^2 + m^2n^2 + n^2l^2) + K_2(lmn)^2 \\ &- (3/2)\lambda_{100}\,\sigma(l^2l_1^2 + m^2m_1^2 + n^2n_1^2 - 1/3) \\ &- 3\lambda_{111}\sigma(lml_1m_1 + mnm_1n_1 + nln_1l_1) \end{aligned} \tag{6.37}$$

When the stress σ is in a particular direction given by l_1, m_1, n_1, then the easy directions of M_S in the domain will correspond with values of l, m, n which make W a minimum in equation (6.37). The spontaneous saturation vector M_S will thus take up an equilibrium position in the domain which is dependent upon the sign and relative magnitudes of K_1, K_2, λ_{100}, λ_{111} and σ. We have already found that in a stress-free domain the easy directions correspond to [100], [110] or [111] crystal axes depending upon the values of K_1 and K_2. The stress σ may disturb these conditions. A solution for the general case is obviously complicated and we shall consider only simple special cases.

The importance of the last two terms in equation (6.37) will depend on the relative values of $\lambda_{100}\sigma$ and $\lambda_{111}\sigma$ on the one hand and the anisotropy constants K_1 and K_2 on the other. In the case of iron suppose we take a stress which is approaching the elastic limit, say $\sigma = 10^8$ N/m². Then with

$$\lambda_{100} = 20 \times 10^{-6}, \lambda_{100}\sigma = 2 \times 10^3 \text{ J/m}^3,$$

and for iron $K_1 = 4{\cdot}4 \times 10^4$ J/m³. Therefore for iron the anisotropy energy predominates over the stress energy even for relatively high stresses. In a stress-free domain there are six equally-easy cube-edge directions for the vector M_S. The effect of a tensile stress, however, is to reduce the number of preferred positions, that is positions of lowest energy, to two in the general case, for M_S will be in a position of minimum energy when it lies very nearly along the cube-edge axis nearest to the direction of the stress, or in the reverse direction. Figure 6.21 illustrates this for the case of iron. If M_S is rotated in a (100) plane and the direction of σ is also in this plane, as shown in Fig. 6.21(*b*), then we may put $l = 0$, $m = \sin\gamma$, $n = \cos\gamma$, $l_1 = 0$, $m_1 = \sin\theta$, $n_1 = \cos\theta$. Equation (6.37), neglecting the constant terms, becomes

$$W = (K_1/8)(1 - \cos 4\gamma) - (3/4)\lambda_{100}\sigma(1 + \cos 2\gamma \cos 2\theta) - (3/4)\lambda_{111}\sigma\,(\sin 2\gamma \sin 2\theta)$$

When there is no stress, and putting in the numerical values for iron, the energy varies with γ as shown by curve *A*, with minima corresponding to the easy directions, where $\gamma = 0°, 90°, 180°, 270°$. If the tensile stress is applied along *OZ*, that is $\theta = 0°$, then the energy with $\sigma = 10^8$ N/m² is represented by curve *B*. The energy minima are now depressed when $\gamma = 0°$ and 180°, and therefore the directions *OZ* and *OZ'* are the preferred positions for M_S. If the stress acts in the direction [011], that is with $\theta = 45°$, curve *C* is the result, again with equal minima at 0°, 90°, 180° and 270°, but when $0° < \theta < 45°$, the energy curve will lie between *B* and *C*, and the easiest directions are again very nearly along *OZ* and *OZ'*.

In the case of polycrystalline iron with no preferred orientation of the crystals the domain vectors will have a random distribution when the material is in the ideal demagnetized state and the magnetostriction λ is then zero. However, if there is a tensile stress in the material, and in the demagnetized state the domain vectors have settled into the cube-edge directions of lowest energy then, although the bulk intensity of magnetization is zero, the magnetostriction is no longer zero because preferred cube-edge directions for the domain vectors have been induced by the stress.

The maximum value of the initial magnetostriction which may be induced in polycrystalline iron by an applied tensile stress may be calculated as follows. In Fig. 6.22 for any domain let the tensile stress σ be in a direction given by l_1, m_1, n_1, where $l_1 = \sin\gamma\cos\phi$, $m_1 = \sin\gamma\sin\phi$ and $n_1 = \cos\gamma$. *OZ* will be the cube-edge direction nearest to the direction of the stress if $n_1 > l_1$, and also if $n_1 > m_1$. In that case M_S will lie along *OZ*, and for the direction of M_S we have $l = m = 0$, $n = 1$. The magnetostriction λ_S in the direction of the stress is then obtained from equation (6.27) and is

$$\lambda_S = (3/2)\lambda_{100}(\cos^2\gamma - 1/3) \qquad (6.38)$$

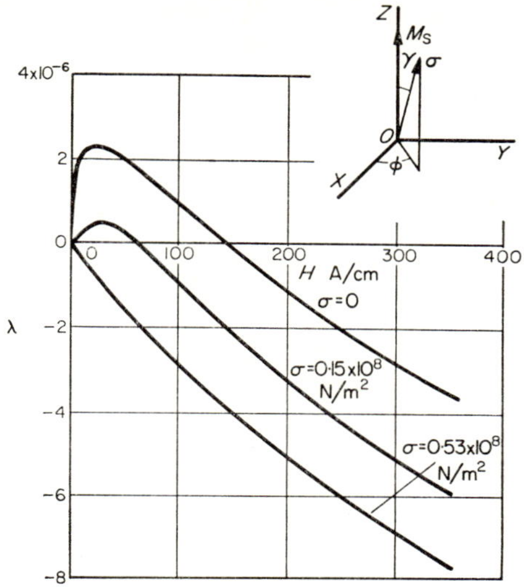

FIG. 6.22. Effect of tensile stress on the magnetostriction of polycrystalline iron.

Consider now the stress vector σ taking up all positions between the limits $\phi = 0$ and $\pi/4$, and for all values of γ which maintain OZ as the nearest cube edge, that is for $n_1 > l_1$. The limits of γ are therefore obtained from $\gamma = 0$ and $\cos\gamma = \sin\gamma\cos\phi$, and are $\gamma = 0$, $\gamma = \tan^{-1}(\sec\phi)$. The mean value of λ_S for all values of ϕ and γ between the limits stated will in fact be the same as the mean value of the initial magnetostriction for the polycrystalline material in the direction of the applied tensile stress. Taking the magnetostriction to be λ_S, as given by equation (6.38), over an element of solid angle of amount $d\gamma . \sin\gamma . d\phi$, then the initial mean value of magnetostriction which we require is given by

$$\lambda_1 = \frac{\displaystyle\int_0^{\phi=\pi/4}\int_0^{\gamma=\tan^{-1}(\sec\phi)} \lambda_S \sin\gamma . d\gamma . d\phi}{\displaystyle\int_0^{\phi=\pi/4}\int_0^{\gamma=\tan^{-1}(\sec\phi)} \sin\gamma . d\gamma . d\phi}$$

$$= (12/\pi)\int_0^{\phi=\pi/4}\int_0^{\gamma=\tan^{-1}(\sec\phi)} (3/2)\lambda_{100}(\cos^2\lambda - 1/3)\sin\gamma . d\gamma . d\phi$$

On carrying out the necessary integrations we find that

$$\lambda_1 = (\sqrt{3}/\pi)\lambda_{100} = 0{\cdot}55\lambda_{100}$$

As a result of the effect just described the observed magnetostriction in relation to an increasing applied field will be less when a tensile stress exists in the direction of the field than in the absence of the stress, since some of the magnetostrictive extension has already occurred before applying the field. This result is indeed observed. Figure 6.22 shows the magnetostriction measured in polycrystalline iron [12] with and without tensile stress, although the reduction in λ due to stress is not as great in these examples as $0{\cdot}55\lambda_{100}$.

If in Fig. 6.22, again in the case of iron, the stress is compressive, that is σ is negative, the direction of lowest energy will be OY provided that OY is the cube edge making the greatest angle with σ. Therefore M_S will lie along OY if $m_1 < l_1$ and $m_1 < n_1$. Since l_1 sin γ cos ϕ, $m_1 = \sin\gamma \sin\phi$, $n_1 = \cos\gamma$ the limits for ϕ are 0 and $\pi/4$, and for γ the limits are 0 and $\tan^{-1}$ (cosec ϕ). For M_S we have $l = n = 0$, $m = 1$ and so from equation (6.27) the magnetostriction in the direction of the compressive stress is

$$\lambda_S = (3/2)\lambda_{100}(\sin^2\gamma \sin^2\phi - 1/3)$$

The initial mean value of the magnetostriction for the polycrystalline material in the direction of the compression is then

$$\lambda_1 = \frac{\displaystyle\int_0^{\phi=\pi/4}\int_0^{\gamma=\tan^{-1}(\operatorname{cosec}\phi)} (3/2)\lambda_{100}(\sin^2\gamma \sin^2\phi - 1/3)\sin\gamma \,.\, d\gamma \,.\, d\phi}{\displaystyle\int_0^{\phi=\pi/4}\int_0^{\gamma=\tan^{-1}(\operatorname{cosec}\phi)} \sin\gamma \,.\, d\gamma \,.\, d\phi}$$

On carrying out the computations we find that $\lambda_1 = -\sqrt{3}(\sqrt{3} - 1)\lambda_{100}/\pi$ or $\lambda_1 = -0{\cdot}40\lambda_{100}$.

Hence we should expect for iron under a compressive stress, when magnetized in the direction of the stress, that the magnetostriction would be greater than if the stress were absent.

The easy direction for nickel is [111] and λ_{111} is negative. Using the same arguments as in the case of iron it is expected that tension in polycrystalline nickel will cause an initial magnetostrictive extension in the direction of the stress, so that the subsequent reduction in length due to magnetization in the direction of the stress is greater with tension than when the nickel is stress-free. This is observed experimentally [13] but a theoretical analysis based on equation (6.37) is difficult because the components of energy due to the magnetic anisotropy and to the stress are now comparable in magnitude. For example we have with nickel $K_1 = -0{\cdot}34 \times 10^4$ J/m^3 and using $\sigma = 10^8$ N/m^2, as we did for iron, with $\lambda_{111} = -25 \times 10^{-6}$, $\lambda_{111}\sigma = -0{\cdot}25 \times 10^4$ J/m^3.

If we consider the simple particular case of a domain in which $\lambda_{100} = \lambda_{111}$ then the saturation value of the magnetostriction in the direction of M_S in the domain will be the same for M_S in any direction, as will be seen from equation (6.29). The magnetostriction in the direction of magnetic saturation in a polycrystalline material made up of such domains will also have the same value, as may also be seen from equation (6.30). Thus this value is $\lambda_m = \lambda_{100} = \lambda_{111}$. In any domain in this case, or in the polycrystalline material, the value of magnetostriction λ_S, under conditions of magnetic saturation, measured at an angle θ to the direction of M_S, is given by equation (6.32) and is $\lambda_S = (3/2)\lambda_m(\cos^2\theta - 1/3)$. If, in the direction in which λ_S is observed, there is a tensile stress σ then the magnetoelastic energy is given by $-\lambda_S\sigma$. Thus for an angle θ between M_S and σ in a domain or a polycrystalline material the variation of magnetoelastic energy with θ, neglecting the constant term is $W_E = -(3/2)\lambda_m\sigma \cos^2 \theta$, where in this special case there is the assumption that $\lambda_m = \lambda_{100} = \lambda_{111}$. This is an equation which sometimes occurs in the literature in discussions of the energy associated with stress and magnetostriction.

6.10 Origin of Magnetic Anisotropy and Magnetostriction

Magnetic anisotropy and magnetostriction in a crystal are considered to have a common origin. Recapitulating briefly we may say that the magnetic moment of a free atom is associated with the orbital and spin motions of electrons in an incompleted sub-shell of the electronic structure of the atom. In crystals the orbital motions are quenched, that is the orbital planes may be considered to be fixed in space relative to the crystal lattice, and in such a way that in bulk the crystal has no resultant moment from this source. Moreover this orbital–lattice coupling is so strong that the application of a magnetic field has little effect upon it. However, in the case of a ferromagnetic crystal made up of domains there are the powerful forces of interaction between spins which produce parallel alignment of their axes and the spontaneous magnetic saturation of the domains. The spin axes are not tightly bound to the lattice as are the orbital axes. There is, however, a coupling between the spin and orbital motions and thereby a comparatively weak coupling between the spins and the crystal lattice. It is this spin–orbit coupling which causes the domain to have directional magnetic properties and provides the forces of magnetic anisotropy. The energy associated with this coupling is a minimum in iron when the spin axes are parallel with [100] directions, and in nickel for [111] directions. A change in energy and a torque on the lattice occur when, under the influence of an applied field, the spin axes are turned away from the easy directions. This explains qualitatively the origin of the magnetic anisotropy of ferromagnetic crystals.

The incompleted sub-shells in the atoms of a crystal are not spherically symmetrical; hence if in a ferromagnetic domain a deflection of the orbital axes occurs when a field is applied, this deflection occurring because of the spin–orbit torque, then a change in the lattice dimensions may be expected. This we may identify with the observed saturation magnetostriction occurring basically in the domains. Magnetostriction and magnetic anisotropy are therefore closely interrelated.

In principle it should be possible to derive mathematically values for the anisotropy constants K_1 and K_2 and for the magnetostriction constants λ_{100} and λ_{111}. The mathematical work has not met with great success so far as numerical values for these constants is concerned, for this work is difficult in the extreme and, to get any solutions, involves the use of drastic approximations. The basic physical mechanisms, however, seem to be clear.

PROBLEMS

6.1. The magnetization curve for a [110] direction of a single-crystal specimen of silicon-iron is given in MKS units by the equation $H = (M^3 - 2M)10^4$. Calculate the saturation intensity of magnetization and the first magnetic anisotropy constant of the material.

6.2. Show that for nickel crystal specimens cut in the [100], [110] and [111] directions, respectively, the theoretical knee points of the magnetization curves are, respectively, $M_S/3^{\frac{1}{2}}$, $M_S(2/3)^{\frac{1}{2}}$ and M_S.

6.3. The magnetization characteristic of a cobalt single-crystal specimen for the axial direction is given by $M = 1{\cdot}8$. For directions in the basal plane saturation is reached when $H = 8 \times 10^5$ A/m and half the saturation value is obtained when $H = 2{\cdot}5 \times 10^5$ A/m. Calculate the magnetic anisotropy constants K_1 and K_2.

6.4. The torque in N.m on a thin, oblate spheroidal single-crystal specimen of silicon-iron, cut in a (110) plane, when rotated in a strong saturating magnetic field is given by

$$L = -0{\cdot}01662 \sin 2\theta - 0{\cdot}0265 \sin 4\theta + 0{\cdot}001876 \sin 6\theta$$

If the volume of the specimen is 2 cm^3, calculate the magnetic anisotropy constants K_1 and K_2.

6.5 The vector M_S is rotated in a (110) plane in a domain in a cubic crystal, and θ is the angle between the vector M_S and [001]. Derive an expression for the magnetostrictive strain in relation to θ

(a) in a [111] direction, and

(b) in an [001] direction.

If $\lambda_{100} = 20 \times 10^{-6}$ and $\lambda_{111} = -20 \times 10^{-6}$ calculate the magnetostrictive strain in Case (a) and in Case (b) when $\theta = 0°, 45°, \tan^{-1}\sqrt{2}$ and $90°$.

SUGGESTED READING

BOZORTH, R. M. *Ferromagnetism*, Chapters 12 and 13, Van Nostrand, New York (1951).

LEE, E. W. 'Magnetostriction and magnetomechanical effects', *Rep. Prog. Phys.* **18,** 184 (1955).

REFERENCES

[1] HONDA, K. and KAYA, S. 'On the magnetization of single crystals of iron', *Sci. Rep. Tohoku. Univ.* **15,** 721 (1926).

[2] KAYA, S. 'On the magnetization of single crystals of nickel', *Sci. Rep. Tohoku. Univ.* **17,** 639 (1928).

[3] KAYA, S. 'On the magnetization of single crystals of cobalt', *Sci. Rep. Tohoku Univ.* **17,** 1157 (1928).

[4] JOULE, J. P. 'On the effects of magnetism upon the dimensions of iron and steel bars', *Phil. Mag.* **30,** 76, 225 (1847).

[5] HONDA, K. and NAGAOKA, H. 'On the magnetostriction of steel, nickel, cobalt and nickel-steels', *Phil. Mag.* **4,** 45 (1902).

[6] MASIYAMA, Y. 'On the magnetostriction of Ni–Co alloys', *Sci. Rep. Tohoku. Univ.* **22,** 338 (1933).

[7] KORNETSKI, M. 'The dependence of volume magnetostriction and of the Weiss Factor on temperature and lattice constant', *Z. Phys.* **98,** 289 (1935).

[8] AZUMI, K. and GOLDMAN, J. E. 'Volume magnetostriction in nickel and the Bethe–Slater interaction curve', *Phys. Rev.* **93,** 630 (1954).

[9] *See* References 13 and 14, Chapter 5.

[10] WEBSTER, W. L. 'Magnetostriction in iron crystals', *Proc. R. Soc.* A, **109,** 570 (1925).

[11] KAYA, S. and TAKAKI, H. 'Hysteresis loops and magnetostriction of iron crystals', *Sci. Rep. Tokoku Univ.* (**Honda**) 314 (1936).

[12] HONDA, K. *Magnetic Properties of Matter*, Syokwabo, Tokyo (1928).

[13] KIRCHNER, H. 'The effect of tension, compression and torsion on longitudinal magnetostriction', *Annln Phys.* **27,** 49 (1936).

Chapter 7

FERROMAGNETIC DOMAINS

7.1 The Development of the Domain Concept

A ferromagnetic material, when compared with one which is paramagnetic, is characterized by the ease with which it may be made to acquire a high intensity of magnetization. Polycrystalline iron, for example, may be magnetized almost to saturation in applied fields of about 400 A/cm, while the knee of the magnetization curve, where M is about $0{\cdot}7M_S$, may be reached in fields of about 4 A/cm. As discussed in Chapter 5, the extraordinary properties found in ferromagnetism were explained by Weiss on the basis of the postulates of a molecular field producing spontaneous saturation in the material and of the subdivision of the self-saturated material into domains.

Any ferromagnetic crystal, in the absence of an externally-applied field, will be made up of domains. The saturation magnetization vectors of the domains will, on account of the crystalline forces of magnetic anisotropy, prefer particular crystallographic directions. In Chapter 6 we derived equations for the upper part of the magnetization curves in different directions in a crystal. Theoretically the magnetization process above the knee point is one of smooth reversible rotation of the M_S vectors relative to the direction of the applied field, and theory and observation in crystals are in reasonably good agreement. So far, however, we have not considered the phenomena occurring during magnetization below the knee of the curve.

Figure 7.1 shows, for example, the initial magnetization curve and two hysteresis loops of a 4% silicon-iron polycrystalline material. In this material $M_S = 1{\cdot}95$ Wb/m^2 and the knee occurs where $M = 1{\cdot}35$ Wb/m^2 approximately. On magnetizing the material from the demagnetized condition at O along the initial curve from O to P in the figure, sudden changes in direction of M_S in the domains through 90° or 180° occur and there are also movements of the boundaries between neighbouring domains. There is a combination of irreversible jumps and reversible boundary movements so that a highly magnified view of the curve at any point might be represented as shown at A. When at point P the applied field is decreased the original magnetization curve is not retraced but the familiar hysteresis loop is obtained. This may be compared with the loop corresponding to a high intensity of magnetization at C. The statement that the upper part of the magnetization curve is reversible

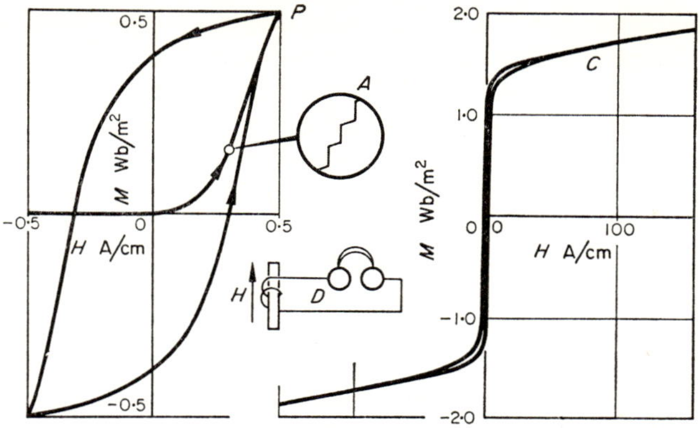

FIG. 7.1. Magnetization curve and hysteresis loops of 4% silicon-iron below and above the knee illustrating Barkhausen jumps.

needs some qualification since, as the figure shows, the ascending and descending sides of the hysteresis loop are not quite coincident. Remembering that hysteresis loss is proportional to the area of the loop, and noting the scale of H in this case, there is in fact in practice, even with a good quality material, a substantial energy loss associated with this upper part of the loop.

The discontinuities in the magnetization curve illustrated at A are due to the presence of domains. If a search coil is wound on the material and connected to headphones, as shown diagrammatically at D, then on increasing the applied field H, however slowly, the sudden jumps in M will induce a rapid succession of impulses of electromotive force in the coil and a rushing noise is heard in the headphones. This experiment was carried out by BARKHAUSEN [1] and indeed provided the first direct experimental support for the existence of discrete self-saturated domains in a ferromagnetic material. With care, and in quiet conditions, the Barkhausen noise can be heard directly with a search coil and headphones, but using an audiofrequency amplifier the noise may be reproduced at any desired volume.

BOZORTH and DILLINGER [2] elaborated the Barkhausen experiment and obtained oscillographic records of the impulses of e.m.f. produced by the discontinuities, after amplification, in the case of iron and a number of other ferromagnetic materials. On the assumption that each impulse was due to the reversal of M_S in a domain it was estimated that on the steepest part of the magnetization curve, where the impulses were largest, and for a variety of materials, the volumes involved in a reversal were of the order of 10^{-8} to 10^{-9} cm^3. These results were at the time thought to represent the sizes of whole domains. We now know, however, that domains in general in soft magnetic materials are very much larger than this and are not fixed in size. Domains may, in fact, grow or contract in volume due to domain boundary movements when the material is acted upon by a changing magnetic field.

There is a well-known technique in engineering whereby fine cracks in steel components may be found by magnetizing the components and sprinkling the surface with a liquid containing a fine magnetic powder. The particles are attracted by stray field to the cracks which are thus made visible. BITTER [3] attempted to map out domain boundaries on ferromagnetic surfaces by a refinement of this technique, and patterns were indeed observed which were at first thought to represent the domains. The method was not entirely successful due to imperfect polishing of the specimens. The Bitter patterns resulted from a strained surface layer produced during polishing and did not, in fact, show the underlying domain arrangements.

NÉEL [4, 5] subsequently, from a theoretical analysis of the energies associated with domains, predicted the shapes and sizes to be expected for them. His predictions were almost immediately confirmed by experiment by WILLIAMS and others [6], more particularly on crystals of silicon-iron. The Bitter process was improved by removing electrolytically the strained surface layer on polished samples. The patterns which may be observed are now well established as representing the domain boundaries where they meet the surface of the specimen. The patterns are relatively simple in cases where the surface coincides with particular crystal planes but otherwise the patterns may be complicated and difficult to interpret. The subject will be discussed in some detail in the following sections.

7.2 Magnetostatic Energy and Domain Subdivision

Let us first consider a crystal of iron in the form of a sphere as shown in Fig. 7.2(*a*). If this were to remain a single domain then the saturation vector M_S would take up one of the cube-edge directions since this would make the magnetic anisotropy energy a minimum. As mentioned in Chapter 1 the demagnetizing factor of a sphere $N_D = 1/3\mu_0$, the demagnetizing field inside the sphere, arising from the free poles on the surface, would be $H_D = N_D M_S$, and the potential energy or magnetostatic energy per unit volume arising from the intensity M_S in the opposed field H_D is $W_M = (1/2)H_D M_S = (1/2)N_D M_S^2 = M_S^2/6\mu_0$. A domain in the form of a cube as shown at (*b*) in

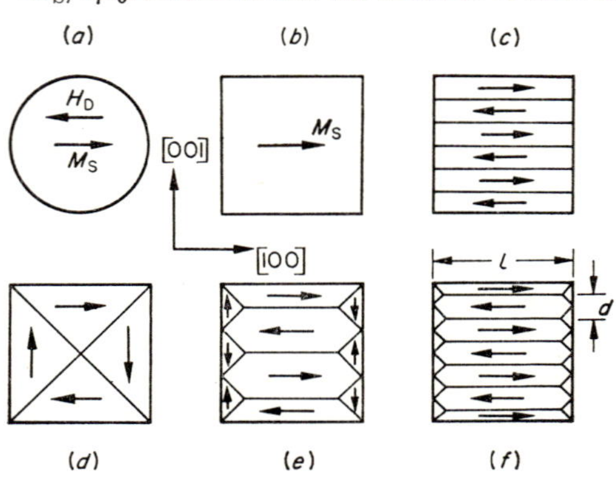

FIG. 7.2. Showing formation of thin lamellar domains in iron.

the figure has theoretically, according to RHODES and ROWLANDS [7], exactly the same magnetostatic energy as a sphere of the same volume. For iron with $M_S = 2{\cdot}14$ Wb/m^2, $W_M = 2{\cdot}14^2 \times 10^7/(6 \times 4\pi) = 6 \times 10^5$ J/m^3. On referring back to Chapter 6 it will be seen that this magnetostatic energy is large compared with the anisotropy and magnetoelastic energies.

It is easy to see from NÉEL's work [8] that the single domain is unstable, except in the special case to be discussed later when the domain volume is very small. The magnetostatic energy may be reduced and indeed eliminated by a breakdown of the large domain into a number of smaller domains. For example, if the block were subdivided into lamellar domains with an antiparallel arrangement of the M_S vectors, as illustrated in Fig. 7.2(*c*), then the internal demagnetizing field is greatly reduced, while the magnetic anisotropy energy is unchanged. Because free poles still appear on the ends of the block there is, however, still some magnetostatic energy. Alternatively if only four domains were formed in the cube, as shown at (*d*), the domain boundaries would be at 45° to the directions of M_S. There is now no free pole on the boundaries or anywhere in the block and the magnetostatic energy has been eliminated, while again the anisotropy energy is a minimum.

We wish to find the most stable domain arrangement, that is one in which the total energy of the system is a minimum. The pattern in Fig. 7.2(*d*) does not provide a minimum in fact because the magnetostriction of the iron domains has to be taken into account. Each of the domains of triangular section in the figure would, if free, have a longitudinal strain of λ_{100} in the direction of M_S and be contracted laterally by $\lambda_{100}/2$. The domains in the arrangement of Fig. 7.2(*d*) obviously place constraints upon each other so that a component of magnetoelastic energy now exists. However, if the structures shown at (*c*) and (*d*) are combined to give the arrangement illustrated at (*e*) we will then have a minimum for the anisotropy energy, zero magnetostatic energy and magnetoelastic energy confined mainly to material near the ends of the block. The small triangular domains with the boundaries at 45° to the directions of M_S shown in Fig. 7.2(*e*) are called closure domains. The boundaries where there is a change of direction of M_S through 90° are known as 90° boundaries, while those between the antiparallel lamellar domains are called 180° boundaries.

The total volume of the closure domains in the block, and hence the magnetoelastic energy, will decrease as the 180° domain boundaries come closer together, as may be seen in Fig. 7.2(*f*). However, as the thickness of these plate-like domains gets less, so more and more domain boundary surfaces are introduced into the block. These boundaries between domains are in fact walls or layers of a finite thickness with which energy also is associated, as described in the next section. The distance between the 180° walls in the block we are considering will finally be determined when the sum of the boundary wall energy and the magnetoelastic energy associated with the closure domains is a minimum.

7.3 Domain Boundary Energy

In Fig. 7.3(*a*) we consider a domain in a single crystal of iron. Let the plane of the paper represent the *YZ* plane or (100) plane of the crystal, and let the arrows correspond with the atomic moments in parallel layers of atoms in *XZ* or (010) planes, that is planes perpendicular to the paper. If the lattice constant is a then these layers in body-centred iron are $a/2$ apart. The domain at (*a*) is maintained in the self-saturated condition by the forces of interaction between neighbouring layers. At (*b*) two domains are represented with a sharp boundary at the broken line where the layers of atomic magnets on either side of the boundary surface point in opposite directions. Because of the positive interaction, work is required to produce the antiparallel arrangement of atomic magnets in the adjacent layers at the boundary, and there is

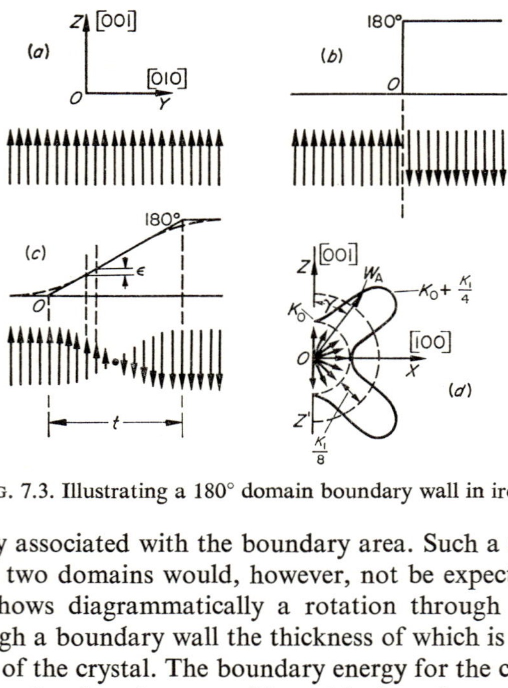

FIG. 7.3. Illustrating a 180° domain boundary wall in iron.

therefore energy associated with the boundary area. Such a sudden division as this between two domains would, however, not be expected in a crystal. Figure 7.3(*c*) shows diagrammatically a rotation through 180° occurring gradually through a boundary wall the thickness of which is many times the lattice constant of the crystal. The boundary energy for the condition shown at (*c*) is less than for that shown at (*b*), and in fact the energy of the wall decreases as the wall thickness increases, as we shall show. There is, however, also magnetic anisotropy energy in the wall which increases with wall thickness. A stable thickness of wall is reached when the sum of the interaction energy and anisotropy energy is a minimum.

Early theoretical work on domain boundaries was done by BLOCH [9], by LANDAU and LIFSHITZ [10], and subsequently by NÉEL [4] and others. The boundaries are often referred to as Bloch walls. To avoid the complications of an exact solution of the problems involved we shall make some simplifying assumptions, but these do not introduce any great error in the determination of wall energies and thicknesses.

In Section 5.6 the energy of interaction between neighbouring spinning electrons was discussed and relations obtained in equations (5.9), (5.10) and (5.11) between the exchange integral, the Curie temperature and the Weiss molecular field. It is convenient to think of the forces of interaction in terms of the Weiss field.

To avoid magnetostatic energy the condition to be met in Figs. 7.3(*b*) and (*c*) is that there shall be no free pole in the region of the boundary, and this is achieved if the atomic magnetic axes remain in the *XZ* or (010) plane. The arrows in Fig. 7.3(*c*) represent this condition.

Let t be the thickness of the Bloch wall. The number of atom layers in the wall is then $n = 2t/a$ where a is the lattice constant. We will assume that the increments of angular rotation between successive layers are the same and equal to ε. In a more rigid solution these increments are not equal, and the angle of rotation of the atomic axes across the wall will vary in a manner indicated by the broken line in Fig. 7.3(*c*). The wall then has an indeterminate thickness. However, we put, for the 180° boundary, $\varepsilon = \pi/n$. The volume associated with a single layer of atoms for a unit area of the boundary is $a/2$. If in Fig. 7.3(*a*) we were to rotate the atomic magnets in one layer through 180°, and the Weiss field is H_W, then the work done for a unit area of layer would be $2H_W M_S(a/2)$. If we suppose the Weiss field to be due to the interaction between nearest neighbours, then the energy per unit area of boundary for the condition shown at (*b*) would be half this amount, that is $H_W M_S a/2$. If we imagine that the wall in Fig. 7.3(*c*) is formed by rotating the right-hand domain in successive increments of ε, bringing in the layers in the wall one by one, then the work done for each layer would be $(1/2)H_W M_S \times (a/2)(1 - \cos \varepsilon)$. The total exchange or interaction energy for n layers, per unit area of boundary, is

$$\begin{aligned} W_E &= (1/4)nH_W M_S a(1 - \cos \varepsilon) \\ &= (1/2)nH_W M_S a \sin^2 (\varepsilon/2) \end{aligned}$$

If ε is small, and putting $H_W = N_W M_S$, where N_W is the Weiss molecular field constant, then

$$\begin{aligned} W_E &= nN_W M_S^2 a\varepsilon^2/8 \\ &= N_W(M_S a\pi/4)^2(1/t) \end{aligned} \tag{7.1}$$

The interaction energy in the boundary therefore decreases as the wall thickness increases. However the anisotropy energy increases with increasing thickness. Figure 7.3(*d*) represents a view looking along the direction OY, that is normal to the boundary. The atomic magnetic axes in the wall take up all positions from OZ through 180° to OZ' as shown. For any inclination of the axes by an angle γ from OZ the magnetic anisotropy energy per unit volume is, from equation (6.2), given by $W_A = K_0 + (K_1/8)(1 - \cos 4\gamma)$.

W_A is plotted in relation to γ in Fig. 7.3(*d*), and the mean value of W_A for the whole boundary wall is $K_0 + K_1/8$. The part K_0 is independent of γ and will not change with the wall thickness, and so may be neglected. The total energy of the wall per unit area of the boundary is therefore $W = W_E + W_A$ or $W = N_W(M_S a\pi/4)^2(1/t) + K_1 t/8$. The wall will assume a thickness which makes W a minimum. Whence putting $dW/dt = 0$ we have

$$-N_W(M_S a\pi/4)^2(1/t^2) + K_1/8 = 0$$

or

$$t = \pi a M_S(N_W/2K_1)^{\frac{1}{2}} \tag{7.2}$$

and W is a minimum when $W_E = W_A$ or $W = 2W_A$.
∴ the wall energy per unit area is

$$W = K_1 t/4 \tag{7.3}$$

The thickness and energy of a 180° boundary wall in iron may be obtained from equations (7.2) and (7.3) by putting in the appropriate numerical values. We have $a = 2{\cdot}86 \times 10^{-8}$ cm, $K_1 = 4{\cdot}4 \times 10^4$ J/m³ and $M_S = 2{\cdot}14$ Wb/m². We may obtain a value for N_W using equation (5.10) and the data given in Section 5.6, though these data are based on assumptions which introduce some uncertainty into the value obtained for N_W. However, put $N_W = H_{W0}/M_0 = 12 \times 10^8/2{\cdot}19 = 5{\cdot}5 \times 10^8$ MKS units. Then from equation (7.2)

$$t/a = \pi \times 2{\cdot}14\surd\,\{5{\cdot}5 \times 10/(2 \times 4{\cdot}4 \times 10)\} = 530$$

The wall thickness is therefore

$$t = 530a = 530 \times 2{\cdot}86 \times 10^{-8} = 1510 \times 10^{-8} \text{ cm}$$

The wall energy is

$$W = 4{\cdot}4 \times 1510 \times 10^4 \times 10^{-10} \text{ J/m}^2 = 1{\cdot}66 \text{ ergs/cm}^2$$

The 180° domain wall illustrated in Fig. 7.3(*c*) may be considered to be made up of two 90° walls each of thickness $t/2$. A closer look at this reveals, as NÉEL [4] has pointed out, that if these halves were drawn apart, so that interposed between them was a domain with M_S normal to the plane of the paper in the figure, there would be no change of energy but the overall length t would become indeterminate. However, the rotation of the atomic magnetic axes in the boundary is accompanied by a local variation of the magnetostriction and this introduces a component of magnetoelastic energy. As a result of this it is found that the 180° boundary wall remains intact, and there will be no great change in the thickness or in the energy we have already calculated.

Different kinds of domain boundary in which there is no free pole on the boundary surfaces are shown for iron in Fig. 7.4. The wall thicknesses and energies for these types of boundary, as calculated by LILLEY for iron [11], are given in Table 7.1. In the case of a 180° boundary in an (001) plane the energy given in the table is 1·24 ergs/cm². This compares with a value of 1·66 ergs/cm² just calculated and with an estimate by BOZORTH [12] of 1·5 ergs/cm². The differences are due to inexact knowledge of N_W or the interaction energy in the wall.

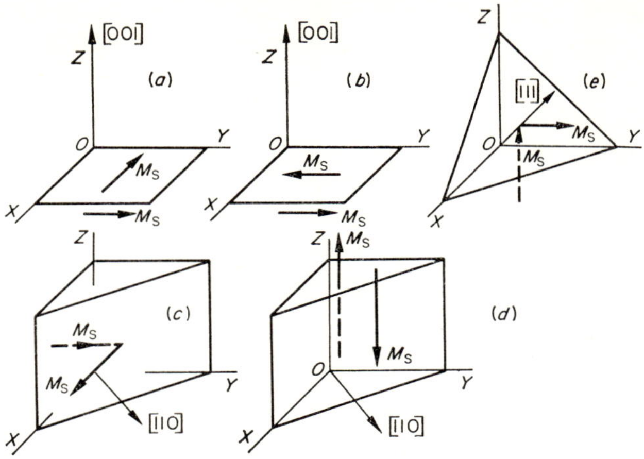

FIG. 7.4. Types of domain boundary in iron: (*a*) 90°, (001) plane: (*b*) 180°, (001) plane: (*c*) 90°, (110) plane: (*d*) 180°, (110) plane: (*e*) 90°, (111) plane.

In nickel the easy directions in the domains correspond with [111] diagonals. In the cubic structure these directions intersect at angles of 70·53° and 109·47°. The kinds of domain boundaries in nickel are then as given in Table 7.1. The energies and wall thicknesses are due to Lilley. In cobalt there are two easy directions, parallel and antiparallel along the *c*-axis. A 180° boundary is therefore the only type to be considered and, according to Lilley, this has a wall energy of 8·2 ergs/cm^2 and a thickness of 157 $\times$ 10^{-8} cm.

TABLE 7.1

Fig. 7.4	Boundary Plane	Type	Wall Energy erg/cm^2	Wall Thickness $10^8 \times t$ cm
		Iron		
a	(001)	90°	0·62	408
b	(001)	180°	1·24	1410
c	(110)	90°	1·07	516
d	(110)	180°	1·71	728
e	(111)	90°	0·74	408
		Nickel		
	(001)	70·53°	0·076	1000
	(001)	109·47°	0·152	∞
	(110)	70·53°	0·0645	1110
	(110)	109·47°	0·192	860
	(110)	180°	0·306	2060
	(111)	109·47°	0·181	1000

7.4 Domain Sizes in Iron

In the last paragraph of Section 7.2 we concluded that the iron crystal shown in Fig. 7.2 would tend to subdivide itself into domains in the form of thin laminae. Subdivision in this way reduces the magnetoelastic energy but introduces an increasing domain wall energy. Having now determined the energy per unit area to be associated with the boundaries we may estimate the ultimate thickness of the thin plate-like domains.

In Fig. 7.2(*f*) let l be the side of the cubical block and d the distance between the boundaries. If d is small compared with l then the total area of the 180° boundaries is l^3/d. The 90° closure domain boundaries are comparatively small and will be neglected. If W is the wall energy per unit area the total wall energy of the block is $W_W = Wl^3/d$.

The total volume of the closure domains in Fig. 7.2(*f*) is $l^2d/2$. If we suppose that magnetostrictive extension in these closure domains is entirely prevented by the restraint imposed by the remainder of the block the stress in them will be $(3/2)\lambda_{100}E_Y$ where E_Y is Young's Modulus. The magnetoelastic energy stored is then

$$W_M = (1/2)(3\lambda_{100}/2)^2E_Y(l^2d/2) = (9/16)\lambda_{100}^2E_Yl^2d$$

The sum of the wall energy and magnetoelastic energy for the whole block is then

$$W_T = W_W + W_M = Wl^3/d + (9/16)\lambda_{100}^2E_Yl^2d$$

The value of d which makes W_T a minimum is obtained by putting $dW_T/dd = 0$. Whence W_T is a minimum when

$$d = (4\lambda_{100}/3)(Wl/E_Y)^{\frac{1}{2}} \tag{7.4}$$

and when $W_W = W_M$.

For iron we may substitute the values $\lambda_{100} = 20 \times 10^{-6}$, $W = 1{\cdot}24$ ergs/cm² $= 1{\cdot}24 \times 10^{-3}$ J/m² and $E_Y = 12{\cdot}3 \times 10^{10}$ N/m². Then from equation (7.4) we have $d = 0{\cdot}007\sqrt{l}$ where d and l are in metres, or $d = 0{\cdot}07\sqrt{l}$ where d and l are in centimetres.

7.5 Observed Domain Patterns

Domain boundaries at the surface of ferromagnetic crystals may be mapped out and observed under the microscope, as already mentioned, and extensive investigations of domain patterns have been made, notably in the early days by Bozorth and his co-workers. Most work has been done on silicon-iron but patterns have also been obtained on nickel and cobalt and on some alloys. By an ingenious technique, using a permanent magnet probe, WILLIAMS, BOZORTH and SHOCKLEY [6] were able also to determine the direction of the magnetic saturation in the domains. Some of the results are given in Figs. 7.5 and 7.6 where the arrows show the directions of M_S.

Patterns for 3·8% silicon-iron crystals on surfaces parallel, or very nearly

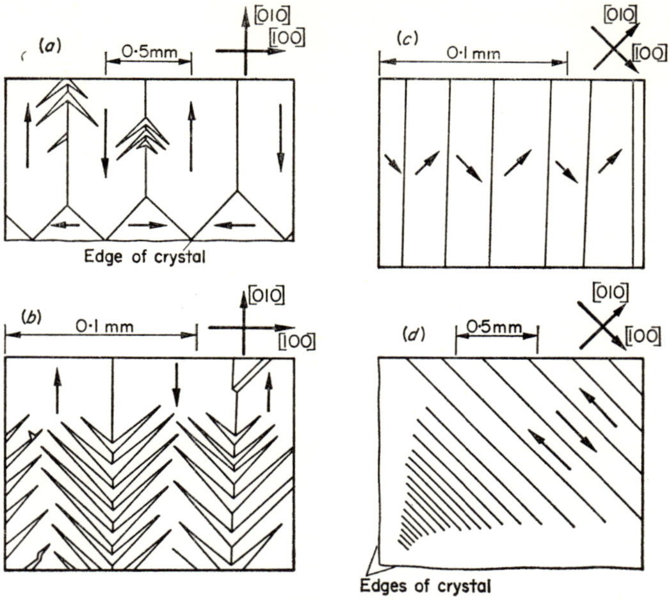

FIG. 7.5. Domain patterns observed on (001) planes of silicon-iron crystals. The directions of the crystal cube-edge axes are as shown.

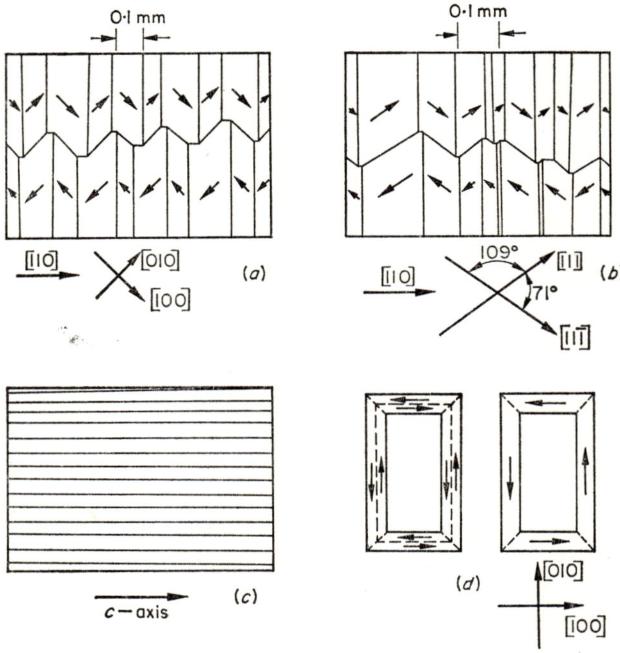

FIG. 7.6. Domain patterns: (a) 3·8% Si–Fe, (001) plane: (b) nickel, (1$\bar{1}$0) plane: (c) cobalt, plane parallel to c-axis: (d) 3·8% Si–Fe, [100] 'picture-frame' specimen.

parallel, to cube-face or (001) crystal planes are shown in Fig. 7.5. At (*a*), near the edge of a crystal, the closure domains are clearly seen. The tree-like patterns in (*a*) and (*b*), often referred to as Néel spikes are a frequently occurring feature. They are considered to be due to a small deviation, one or two degrees, of the specimen surface from the true (001) plane of the crystal. The M_s vectors will in this case break the surface at a small angle, thus producing some free pole and introducing magnetostatic energy. The spiked domains are in fact special, shallow, surface closure domains carrying magnetization, in the horizontal direction in the figures, between the main domains. They have the effect of reducing the weak magnetostatic energy at the expense of adding some energy due to the walls of the spikes. The spikes vanish on surfaces accurately parallel to (001) planes, but reappear in the neighbourhood of cavities or inclusions in the crystal, and this will be discussed again in Section 8.4.

90° boundary walls which are parallel to (110) planes are shown in Fig. 7.5(*c*), and again in Fig. 7.6(*a*). In this case, as in all these diagrams for silicon iron, the saturation vectors lie along cube-edge directions. Figure 7.5(*d*) shows 180° boundaries in (010) planes near the corner of a crystal, on the (001) surface of the specimen. It will be seen that the domain walls come closer together as the length of the domains decreases. This is perhaps some confirmation of the theory developed in Section 7.4 where the domain width was found to be proportional to the square root of the length.

Figure 7.6(*a*) shows, for a 3·8% silicon-iron crystal, a pattern observed on an (001) surface, where the 90° boundaries are parallel to (110) planes, as already mentioned. The mean intensity of magnetization in the upper part of the diagram is $M_S/2^{\frac{1}{2}}$ to the right, while in the lower part there is a corresponding intensity to the left. The diagram is divided by the zig-zag boundaries which are mainly of the 180° type. It was found that, on applying a field H horizontally, that is in the [110] direction, these boundaries moved downwards. Thus in this case the resultant intensity of magnetization in the field direction increased by a growth in volume of the upper domains at the expense of the lower. This is precisely the effect to be expected.

The pattern shown in Fig. 7.6(*b*) is for a nickel crystal and is due to Williams and Walker [13]. The specimen surface corresponds with a $(1\bar{1}0)$ plane and the vertical lines to domain boundaries in (110) planes. The easy direction of magnetization in nickel is the body diagonal or [111] direction and the angles between these diagonals, as previously mentioned, are approximately 71° and 109°, as shown in the figure. The M_S vectors in the diagram are then parallel or antiparallel to [111] or to $[11\bar{1}]$ and the principal boundaries are of the (110), 109° type (see Table 7.1). The zig-zag boundaries are of the 180° type, and application of a magnetic field in the [110] direction, horizontally in the figure, caused these boundaries to move downwards as expected.

In cobalt the easy directions are parallel or antiparallel to the *c*-axis and only 180° boundaries, parallel to this axis, occur. Figure 7.6(*c*) shows a pattern which has been observed in this case.

WILLIAMS and SHOCKLEY [14] prepared single crystal specimens from 3·8% silicon-iron in the form of hollow rectangles or 'picture frames'. One of their specimens is shown in Fig. 7·6(*d*), although not to scale. The overall dimensions of the specimen were 1·9 × 1·3 cm, the width of the four limbs was 0·1 cm and the thickness 0·07 cm. The rectangle was cut so that each side was parallel to a cube-edge crystal direction. When this specimen was heated above the Curie temperature, thus eliminating any domain structure, and then allowed to cool slowly it was found to consist of four domains, as shown in the right-hand part of the figure, with the 90° boundaries in the corners. This is a condition of lowest energy since there is no magnetostatic energy, anisotropy energy is a minimum, and magnetoelastic energy will be small. There remains the small boundary energy. In the demagnetized state eight domains were formed, as shown in the left-hand figure, with domain boundaries represented by the broken lines. Additional boundary energy has now been introduced. Hysteresis losses accompanied any subsequent movement of the boundaries. These particular domain patterns seem to be the simplest conceivable in a whole specimen.

7.6 Single Domain Particles

It was shown in Section 7.3 that for a 180° domain boundary in iron the wall thickness was about 530a or 1510 × 10^{-8} cm in thickness. It is of some interest and practical importance to consider the conditions in small ferromagnetic particles of this order of size. It has, in fact, been shown by NÉEL [15] and independently by STONER and WOHLFARTH [16] that when the particles are small enough they remain as single domains, domain boundaries being energetically unable to form in them.

It is convenient to consider single crystal particles, for example of iron, in the form of prolate ellipsoids of revolution of length l and equatorial diameter d as shown in Fig. 7.7. Let N_D be the demagnetizing coefficient of a particle in the direction of the polar axis. Then N_D depends on the ratio l/d and $\mu_0 N_D$, where $\mu_0 = 4\pi \times 10^{-7}$ H/m, is plotted in the figure in relation to l/d. For a sphere, where $l = d$, $\mu_0 N_D = 1/3$. If now the dimensions of the particle are many times the thickness of a Bloch wall then we should expect the particle to be subdivided by domain walls into a number of domains. Magnetostatic energy could thus be eliminated or made small but domain wall energy would be introduced. However, the wall energy is a function of the thickness of the wall and increases with decreasing thickness. The interaction energy is given in equation (7.1). Hence if the particle size becomes small compared with the normal wall thickness the wall energy may now rise to a comparatively high value, so that walls cannot form in the small particle except by introducing wall energy greater than the magnetostatic energy of the single domain particle. In these circumstances we should expect the particle to remain as a single domain with M_S along the polar axis, that is in the direction having the lowest demagnetizing factor.

The critical diameter below which the particle would remain a single domain may be estimated by calculating the work required to demagnetize the particle and thus eliminate the magnetostatic energy.

The single domain particle shown in Fig. 7.7(*a*) could be demagnetized by rotating the atomic magnets layer by layer through a total angle of 360° across the diameter as shown at (*b*). Rotation through 180° would not be sufficient since this would not reduce the magnetization to zero in all directions but would leave a component in a direction perpendicular to the paper.

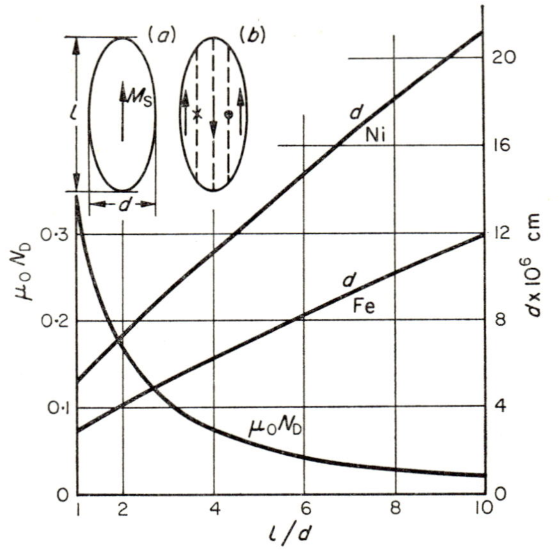

FIG. 7.7. Critical sizes of single-domain particles of iron and nickel; prolate spheroids.

The demagnetized state at (*b*) therefore may be regarded as made up of two 180° boundaries each of thickness $t = d/2$. The wall energy per unit area for a 180° boundary of thickness t due to interaction is given by equation (7.1). Since a unit area of wall thickness t has a volume t, then the interaction energy per unit volume is $W_E = N_W(M_S a\pi/4t)^2$. There will also be anisotropy energy which is, as in Section 7.3, $W_A = K_1/8$ per unit volume. On the other hand the magnetostatic energy per unit volume of the single domain shown in Fig. 7.7(*a*) is $W_M = N_D M_S^2/2$. Equating the energies for the two cases and putting $t = d/2$ we have.

$$N_D M_S^2/2 = N_W(M_S a\pi/2d)^2 + K_1/8$$

whence

$$d = \pi a(N_W/2N_D)^{\frac{1}{2}}\{1 - K_1/(4N_D M_S^2)\}^{-\frac{1}{2}} \tag{7.5}$$

The term in this equation containing K_1 turns out to be small compared with unity, for example in the case of a spherical iron particle we may put $K_1 = 4{\cdot}4 \times 10^4$ J/m³, $M_S = 2{\cdot}14$ Wb/m² and $N_D = 1/3\mu_0 = 10^7/12\pi$, and then $K_1/(4N_D M_S^2) = 0{\cdot}01$. By neglecting the K_1 term, equation (7.5) becomes

$$d = \pi a(N_W/2N_D)^{\frac{1}{2}} \tag{7.6}$$

Equation (7.6) gives a critical value of d. When d is less than this, the magnetostatic energy is less than the energy of the particle in the demagnetized state represented in Fig. 7.7(*b*). Moreover domain boundaries of small thickness could not form inside the particle since the energy associated with them would be greater than in Fig. 7.7(*b*). Therefore when d is less than the value given by equation (7.6) the particle will remain as a single domain.

For a spherical particle of iron, putting $N_W = 5{\cdot}5 \times 10^8$ and $a = 2{\cdot}86 \times 10^{-8}$ cm, equation (7.6) gives $d = 101\ a = 2{\cdot}9 \times 10^{-6}$ cm. If the particle is in the form of a prolate spheroid, for example with $l/d = 10$, then $N_D = 0{\cdot}02/\mu_0$, the anisotropy energy is still not very important, and the equatorial diameter of the particle is $d = 413a = 12 \times 10^{-6}$ cm. A relation between d and l/d for iron is given in Fig. 7.7.

For nickel, critical sizes of the single domain particles are greater than for iron. From Section 5.6 we have $N_W = k\theta_C/M_0\mu_B$. For nickel $\theta_C = 631$°K, $M_S = 0{\cdot}641$ Wb/m². Boltzmann's constant $k = 1{\cdot}38 \times 10^{-23}$ J/°K and the Bohr magneton $\mu_B = 1{\cdot}165 \times 10^{-29}$ Wb.m. Whence $N_W = 11{\cdot}7 \times 10^8$ MKS units. Putting the lattice constant for nickel $a = 3{\cdot}517 \times 10^{-8}$ cm, the critical diameter for a spherical particle from equation (7.6) is $d = 147a = 5{\cdot}2 \times 10^{-6}$ cm. When $l/d = 10$, then $d = 600a = 21 \times 10^{-6}$ cm. A relation between d and l/d for nickel is also shown in Fig. 7.7.

The foregoing analysis and the results plotted in Fig. 7.7 give the critical dimensions of single domain particles of iron and nickel of different ratios of length to diameter. The figures obtained can, however, only be regarded as approximate, and as indicating the order of size of these particles, because of the uncertainty in the values to be assigned to N_W and the simplifying assumptions made in the analysis. However, an important characteristic of these single domain particles, when they depart from the spherical form, is the strong magnetic anisotropy arising from the particle shape. The state of lowest energy is when M_S lies along the axis corresponding to the lowest demagnetizing coefficient, that is along the polar axis of the prolate spheroids. To rotate M_S away from this direction may require a considerable amount of energy. The matter is thought to be important in permanent magnet materials of high coercivity and is discussed further in Chapter 8.

7.7 Single Domains in Thin Films

It has been found experimentally that very thin ferromagnetic films tend to become single domains with the saturation magnetization vector lying parallel to the surface of the film. Films of thickness less than about 2000 Å (2×10^{-5} cm) are considered thin in this connection. The phenomenon occurs, not only in single crystal films, but also in films which are polycrystalline, so that the domain may in fact spread over many crystals. The single crystal films are found to have a strong uniaxial magnetic anisotropy superimposed upon the normal crystal anisotropy. The polycrystalline films can also be made to have a uniaxial anisotropy. In either case the films, as well as being single domains, have one preferred axis in them for M_S. The magnetic state is thus bistable in the sense that on the application of a reverse magnetic field of sufficient magnitude the direction of M_S may be reversed. They thus behave like permanent magnets capable of reversal with two directions, forward and reverse, for the magnetic vectors. Reasons why thin ferromagnetic films should tend to become single domains can be given, but the origin of the uniaxial behaviour is obscure.

That thin films might be single domains was anticipated on theoretical grounds by Kittel. An analysis of the phenomenon will be made with reference

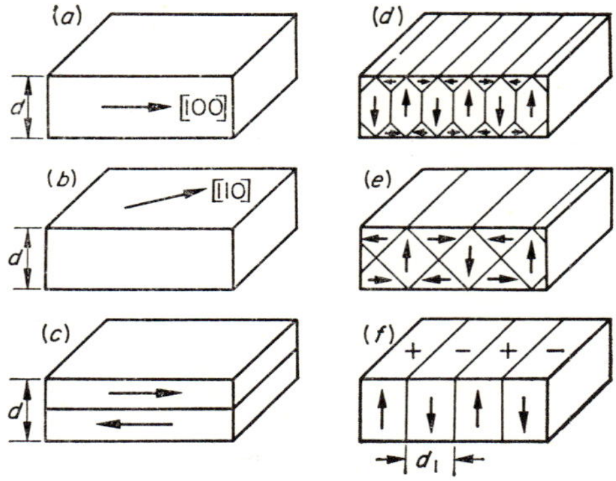

FIG. 7.8. Hypothetical domain arrangements in a thin film of iron.

to Fig. 7.8 but in a slightly different way from that given by KITTEL [17]. The figure shows a single crystal film of iron of thickness d with several hypothetical domain arrangements. We assume the film surface is parallel to an (001) plane and in Fig. 7.8(a) the film is a single domain with M_S along the [100] cube-edge direction. If d is small compared with the surface dimensions of the film, the demagnetizing factor of the film in any direction parallel to the surface will be very small. The magnetostatic energy is thus negligible and with M_S parallel to [100] the anisotropy energy is also small. We may therefore expect the film in condition (a) to be a single domain with energy $W_a \to 0$. However, in polycrystalline films which are found to be single domain, M_S in some of the constituent crystals will lie in crystal directions where the anisotropy energy is not zero. We therefore consider Case (b) in the figure.

In Case (b) let M_S be parallel to [110]. The anisotropy energy per unit volume is now $K_1/4$ (see Table 6.1) and the total energy per unit area of the film is $W_b = K_1 d/4$. The [110] axis is not the direction of highest anisotropy energy in iron. If in fact we had a crystal in which [111] was parallel to the film surface then the energy would be $(K_1/3 + K_2/27)$.

Case (*c*) in Fig. 7.8 shows the film divided into two domains with a 180° domain boundary and M_S in an easy direction in each domain. The energy existing now is the domain wall energy discussed in Section 7.3 so that the energy per unit area of film is $W_c = W_W$ where W_W is the 180° wall energy. Alternatively the boundary in this case might have been of the 90° type and in this case the wall energy would be halved.

Another arrangement of domains in which the magnetostatic energy and the anisotropy energy have been eliminated is shown at (*d*). The wall energy in this case would be reduced by increasing the distance between the 180° walls until the condition at (*e*) was obtained. In Case (*e*) the total area of the 90° walls is $2\sqrt{2}$ times the area of the 180° walls in Case (*c*). As just stated a 90° wall has only half the wall energy of a 180° wall of the same area (see Table 7.1), but Case (*c*) is still energetically a more favourable condition than Case (*e*).

Figure 7.8(*f*) shows another possibility with domains in parallel strips separated by 180° boundaries. Because of the surface distribution of free poles in this case there is magnetostatic energy as well as the wall energy. If d_1 is the distance between the walls the total wall energy per unit area of film is $W_W d/d_1$. According to Kittel the magnetostatic energy per unit area of film is given by $1{\cdot}7M_S^2 d_1$ in CGS units, which in MKS units becomes $1{\cdot}07 \times 10^5 M_S^2 d_1$. The total energy is then

$$W = W_W d/d_1 + 1{\cdot}07 \times 10^5 M_S^2 d_1$$

W is a minimum when $dW/dd_1 = 0$, and then $d_1 = (W_W d/1{\cdot}07 \times 10^5 M_S^2)^{\frac{1}{2}}$ and in this case $W_f = 645 M_S (W_W d)^{\frac{1}{2}}$.

For iron we may put $M_S = 2{\cdot}14$ Wb/m², $K_1 = 4{\cdot}4 \times 10^4$ J/m³, $K_2 = 2{\cdot}1 \times 10^4$ J/m³ and $W_W = 1{\cdot}66 \times 10^{-3}$ J/m². In Fig. 7.9 W_b, W_c and W_f are plotted in relation to the film thickness d.

So far as a single crystal film is concerned, the condition in Fig. 7.8(*a*) is the state of lowest energy with $W_a \to 0$, but the single domain state at (*b*) is also of lower energy than any of the multi-domain conditions in the figure provided that the thickness of film d is less than about 10^{-5} cm. This will be clear from Fig. 7.9. It is therefore to be expected that polycrystalline films below this critical thickness, and in which M_S in the constituent crystals may have any direction relative to the crystal axes, will also be single domain. It will be noticed by reference to Table 7.1 that the critical film thickness is of the same order of magnitude as the calculated thickness of a domain wall between domains in thicker material.

The usual method of making the films is to melt the metal in a vacuum and to condense the metal vapour on to a flat, clean substrate which may be of glass. The polycrystalline films thus formed are, however, somewhat imperfect. They are not uniform in thickness but appear to be an aggregation of

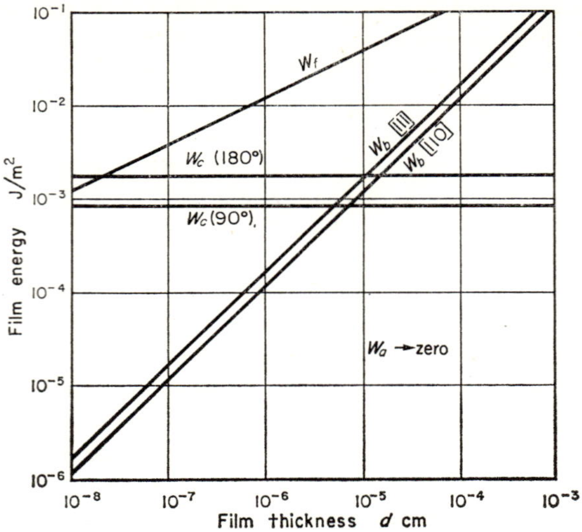

FIG. 7.9. Energy of a thin iron film in relation to thickness for different domain conditions.

small particles. The films are found to have the same M_S–T characteristics as the bulk metal, at least in thicknesses down to about 20 Å, and to have single domain properties. They are, however, magnetically isotropic unless specially treated.

Although the phenomenon of uniaxial anisotropy is not well understood, it can be induced in polycrystalline films by cooling them from above the Curie temperature in a magnetic field, or better still by vacuum deposition of the film in the presence of a steady magnetic field of a few hundred A/cm. The direction of the field during treatment fixes the subsequent easy direction of the film, and its anisotropy energy may be written $W_\mu = K_\mu \sin^2 \theta$ where K_μ is the uniaxial anisotropy constant and θ is the angle between M_S and the easy direction. The equivalent internal anisotropy field acting to hold M_S in the easy direction is $H_\mu = 2K_\mu/M_S$. This case may be compared with that of a bulk cobalt crystal by referring to equation (6.3) and Table 6.2. In the case of nickel-iron films, 80% Ni, 20% Fe, deposited on glass, K_μ may be about 300 J/m³ and H_μ about 8 A/cm. Reversal of M_S in the film occurs for a reverse field of only about $0{\cdot}1H_\mu$.

When a rock salt crystal is used as substrate films of iron, nickel or their alloys may be formed which are themselves single crystals. These single crystal films have a uniaxial anisotropy $W_\mu = K_\mu \sin^2 \theta$ superimposed upon their normal cubic anisotropy even when no external field is applied during deposition. The reason for this is obscure also but K_μ may be about 10^3 J/m³.

PROBLEMS

7.1. A soft-iron sphere of radius 1 cm is placed in a magnetic field $H = 2500$ A/cm. Calculate

(a) the magnetostatic energy of the sphere,

(b) the magnetic potential energy of the system, and

(c) the work required to remove the sphere from the field.

7.2. If the sphere of Problem 7.1 is a single crystal with $M_S = 2{\cdot}14$ Wb/m^2 and $K_1 = 4{\cdot}4 \times 10^4$ J/m^3 calculate the external field required to saturate in a [110] direction, the magnetostatic energy and the magnetic anisotropy energy.

7.3. If the thickness of a 180° domain wall in an (001) plane is 1200 Å when $K_1 = 4 \times 10^4$ J/m^3 estimate the magnetic anisotropy energy and the total energy per unit area of the wall.

7.4. If a sample of silicon-iron has a Curie temperature of 720°C, a saturation value $M_S = 1{\cdot}9$ Wb/m^2 and a lattice constant $a = 2{\cdot}85$ Å, estimate the critical diameter for a spherical particle to be a single domain.

SUGGESTED READING

Stewart, K. H. *Ferromagnetic Domains*, Cambridge University Press, London, (1954).

Anderson, J. C. 'Thin magnetic films', *Electron. Technol.* **39,** 230 (1962).

REFERENCES

[1] Barkhausen, H. 'Two phenomena uncovered with the help of the new amplifiers', *Phys. Z.* **20,** 401 (1919).

[2] Bozorth, R. M. and Dillinger, J. F. 'Barkhausen Effect II. Determination of the average size of the discontinuities in magnetization', *Phys. Rev.* **35,** 733 (1930).

[3] Bitter, F. 'On the inhomogeneities in the magnetization of ferromagnetic materials', *Phys. Rev.* **38,** 1903 (1931).

[4] Néel, L. 'Some properties of boundaries between ferromagnetic domains', *Cah. Phys.* **25,** 1, 21 (1944).

[5] Néel, L. 'Laws of magnetization and subdivision of elementary domains of iron', *J. Phys. Radium, Paris,* **5,** 241, 265 (1944).

[6] Williams, H. J., Bozorth, R. M. and Shockley, W. 'Magnetic domain patterns on single crystals of iron', *Phys. Rev.* **75,** 155 (1949).

[7] Rhodes, P. and Rowlands, G. 'Demagnetizing energies and domain structures in ferromagnetics', *Proc. Leeds phil. lit. Soc.* **6,** 191 (1954).

[8] Néel, L. 'Principles of a new general theory of the coercive field', *Annls Univ. Grenoble,* **22,** 299 (1946).

[9] Bloch, F. 'Theory of the exchange problem and of residual ferromagnetism', *Z. Phys.* **74,** 295 (1932).

[10] Landau, L. and Lifshitz, E. 'Theory of dispersion of magnetic permeability in ferromagnetic bodies', *Phys. Z. Sowju.* **8,** 153 (1935).

[11] Lilley, B. A. 'Energies and widths of domain boundaries in ferromagnetics', *Phil. Mag.* **41,** 792 (1950).

[12] Bozorth, R. M. *Ferromagnetism*, p. 817, Van Nostrand, New York (1951).

[13] Williams, H. J. and Walker, J. G. 'Domain patterns on nickel', *Phys. Rev.* **83,** 634 (1951).

[14] Williams, H. J. and Shockley, W. 'Simple domain structure in an iron crystal showing direct correlation with magnetization', *Phys. Rev.* **75,** 178 (1949).

[15] Néel, L. 'Properties of a cubic ferromagnetic having small grains, *C. r. Acad. Sci., Paris,* **224,** 1488, 1550 (1947).

[16] Stoner, E. C. and Wohlfarth, E. P. 'A mechanism of magnetic hysteresis in heterogeneous alloys', *Phil. Trans. R. Soc.* A, **240**, 599 (1948).

[17] Kittel C. 'Theory of structure of ferromagnetic domains', *Phys. Rev.* **70,** 965, (1946).

Chapter 8
SOME MAGNETIZATION PROCESSES

8.1 The Magnetization Curve of a Ferromagnetic Substance

In Chapter 6 equations were obtained for the magnetization curves in particular crystal directions of single crystals of cubic and of hexagonal structure. In deriving these equations the principal factor considered was the crystal anisotropy and the rotational process in which the domain saturation vector M_S was rotated by the applied field from an easy direction. In the case of crystals of cubic structure only the upper part above the knee of the magnetization curve was involved, and the kind of agreement between theory and observation found in these cases is illustrated in Figs. 6.6 and 6.7, and in Fig. 6.11 for hexagonal cobalt.

It is assumed in the foregoing simple theory that the magnetization at the knee point is attained in a negligibly small applied field and in fact, in practice, for a soft magnetic material, this field will be less than, say, 10 A/cm. Such a small field will not in general cause the domain M_S vectors to depart appreciably from the easy crystal direction and, at the knee, it is supposed that M_S in all the domains has taken up the nearest easy direction to the direction of the applied field.

A useful theoretical, quantitative analysis of the relation between M and H below the knee point presents formidable difficulties and has not yet been accomplished. A brief mention of the phenomena occurring has already been given in Section 7.1. The arrangements of the domains in simple, single crystal cases when the crystals are in the demagnetized state, or when the resultant M is relatively small, are also shown, for example, in Figs. 7.2, 7.5 and 7.6. On this part of the magnetization curve the applied magnetic field is relatively low and the rotational process does not occur. The magnetization in the direction of an increasing field can only increase by an increase in the volume of the domains directed favourably with respect to the field. There will be a corresponding decrease in the volume of the remaining domains. This increase in volume will occur by movements, smooth or in Barkhausen jumps, of the domain boundary walls. For any value of H, and the corresponding value of M, a state of equilibrium occurs when the total energy of the system has attained a minimum, and the components of energy are those already discussed in Chapter 7, namely magnetostatic, domain boundary and magnetoelastic energy. Damping of domain boundary movements, involving losses in the form of heat, will also occur. A further complication is that in the general case of a polycrystalline material no precise knowledge of the domain shapes and sizes inside individual crystals will be available, especially since in these crystals any crystal plane may be parallel with the surface of the specimen.

Although a general theory of the magnetization curve is not available because of these complexities, it is of interest to investigate further some of the factors which must be involved particularly in domain boundary movements. First, however, we will consider certain phenomena observable in paramagnetic materials, since phenomena of a similar character may, perhaps, also occur in ferromagnetic materials.

8.2 Relaxation Effects in Paramagnetic Materials

We will consider a solid paramagnetic material such as a paramagnetic salt in which orbital motions are quenched and only electron spins contribute to the bulk magnetization of the material.

If a steady magnetic field applied to the material produces a steady intensity of magnetization M_A and then the field is suddenly removed, the magnetization M at time t decreases to zero, not instantly, but to a first approximation, exponentially with time. Then $M = M_A \varepsilon^{-t/\tau}$ where τ is known as the relaxation time. By differentiating and rearranging this becomes

$$\tau\, dM/dt + M = 0 \tag{8.1}$$

The equation represents the condition of the material in the absence of an applied field. If a field H varying sinusoidally with time is applied and its periodic time is long compared with τ we may put

$$M = \kappa_0 H \tag{8.2}$$

where κ_0 is the volume susceptibility of the material corresponding to a low frequency, and in this case M and H would be sinusoidal quantities in phase.

However, in the general case, with no restrictions on the relative values of τ and the periodic time we may use equations (8.1) and (8.2) to obtain

$$\tau\, dM/dt + M = \kappa_0 H \tag{8.3}$$

The steady state solution of equation (8.3) shows that at any frequency except zero M will lag behind H. It is convenient for analytical purposes to make use of complex quantities. If H_1 is the root mean square value of the alternating field H we may put $H = H_1 \varepsilon^{j\omega t}$ where $\omega = 2\pi f$ and f is the frequency. Substituting this value of H in equation (8.3), solving and rearranging gives

$$M = \kappa_0(1 - j\omega\tau)H/\{1 + (\omega\tau)^2\} \tag{8.4}$$

The complex susceptibility κ at any frequency is then given by

$$\kappa = M/H = \kappa_0(1 - j\omega\tau)/\{1 + (\omega\tau)^2\}$$

Putting $\kappa = \kappa' - j\kappa''$ we then have

$$\kappa' = \kappa_0/\{1 + (\omega\tau)^2\} \tag{8.5}$$

$$\kappa'' = \kappa_0\omega\tau/\{1 + (\omega\tau)^2\} \tag{8.6}$$

The significance of the coefficients κ' and κ'' of the real and imaginary parts of the complex susceptibility κ may be understood by reference to Fig. 8.1. In Fig. 8.1(*a*) material of unit length and unit cross-sectional area is magnetized by a single turn by an applied voltage V_1 and current I_1. Remembering that $B = \mu_0 H + M$ the vector diagram is as shown in Fig. 8.1(*b*) where all quantities are r.m.s. values. Representing the vectors as complex quantities in the Argand diagram in Fig. 8.1(*c*) we have $M = \kappa H$ and $\kappa' H$ is the component of the magnetization in phase with the applied field, while $-j\kappa'' H$ is the component in quadrature.

Because of the time lag between M and H associated with the relaxation

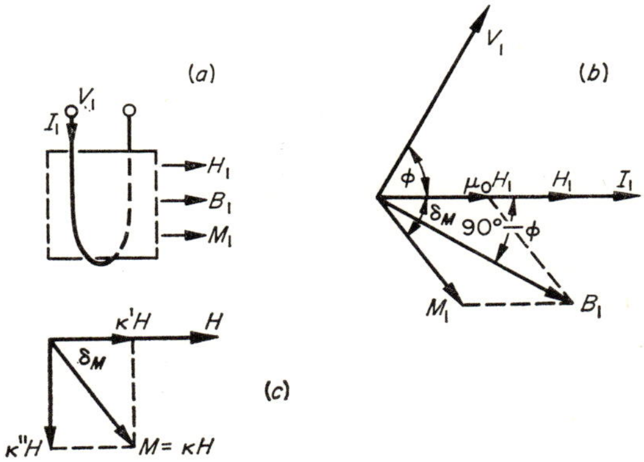

FIG. 8.1. Illustrating complex susceptibility in a magnetic material.

time there will be an energy loss appearing as heat in the material. The rate at which this energy is produced, or the power input per unit volume, is given from theory well known in electrical engineering as $P = V_1 I_1 \cos\phi$, where $\cos\phi$ is the power factor. Now $V_1 = \omega B$, $I_1 = H_1$ and in Fig. 8.1(*b*) $M_1 \sin\delta_M = B_1 \cos\phi$, where δ_M is the loss angle and $\tan\delta_M = \kappa''/\kappa'$. Whence

$$P = \omega H_1 M_1 \sin\delta_M \tag{8.7}$$

Now $|\kappa| = \{(\kappa')^2 + (\kappa'')^2\}^{\frac{1}{2}}$, $\sin\delta_M = \kappa''/|\kappa|$ and $M_1 = |\kappa| H_1$. Whence, by using also equations (8.5) and (8.6), equation (8.7) may be written

$$P = \omega H_1^2 \kappa'' = (\kappa_0/\tau) H_1^2 (\omega\tau)^2/\{1 + (\omega\tau)^2\} \tag{8.8}$$

or alternatively

$$P = (\tau/\kappa_0)\omega^2 M_1^2 \tag{8.9}$$

κ'/κ_0 and κ''/κ_0 are plotted in relation to $\omega\tau$ in Fig. 8.2. κ'/κ_0 reaches a maximum equal to 0·5 when $\omega\tau = 1$. The relation between P and $\omega\tau$ when

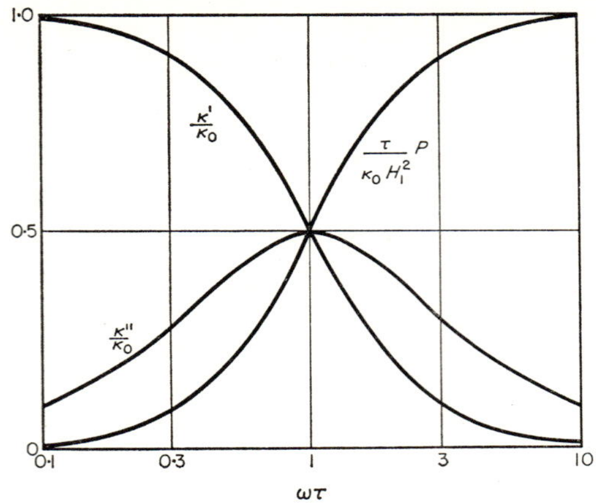

FIG. 8.2. Showing components of complex susceptibility and the power loss for constant alternating field strength in relation to frequency and relaxation time for a paramagnetic.

the applied field H_1 is maintained constant is obtainable from equation (8.8) and is also shown in the figure. The variation of κ' with frequency is called paramagnetic dispersion and of κ'' with frequency the paramagnetic absorption.

An experimental way of determining the value of τ for a particular material is by noting the frequency at which κ'' reaches a maximum. The first work on paramagnetic relaxation was by GORTER [1, 2] on ferric ammonium alum. The position of the maximum value of κ'' was found to depend on the field strength but occurred in the neighbourhood of $f = 10^6$ c/s. This frequency gives a value $\tau = 1/(2\pi \times 10^6) = 1{\cdot}6 \times 10^{-7}$ s.

To consider the origin of the delay in magnetization, which is represented by the relaxation time, suppose a magnetic field is suddenly applied to a paramagnetic. The immediate effect is that there is no increase in magnetization in the field direction because the axes of the spinning electrons, parallel or antiparallel to the field, will simply precess about the field direction. An increment of magnetization occurs when an antiparallel spin reverses and this involves a reduction in magnetic potential energy of $2H\mu_B$. In some way this energy must be transferred to the body of the crystal, that is it has to be added to the thermal energy of the crystal ions. The spinning electron may be considered to be connected to the lattice ions through the spin–orbit coupling discussed in Section 2.5, and this may provide the mechanism whereby the magnetic potential energy in the reversal is transferred to the crystal, producing in fact in the process a rise of temperature. The resultant delay associated with this process gives what is called a spin-lattice relaxation time which is temperature dependent, increasing with decreasing temperature to about 10^{-2} second at liquid helium temperature (4·2°K), and falling to 10^{-7} second or lower at room temperature.

Measurements of the iron losses in thin ferromagnetic laminations at low frequencies, say below 10^3 c/s, and also the experimental results given in Section 8.3, seem to show that a relaxation time effect may also be operating in these cases, but with a relaxation time at room temperature considerably greater than those given above for the paramagnetics. If indeed such an effect is occurring, the appropriate equation for the alternating energy loss would be equation (8.9) where the power is proportional to the square of both the frequency and the intensity of magnetization. Unfortunately in the ferromagnetic metals eddy currents, or Foucault currents, are also present and the eddy-current component of power loss is, according to classical theory, also proportional to the square of the frequency and the flux density. The flux density B and the intensity of magnetization M are in this case nearly the same. It therefore becomes difficult, if not impossible, experimentally to separate the true eddy-current loss from a relaxation loss if it is present. These losses are a matter of some importance in power supply transformers and other electrical equipment and the subject will be referred to again in this chapter.

8.3 Velocity of Movement of Domain Boundaries

Magnetization proceeds in a paramagnetic by reversal of electron spins with some time delay. In a ferromagnetic the resultant magnetization below the knee of the magnetization curve increases by domain boundary movements. The applied field in this case produces reorientation of the electron spins only within the width of the boundary walls as these pass across the domains. Again the spin axes can reorient themselves only by some mechanism operating at a finite speed. A restraint on the movement of the walls in a changing applied field, accompanied by energy losses, will therefore occur on this account. Induced electromotive forces producing eddy currents near the boundaries will cause a time delay and losses. Direct examination of these phenomena in a normal specimen with small domains and small boundary movements are hardly possible, but specimens can be produced involving a single domain wall moving over large distances.

In a classic experiment SIXTUS and TONKS [3] showed that hard-drawn nickel-iron wires, the material having between 10% and 20% nickel, positive magnetostriction and relatively low magnetic anisotropy, when magnetized and put into strong tension, would form a single domain with the saturation magnetization vector along the axis of the wire. Measurements were made on a wire 0·019 cm radius containing 15% nickel, stretched along the axis of a main magnetizing coil 65·5 cm long. High tensile stresses, up to 92 kg/mm^2 (58 tons/in^2), were used. Reversal of the magnetizing field produced no effect on the single domain until a critical value H_0 was exceeded. For greater fields it was possible to start a reversal of the domain magnetization at one end of the wire by means of a local magnetizing coil. Figure 8.3(a) shows the experimental scheme diagrammatically. D is the starting coil while the search

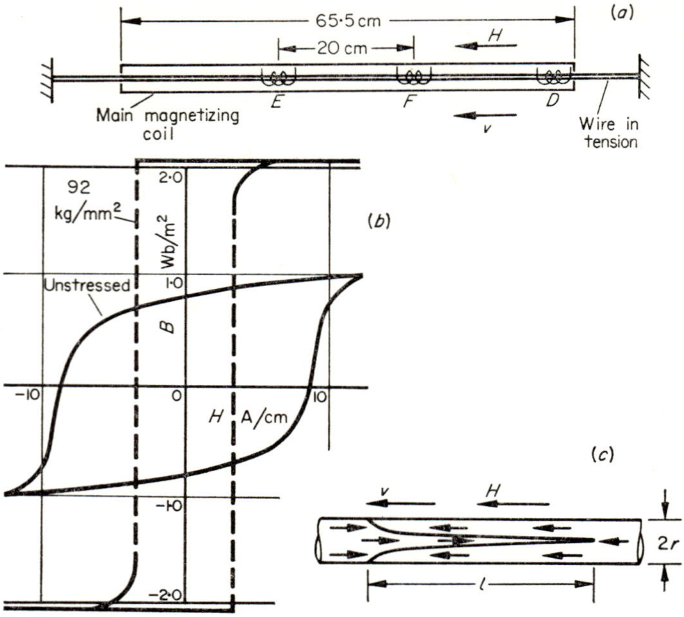

FIG. 8.3. Sixtus and Tonks' experiments on moving domain boundaries in 15% Ni–Fe wire.

coils E and F, 20 cm apart, were used to determine the velocity with which the domain boundary passed along the wire, and also to give some information about the shape of the advancing domain boundary.

Figure 8.3(*b*) shows a hysteresis loop of the unstressed multi-domain 15% nickel-iron wire, and also a loop when the stress was 92 kg/mm². The broken lines correspond with the reversal of magnetization produced by the movement of the boundary from one end of the wire to the other. The travelling domain boundary was deduced to have the long trumpet-like shape illustrated in Fig. 8.3(*c*). This diagram is not to scale for the overall length of the boundary region was about 100 cm.

Figure 8.4(*a*) shows the boundary velocity in relation to the field H applied by the main magnetizing coil for the 15% nickel-iron wire for several values of the tensile stress. The relation between the velocity and field strength may be written

$$v = A(H - H_0) \tag{8.10}$$

The critical field H_0 depends on the stress in the wire but the constant A was found, in a number of experiments, to be independent of the stress, the nickel content or the radius of the wire. Velocities ranging from 500 to 40,000 cm/s were observed. The value found for A was 25,000 cm/s per oersted or 20,000 cm/s per A/cm.

The field H_0, corresponding to an infinitely slow velocity of the boundary, may be classified as a direct hysteresis term. This field is required in order to

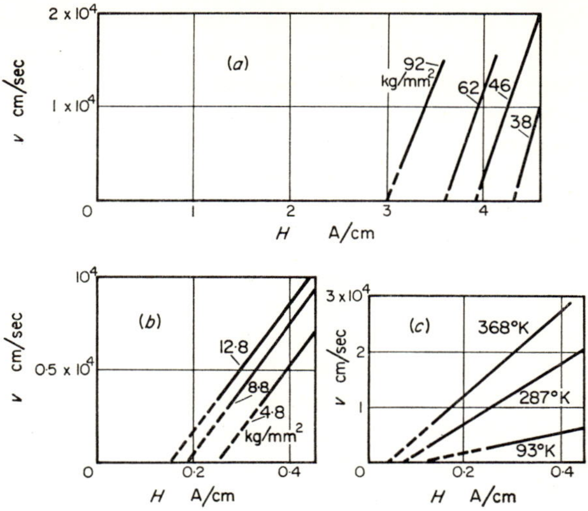

FIG. 8.4. Domain-boundary velocities: (*a*) Sixtus and Tonks, hard-drawn 15% Ni–Fe wire, $r = 0{\cdot}019$ cm: (*b*) Dijkstra and Snoek, annealed 60–40 Ni–Fe wire, $r = 0{\cdot}01$ cm: (*c*) Same as (*b*) but $r = 0{\cdot}005$ cm.

drag the boundary past imperfections in the metal such as inclusions, cavities and local strains according to the ideas discussed in the following sections. An additional field $H - H_0$ is required to produce the velocity v. Reversal of the magnetization at the advancing domain front will set up eddy currents and an effective eddy current field H_E opposing the applied component $H - H_0$. An obvious and attractive interpretation of the observed results would be to put

$$H - H_0 = A^{-1}v = Cv = H_E$$

that is to suppose that the propagation velocity is determined entirely by the eddy currents. Sixtus and Tonks, in these experiments, investigated this possibility theoretically. They concluded that the opposing field due to eddy currents was much less than $H - H_0$ so that some other mechanism of energy loss in the moving boundary had to be invoked. We might therefore put

$$H - H_0 = H_E + H_R = (C_E + C_R)v \tag{8.11}$$

where $H_E = C_E v$ is the eddy current field and $H_R = C_R v$ is an effective, equivalent field due to another effect.

LANDAU and LIFSHITZ [4] consider that the spinning electrons in the moving boundary wall precess under the action of an effective field arising from the exchange forces and the anisotropy and that this precession is damped by some mechanism not specified. The term $H_R = C_R v$ and its associated losses would be thus accounted for. We may compare this with the relaxation due to spin–orbit coupling already discussed for paramagnetics in Section 8.2, but the relaxation time in this case at room temperature would have to be much greater than the 10^{-7} seconds found there.

DIJKSTRA and SNOEK [5] have considered this matter further and extended the experiments of Sixtus and Tonks. Their specimens were nickel-iron wires over 250 cm long of 60% Ni, 40% Fe and 50% Ni, 50% Fe, in both the annealed and the cold-worked conditions. The wires varied in radius from about 0·001 cm to 0·01 cm. When the radius exceeded 0·02 cm a well-defined boundary could not be realized. The shape of the moving boundary was again estimated to be as Fig. 8.3(*c*) with a ratio l/r of about 1000 in all cases. Figure 8.4(*b*) shows the results for the 60–40 Ni-Fe wire of radius $r = 0{\cdot}01$ cm in the annealed state under various tensile stresses. In this case again $v = A(H - H_0)$ and $A = 24{,}000$ cm/s per A/cm approximately. It will be noted that the stresses and field strengths are much lower than for the Sixtus and Tonks hard-drawn specimen. Figure 8.4(*c*) gives results for the same annealed material of radius $r = 0{\cdot}005$ cm measured at different temperatures. In this case A was found to be directly proportional to the absolute temperature.

Dijkstra and Snoek also concluded that eddy current damping on the propagation velocities was quite insufficient to account for the experimental observations. According to their theoretical investigation of the eddy-current damping they found, for the 0·01 cm nickel-iron wire, that the value of C_{E} in equation (8.11) was approximately $1{\cdot}3 \times 10^{-5}$ A cm^{-2} s, while the observed value $A^{-1} = C_{\mathrm{E}} + C_{\mathrm{R}} = (24{,}000)^{-1} = 4 \times 10^{-5}$ A cm^{-2} s. Hence the eddy-current damping was only about one-third of the total experienced by the moving boundary. We may thus write an eddy-current anomaly factor $\eta' = (C_{\mathrm{E}} + C_{\mathrm{R}})/C_{\mathrm{E}} = 3$ in this case.

The results in Fig. 8.4(*c*) show that A^{-1} is inversely proportional to temperature, so that the damping effect increases with decreasing temperature as in the case of paramagnetic relaxation. This lends support to the idea that a spin-lattice relaxation mechanism is involved in this case also.

8.4 Inclusion Theories of Hysteresis and Coercivity

The experimental results in the preceding section show that, even when the velocity of motion of a domain boundary approaches zero, a field H_0 is still required. As already suggested, H_0 may be associated with the static hysteresis and may be of the order of magnitude of the coercivity H_{C} in a hysteresis loop. In different materials H_{C} may vary in magnitude over a very wide range. For example, the single crystal 'picture-frame' specimen shown in Fig. 7.6(*d*) of relatively pure, carefully heat-treated silicon-iron, was found to have a nearly rectangular hysteresis loop. An applied field caused the specimen to change from the demagnetized state shown in the figure to saturation by an inward movement of the four domain boundaries. It was reported that this movement was accompanied by Barkhausen discontinuities and a coercive field of about 0·02 A/cm was required. On the other hand the results given in Fig. 8.4(*b*) for an annealed nickel-iron wire show that H_0 was about 0·2 A/cm, and in Fig. 8.4(*a*) for the hard-drawn wire about 4 A/cm. Very much greater values than this can occur; for example, in permanent magnet alloys, where a high coercive force is desired, H_C may reach 500 A/cm and more. It does not seem possible to account for this wide range of values by any one mechanism.

The best magnetic properties, that is high permeability and low hysteresis loss, accompanied by low coercive force, are obtained in ferromagnetic metals in which the crystals are as free as possible from imperfections. The properties are impaired by lattice distortions due to internal strains set up by impurities, cavities, cold-work or to other effects. KERSTEN [6] has put forward an inclusion or foreign-body theory.

In Fig. 8.5(a) a spherical non-ferromagnetic particle or a cavity of radius r is intersected by a 180° domain boundary at distance x from the centre of the particle. The domain boundary has wall energy W per unit area as described in Section 7.3. There is, however, a circular hole in the wall due to the particle which will reduce the wall energy by $W\pi(r^2 - x^2)$. The wall energy is

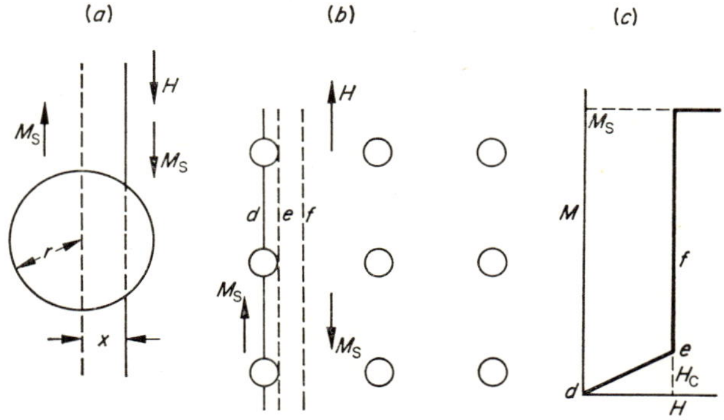

FIG. 8.5. Kersten's foreign-body theory of coercivity.

therefore a minimum when the boundary bisects the particle, and a maximum when $x = r$ with the boundary just leaving the particle. On this account the boundary should tend to cling to the particle or cavity, and energy is required to drag it away. This is the basis of Kersten's theory.

Assuming with Kersten that there are n small, equal particles per unit volume arranged on a cubic system in a crystal, as shown in Fig. 8.5(b), and that a 180° boundary intersects one plane of particles. In this plane there will be $n^{\frac{2}{3}}$ particles per unit area. If the wall is displaced by an applied field H through a distance x from the centres of the particles then the increase in magnetization due to a movement x per unit area of the wall is $2M_Sx$ approximately, if the particles are relatively small, and the magnetic potential energy for the movement x is $-2HM_Sx$. The energy of the perforated wall per unit area is $W\{1 - n^{\frac{2}{3}}\pi(r^2 - x^2)\}$. The total energy for displacement x is

$$W_T = W\{1 - n^{\frac{2}{3}}\pi(r^2 - x^2)\} - 2HM_Sx$$

The equilibrium position of the wall for the field H occurs when $dW_T/dx = 0$ and this leads to

$$H = \pi n^{\frac{2}{3}} Wx/M_S \tag{8.12}$$

Putting $M = 2M_S x$ the susceptibility corresponding to the movement of the boundary across one array of particles, that is when $x < r$, is

$$\kappa = M/H = 2M_S^2/(\pi n^{\frac{2}{3}} W)$$

From equation (8.12), H will reach a maximum when $x = r$ and then there would be nothing to prevent the boundary running forward to give saturation in the crystal. The magnetization curve, corresponding to positions *d*, *e* and *f* of the boundary, would be as shown in Fig. 8.5(*c*), and the coercivity is obtained by putting $x = r$ in equation (8.12) to give

$$H_C = \pi n^{\frac{2}{3}} Wr/M_S \tag{8.13}$$

If the fractional volume of impurities in the crystal is v then $v = n4\pi r^3/3$, and substituting for n in equation (8.13) gives

$$H_C = (9\pi/16)^{\frac{1}{3}} W v^{\frac{2}{3}}/rM_S \tag{8.14}$$

Equation (8.14) can be used quantitatively to determine H_C which, for a total given volume of impurity, will be greatest for the smallest particle size. It is implicit in the theory that the particle diameter is greater than the wall thickness. However, taking the case of iron, having a 180° wall thickness, from Table 7.1, of $1{\cdot}4 \times 10^{-5}$ cm, suppose we put $r = 0{\cdot}7 \times 10^{-5}$ cm with $W = 1{\cdot}2$ erg/cm² and $M_S = 2{\cdot}14$ Wb/m² then $H_C = 9700\ v^{\frac{2}{3}}$ A/m $= 97v^{\frac{2}{3}}$ A/cm. Thus if the iron contained 1% of impurity, $v = 0{\cdot}01$, and $H_C = 4{\cdot}5$ A/cm, or for 10% of impurity, $H_C = 21$ A/cm.

Although Kersten found a good deal of experimental evidence in favour of his theory there are objections to the premises adopted in its derivation. In the first place if the particles are not on a regular lattice, as assumed in the theory, but are distributed in a random way, then it is apparent that a plane domain boundary will intersect fewer of them at any time and the expected value of H would be much smaller than the derived value. A more important objection has been pointed out by NÉEL [7].

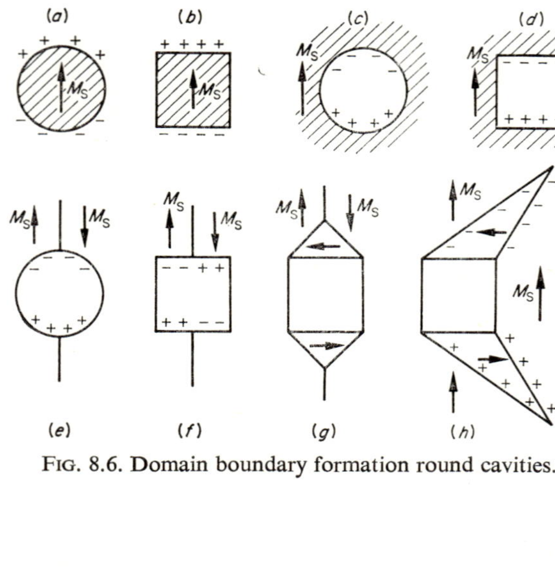

FIG. 8.6. Domain boundary formation round cavities.

In Fig. 8.6(*a*) a spherical ferromagnetic particle magnetized to saturation is shown. There will be free poles on the outer surface of the sphere, represented by the plus and minus signs, and the particle will have magnetostatic energy. A similar particle in the form of a cube is shown in Fig. 8.6(*b*). As already mentioned in Section 7.2 the magnetostatic energies per unit volume of the sphere and the cube are exactly the same and, in the case of iron, this energy is $W_M = 6 \times 10^5$ J/m³. In the figure a spherical cavity at (*c*) and a cubical cavity at (*d*) are shown embedded in a saturated ferromagnetic material. If M_S is the same throughout then the energies per unit volume of the particles (*a*) and (*b*), and of the cavities (*c*) and (*d*), are the same.

Consider now the spherical cavity at (*e*) bisected by a domain boundary as suggested by Kersten. The magnetostatic energy is, according to Néel, approximately one half of that of the isolated cavity. Hence in the case of iron the energy required to pull the boundary clear of the cavity, on account of the free poles, would be about 3×10^5 J/m³, or $(4\pi r^3/3)3 \times 10^5$ J, if r is the radius of the cavity in metres. On the other hand the energy envisaged by Kersten to do this would be only that due to the hole in the boundary $= \pi r^2 W = (\pi r^2)1{\cdot}2 \times 10^{-3}$ J. The ratio of these energies is about $3{\cdot}3 \times 10^8 \times r$. Thus even for the smallest particles where, say, $r = 0{\cdot}7 \times 10^{-5}$ cm, the component of energy due to the free poles is about 20 times that due to the perforation in the boundary, and this factor increases with increasing particle size. Kersten has, in fact, neglected the magnetostatic energy and this neglect is clearly inadmissible.

Néel has pointed out that the energy associated with a cavity in a domain may be greatly reduced by suitable domain configurations. These configurations for a spherical cavity are not easy to visualize but for a cubical cavity the relatively high magnetostatic energy, for the example shown in Fig. 8.6(*f*), would be eliminated by the appearance of the two closure domains with the 90° boundaries shown at (*g*). The additional wall energy in this case would be small compared with the eliminated free pole energy. The magnetostatic energy of an isolated cavity as shown at (*d*) could be reduced by the formation of the spiked domains shown at (*h*). Since the domain walls are now not at 45° to M_S a weak free pole strength will exist over the boundary surfaces of magnitude $M = M_S \cos(45° - \theta) - M_S \cos(45° + \theta) = M\sqrt{2} \sin\theta$ where 2θ is the acute angle in the spike. In this case there is the additional wall energy and some magnetostatic energy due to the weak free poles, and the length of the spike would adjust itself to make the total energy a minimum. These spiked domains, predicted theoretically by Néel, are a common feature in observed domain patterns. Changes in the domain walls as a boundary moves past an inclusion might be as shown in Fig. 8.7.

Néel has further suggested that local inhomogeneities in a crystal due to nonuniformity of chemical composition, or other effects which caused the saturation magnetization M_S to vary, would result in the appearance of diffused free poles. Magnetostatic energy changes would occur as a boundary

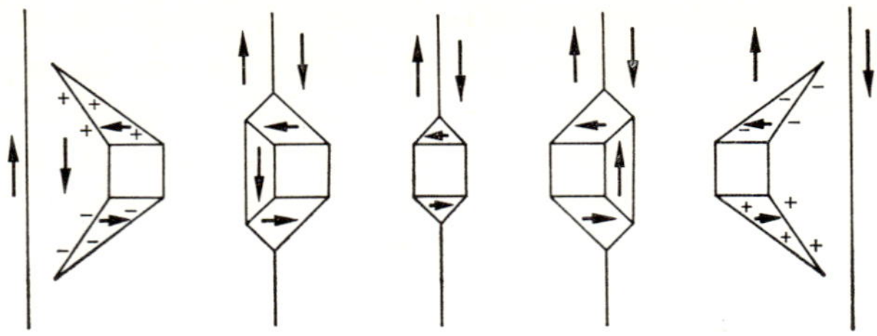

FIG. 8.7. Domain formations as a boundary passes an inclusion.

moved across these areas and this would affect the permeability and coercivity of the material. However, in spite of the evident general correctness of Néel's ideas, quantitatively the theory is not complete.

8.5 The Strain-Magnetostriction Theory of Permeability

The crystal lattice of a ferromagnetic domain will in general be deformed because of the spontaneous saturation in it and the accompanying magnetostriction. The magnetostrictive strain in a particular direction must change when the saturation vector changes direction. These changes in lattice dimensions cannot occur with complete freedom during magnetization of a solid specimen because of the restraints imposed on a domain by the neighbouring material. The magnetic properties must in that case be affected by magnetoelastic energy changes of a complicated character. BECKER [8], however, has shown, for a specimen in highly idealized conditions, how it is possible to derive a relation between permeability, magnetostriction and the internal stresses.

In Fig. 8.8 we assume we have a long strip of unit cross-section in a [100]

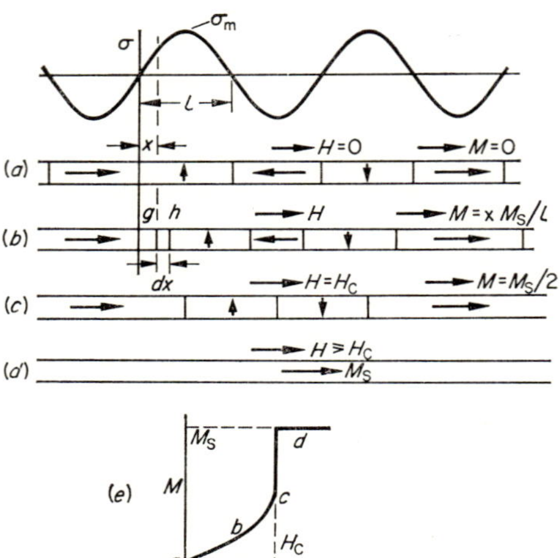

FIG. 8.8. Becker's strain-magnetostriction theory.

direction of a crystal having positive saturation magnetostriction λ_{100}. Due to the general imperfections of the crystal there are internal tensile and compressive stresses. The stress acting along the length of the strip is represented by $\sigma = \sigma_m \sin(\pi x/l)$ as shown in the figure, where positive values of σ indicate compressive stress and negative values tension. The strip may then be divided up into domains each of length l. Those in tension will have M_S parallel to [100], forward or reverse, because of the positive magnetostriction, as already described in Section 6.9, while the domains in compression will have M_S at right-angles to [100]. When the applied field H along [100] is zero, the arrows representing the M_S vectors in the domains may then be as shown in Fig. 8.8(*a*), and the resultant magnetization $M = 0$. On applying a small field H the domain boundaries will move through equal distances x to give a resultant magnetization in the direction of H given by $M = xM_S/l$ as shown at (*b*). In making these boundary movements there are changes in magnetoelastic energy and in magnetic potential energy, equilibrium being reached when the total energy is a minimum. The condition for equilibrium may be found thus:

Suppose the boundary at g moves through a distance dx to h. In the volume dx, before the boundary moved the longitudinal strain was $-\lambda_{100}/2$, while after the movement the strain becomes λ_{100}. Hence the lattice in the volume dx extends by an amount $3\lambda_{100}/2$ against a compressive stress σ and hence magnetoelastic energy is stored given by $(3\lambda_{100}/2)\sigma\, dx$. The corresponding change in magnetic potential energy is $-HM_S\, dx$. The total change in energy is

$$dW = (3\lambda_{100}/2)\sigma\, dx - HM_S\, dx$$

We have equilibrium when $dW/dx = 0$, or when $HM_S = (3\lambda_{100}/2)\sigma$ or

$$H = \{3\lambda_{100}\sigma_m \sin(\pi x/l)\}/2M_S \tag{8.15}$$

The relative permeability $\mu_R = B/\mu_0 H = 1 + M/\mu_0 H$. Since $M = xM_S/l$, and μ_R will in general be great compared with unity, then

$$\mu_R = 2M_S^2(x/l)/\{3\mu_0\lambda_{100}\sigma_m \sin(\pi x/l)\}$$

If x is small then $(\pi x/l)/\sin(\pi x/l) = 1$ and μ_R becomes the initial relative permeability μ_{RI}. Whence

$$\mu_{RI} = (2/3\pi\mu_0)M_S^2/(\lambda_{100}\sigma_m) \tag{8.16}$$

Equation (8.16) shows that μ_{RI} is large when λ_{100} is small and when the internal stress σ_m is small. The value to be assigned to σ_m will not in general be known. Its value should be a minimum in a material which had been purified and carefully heat-treated to remove as far as possible all internal strains and stresses. Even in this case cooling of the material from above the Curie temperature will result in internal stresses being set up due to the magnetostriction induced when the domains come into existence. The internal stress from this cause may be expected to be of the order of $\lambda_{100}E_Y$ where E_Y is

Young's Modulus. If this value is substituted for σ_m in equation (8.16) we have

$$\mu_{RI} = (2/3\pi\mu_0)(M_S^2/\lambda_{100}^2 E_Y) \quad (8.17)$$

Equation (8.17) represents the highest initial relative permeability that might be expected in a very good specimen, and contains only known physical constants of the material. If for iron we substitute the values $M_S = 2{\cdot}14$ Wb/m², $\lambda_{100} = 20 \times 10^{-6}$ and $E_Y = 20 \times 10^{10}$ N/m² we have $\mu_{RI} = 9700$. This is in reasonable agreement with a value found by CIOFFI [9] for purified iron, although ordinary low impurity dynamo iron has a value for μ_{RI} of only about 250.

Some further support for the general correctness of Becker's ideas on the dependence of permeability on magnetostriction is afforded by comparing the calculations of μ_{RI} from equation (8.17) with observed values for the nickel-iron alloys. The saturation magnetostriction of these alloys is plotted in Fig. 6.19(*c*) and will be seen to pass through zero for about 81% nickel.

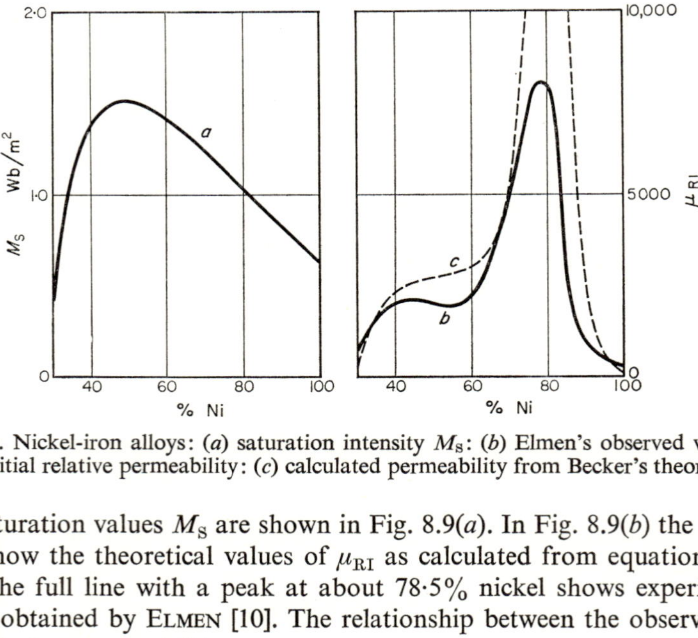

FIG. 8.9. Nickel-iron alloys: (*a*) saturation intensity M_S: (*b*) Elmen's observed values of initial relative permeability: (*c*) calculated permeability from Becker's theory.

The saturation values M_S are shown in Fig. 8.9(*a*). In Fig. 8.9(*b*) the broken lines show the theoretical values of μ_{RI} as calculated from equation (8.17) while the full line with a peak at about 78·5% nickel shows experimental results obtained by ELMEN [10]. The relationship between the observed and theoretical curves is striking.

A value of the coercivity H_C may also be obtained for the idealized model in Fig. 8.8, for $H = H_C$ when the boundaries have moved to the points of maximum stress as in Fig. 8.8(*c*). Then $M = M_S/2$ and $\sin(\pi x/l) = 1$ so that from equation (8.15)

$$H_C = 3\lambda_{100}\sigma_m/2M_S \quad (8.18)$$

When $H > H_C$, $M = M_S$ as shown at (*d*) and the whole magnetization curve is shown at (*e*). If we put $\sigma_m = \lambda_{100}E_Y$ in equation (8.18) and substitute values for iron we have $H = 0{\cdot}5$ A/cm, which is about right for dynamo iron but high for a well-purified specimen.

On the other hand using equation (8.18) for the case, for example, of Alcomax, a permanent magnet material, for which $H_C = 480$ A/cm and $M_S = 1{\cdot}35$ Wb/m^2, and putting $\lambda_S = 20 \times 10^{-6}$ for λ_{100} we have $\sigma_m = 2H_C M_S/3\lambda_{100} = 2{\cdot}16 \times 10^9$ N/m^2. This is a very high stress (about 142 tons/in^2), greater than the breaking strength of the material. There are also some materials of high coercivity which are mechanically relatively soft and which would not have very high internal stresses. A high coercivity can indeed also be obtained in materials compacted from fine soft iron powders. The strain theory of high coercivity is therefore not adequate in these cases.

8.6 The Small-Particle Theory of High Coercivity

It was shown in Section 7.6 that single crystal ferromagnetic particles below a critical size would have no domain boundary in them but would be single domains. It was estimated that for spherical iron particles the critical diameter was about $2{\cdot}9 \times 10^{-6}$ cm. Such a particle uniformly magnetized to the saturation value M_S has a relatively high magnetostatic energy. If the particle is not spherical but, for example, ellipsoidal in shape, the magnetostatic energy will depend on the direction of M_S relative to the axes of the ellipsoid. The energy will be a minimum when M_S lies along the major axis, since the demagnetizing factor of the particle is a minimum in this direction. The direction of M_S can then only be changed, if a field H is applied, by doing work to increase the magnetostatic energy. STONER and WOHLFARTH [11] have shown that, on this account, a single domain particle, even when deformed only slightly from the spherical shape, can have a relatively high coercivity and, if the particle is greatly elongated, very high values of coercivity are theoretically possible. In fact any value of coercivity yet observed in practice might be accounted for by this simple mechanism which, it may be noted, does not depend upon strains and stresses in the material.

The demagnetizing factors of ellipsoidal specimens have been discussed in Chapter 1. In the case of the general ellipsoid with semi-axes $a > b > c$ the demagnetizing factors in the directions of the axes are $N_a < N_b < N_c$ and $\mu_0(N_a + N_b + N_c) = 1$. However, we shall consider prolate spheroids, that is prolate ellipsoids of revolution, in which $a > b = c$ so that $N_a < N_b = N_c$, and $\mu_0(N_a + 2N_b) = 1$, where a is the polar semi-axis and b the equatorial radius. Obviously if the particle is spherical $N_a = N_b = N_c = 1/(3\mu_0)$ and if the particle is very greatly elongated we may put $N_a = 0$, $N_b = 1/(2\mu_0)$.

Consider the single-domain prolate spheroid shown in Fig. 8.10(*a*) in which the saturation vector M_S is initially pointing downwards along the polar axis and to which the field H is applied in the reverse direction. If M_S is turned through an angle ψ as shown we then have, per unit volume of the particle, magnetic potential energy $HM_S \cos \psi$. The components of M_S axially and in the direction of an equatorial diameter are, respectively,

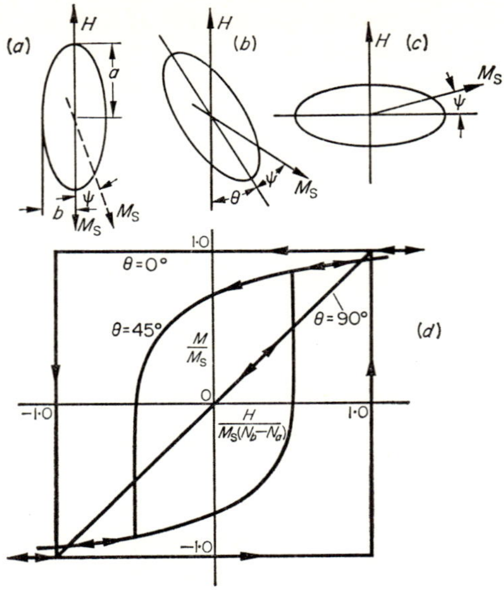

FIG. 8.10. Stoner and Wohlfarth's small-particle theory of high coercivity.

$M_S \cos \psi$ and $M_S \sin \psi$, and the corresponding components of magnetostatic energy are $(1/2)N_a(M_S \cos \psi)^2$ and $(1/2)N_b(M_S \sin \psi)^2$. The total energy is then

$$W = HM_S \cos \psi + (1/2)N_a M_S^2 \cos^2 \psi + (1/2)N_b M_S^2 \sin^2 \psi$$

Equilibrium occurs when the vector M_S lies in a trough of the total energy, that is when $dW/d\psi = 0$ and $d^2W/d\psi^2$ is positive. Differentiating we have

$$dW/d\psi = -HM_S \sin \psi + (M_S^2/2)(N_b - N_a) \sin 2\psi$$

and $$d^2W/d\psi^2 = -HM_S \cos \psi + M_S^2(N_b - N_a) \cos 2\psi$$

$dW/d\psi = 0$ when $\psi = 0$ and then $d^2W/d\psi^2 = -HM_S + M_S^2(N_b - N_a)$. Then $\psi = 0$ represents a stable position for M_S so long as $d^2W/d\psi^2$ is positive, that is $H < M_S(N_b - N_a)$. When H exceeds this value it is apparent that M_S will reverse direction, and pull into line with H, making both the magnetic potential energy and the magnetostatic energy a minimum. The particle for this alignment in the field will therefore have a coercive force $H_C = M_S(N_b - N_a)$. The hysteresis loop for the particle, if the field is alternating in direction and has forward and reverse peak values exceeding H_C, is therefore of the rectangular form shown in Fig. 8.10(*d*).

In the more general case when the polar axis is inclined at an angle θ to the direction of H, as shown in Fig. 8.10(*b*) the total energy is

$$W = HM_S \cos(\theta + \psi) + (1/2)N_a M_S^2 \cos^2 \psi + (1/2)N_b M_S^2 \sin^2 \psi$$

and $dW/d\psi = -HM_S \sin(\theta + \psi) + (M_S^2/2)(N_b - N_a) \sin 2\psi$

Again a condition of equilibrium is when $dW/d\psi = 0$ or

$$H \sin(\theta + \psi) = (M_S/2)(N_b - N_a) \sin 2\psi \tag{8.19}$$

We are interested in M, the component of magnetization resolved in the field direction, and this is given by

$$M = -M_S \cos (\theta + \psi) \tag{8.20}$$

Eliminating ψ between equations (8.19) and (8.20) and putting $m = M/M_S$ and $h = H/\{M_S(N_b - N_a)\}$ leads to the equation

$$4(1 - m^2)(h^2 + 2mh \cos 2\theta + m^2) = \sin^2 2\theta \tag{8.21}$$

Equation (8.21) is a relation between H and M and from it the hysteresis loop of the particle corresponding to any inclination θ of the polar axis to the field can be determined. An easy example to plot is when $\theta = 45°$ for then cos $2\theta = 0$ and sin $2\theta = 1$, and equation (8.21) becomes

$$4(1 - m^2)(h^2 + m^2) = 1 \tag{8.22}$$

Values which satisfy equation (8.22) are given in Table 8.1 and the loop is plotted in Fig. 8.10(*d*). It is interesting to note from the figures for $\theta + \psi$ given in Table 8.1 that, when H is alternating in a fixed direction, M_S is a rotating vector making one complete revolution for each cycle, with sudden irreversible jumps (jumps of 120° when $\theta = 45°$). This rotation of M_S in one direction will occur for any value of θ except perhaps $\theta = 0°$.

TABLE 8.1

SINGLE SMALL-PARTICLE HYSTERESIS LOOP: $\theta = 45°$

Ascending Loop			Descending Loop		
$H/\{M_S(N_b - N_a)\}$	M/M_S	$\theta + \psi$	$H/\{M_S(N_b - N_a)\}$	M/M_S	$\theta + \psi$
$-\infty$	-1	0	∞	1	180°
-1	$-0{\cdot}93$	21·5°	1	0·93	201·5°
$-0{\cdot}75$	$-0{\cdot}90$	25·8°	0·75	0·90	205·8°
$-0{\cdot}5$	$-0{\cdot}866$	30°	0·5	0·866	210°
$-0{\cdot}25$	$-0{\cdot}804$	36·5°	0·25	0·804	216·5°
0	$-0{\cdot}707$	45°	0	0·707	225°
0·25	$-0{\cdot}538$	57·5°	$-0{\cdot}25$	0·538	237·5°
0·5	{0 {0·866	90° 210°	$-0{\cdot}5$	{0 {$-0{\cdot}866$	270° 30°
0·75	0·90	205·8°	$-0{\cdot}75$	$-0{\cdot}90$	25·8°
1	0·93	201·5°	-1	$-0{\cdot}93$	21·5°
∞	1	180°	$-\infty$	-1	0°

Another easy example to calculate is for the case shown in Fig. 8.10(*c*) where $\theta = 90°$. Then cos $2\theta = -1$ and sin $2\theta = 0$. Whence using equation (8.21) the relation between m and h becomes the straight line $m = h$, and this also is shown in Fig. 8.10(*d*).

Hysteresis loops for other values of θ have been evaluated by Stoner and Wohlfarth. The coercivity of a particle for any angle θ corresponds to

$M = 0$ in the loops and, when $\theta \geqslant 45°$, may be obtained by putting $m = 0$ in equation (8.21). Then $4h_C^2 = \sin^2 2\theta$ and hence $H_C/M_S(N_b - N_a) = (\sin 2\theta)/2$. The coercive force in relation to θ is plotted in Fig. 8.11.

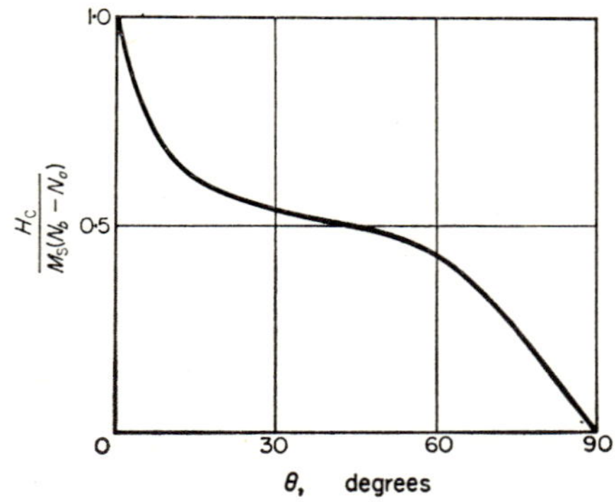

FIG. 8.11. Coercivity of single particles.

If a material is made up entirely of similar single-domain spheroids, with their polar axes orientated at random, then the material will be magnetically isotropic. Stoner and Wohlfarth have calculated the hysteresis loop in this case, and their result for prolate spheroids is plotted in Fig. 8.12. For this

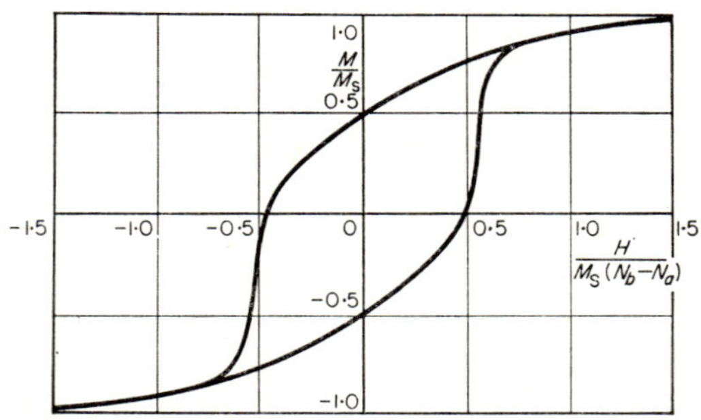

FIG. 8.12. Hysteresis loop for single-domain particles orientated at random.

assembly of random particles the remanent magnetization or retentivity is given by $M_R = M_S/2$ and the coercivity by

$$H_C = 0{\cdot}479 M_S(N_b - N_a) \tag{8.23}$$

Permanent magnet materials made of soft iron powder with a particle size below about 10^{-5} cm have been produced with coercivities exceeding 400 A/cm [12]. This result can be explained by the foregoing theory if it is supposed that the particles are deformed only a little from the spherical shape. For a prolate spheroid the demagnetizing factor in the direction of the polar axis is given by

$$\mu_0 N_a = (1/e^2 - 1)[(1/2e) \log \{(1 + e)/(1 - e)\} - 1] \tag{8.24}$$

where the eccentricity of the ellipse is $e = \{1 - (b/a)^2\}^{\frac{1}{2}}$. Equation (8.24) may be expanded into the series

$$\mu_0 N_a = (b/a)^2(1/3 + e^2/5 + e^4/7 + \ldots) \tag{8.25}$$

If the iron particles are spheroids with, for example, $b/a = 0{\cdot}9$, a relatively small departure from the spherical shape, then $e^2 = 0{\cdot}19$ and, using equation (8.25), $\mu_0 N_a = 0{\cdot}305$ and since in this case $\mu_0(N_a + 2N_b) = 1$, $\mu_0 N_b = 0{\cdot}347$. Assuming the polar axes of the particles are randomly orientated then the coercive force from equation (8.23) is $H_C = 0{\cdot}479 M_S(N_b - N_a) = 0{\cdot}479 \times 2{\cdot}14 \times 0{\cdot}042 \times 10^7/(4\pi)$ A/m $= 343$ A/cm.

This appears to give a very satisfactory explanation of the observed high coercivities of soft iron powder magnets. Moreover very much higher values are obviously possible as the particles become more elongated, that is as b/a becomes smaller. For needle-like particles we may put $N_a = 0$ and $\mu_0 N_b = 1/2$ so that for a random disposition of the needles $H_C = 0{\cdot}479 M_S/(2\mu_0)$. If the needles were in parallel alignment the materials would be strongly anisotropic and, in the direction of the axes of the needles, $H_C = M_S/(2\mu_0)$ while at right-angles to this direction the coercivity would be zero.

Some theoretical values of H_C for material made of iron particles are given in Table 8.2 and plotted in Fig. 8.13. Values of H_C in excess of the values

TABLE 8.2

b/a	$\mu_0 N_a$	$\mu_0 N_b$	H_C A/cm random	H_C A/cm polar axes parallel
1	0·333	0·333	0	0
0·9	0·305	0·347	343	715
0·7	0·244	0·378	1090	2280
0·5	0·173	0·412	1950	4070
0·2	0·055	0·471	3400	7090
0	0	0·5	4080	8520

given in the table do not appear to have been observed. Permanent magnet alloys in general are solid crystalline metals and are not made up of discrete particles, but the small-particle theory of high coercivity may still possibly be applied to them. Such alloys may consist of two metallurgical phases, one being strongly ferromagnetic and finely subdivided in the form of small islands in the weakly ferromagnetic matrix which is the second phase. If the ferromagnetic particles are small and nonspherical the observed high coercivities can then readily be accounted for.

It should be pointed out here that the coercivity H_C used in this section is the field required to reduce the intensity of magnetizaton M to zero. Since $B = \mu_0 H + M$ the coercive field to reduce the flux density B to zero will be

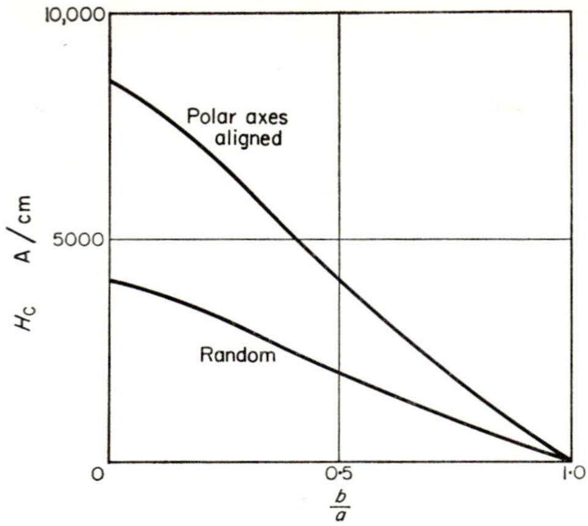

FIG. 8.13. Coercivity of material made of prolate spheroids.

less. The difference only becomes important in practice in the case of materials of very high coercivity, for example $H_C > 500$ A/cm and more especially if the saturation value of the material is low.

In the foregoing only the anisotropy due to the shape of the particle has been considered but, in fact, the magnetic anisotropy associated with the crystal structure must have an additional effect on the coercivity of the particle. Consider first the case of a spherical single-domain particle of cobalt, an element which has uniaxial magnetic anisotropy. Suppose a field H is applied along the c-axis of the hexagonal cobalt and let M_S be deflected by an angle ψ from the c-axis, so that in the direction of the field $M = -M_S \cos \psi$. Then using equation (6.3) and taking only the second term in the equation, the magnetic anisotropy energy of the particle per unit volume is $W_A = K_1 \sin^2 \psi$. The magnetic potential energy is $HM_S \cos \psi$ and the total energy is

$$W = K_1 \sin^2 \psi + HM_S \cos \psi$$

whence $$dW/d\psi = 2K_1 \sin \psi \cos \psi - HM_S \sin \psi$$

and $$d^2W/d\psi^2 = 2K_1 \cos 2\psi - HM_S \cos \psi$$

$dW/d\psi$ is zero when $\sin \psi = 0$, and $d^2W/d\psi^2$ is positive when $\sin \psi = 0$, $\psi = 0$, if $HM_S < 2K_1$. If $HM_S \gg 2K_1$ the position of the vector M_S with $\psi = 0$ becomes unstable and hence the coercivity of the particle is $H_C = 2K_1/M_S$. From Tables 5.1 and 6.3, putting for cobalt $M_S = 1{\cdot}76$ Wb/m^2 and $K_1 = 40 \times 10^4$ J/m^3, the coercivity of the particle with H aligned along the c-axis is $H_C = 2 \times 40 \times 10^4/1{\cdot}76$ A/m $= 4550$ A/cm. The coercivity of an assembly of spherical particles of cobalt with a random orientation of the c-axes would then be about 2200 A/cm. In the particular and hypothetical case of spheroids of cobalt in which the c-axes of the crystals coincided with the polar axes of the spheroids values of H_C somewhat greater than those deduced from the shape anisotropy formula $H_C = 0{\cdot}479 M_S(N_b - N_a)$ would be theoretically possible.

The case of spherical particles of iron or nickel with cubic structure is rather more complicated, but Stoner and Wohlfarth have estimated that orientated spherical particles would have coercivities of about 320 A/cm for iron and about 160 A/cm for nickel. The coercivities for random orientations would be much less than this.

PROBLEMS

8.1. A spherical cavity of radius r in iron is assumed to be bisected by a 180° domain wall. Estimate the energy required to move the wall clear of the cavity when $r = 10^{-4}$ cm, and when $r = 10^{-3}$ cm,

(a) on Kersten's theory which neglects magnetostatic energy,

(b) taking account of the magnetostatic energy.

Assume the wall energy to be 1 erg/cm² and $M_S = 2{\cdot}14$ Wb/m².

8.2 Using Becker's theory estimate the highest value of initial relative permeability, and the lowest value of coercivity, to be expected for an alloy having the values $M_S = 1{\cdot}4$ Wb/m², $\lambda_{100} = 25 \times 10^{-6}$, $E_Y' = 18 \times 10^{10}$ N/m².

8.3. A single-domain particle of iron is in the form of a prolate spheroid. The ratio of the semi-axes is $b/a = 0{\cdot}5$ and a field H is applied at an angle

(a) of 30° to the polar axis, and

(b) of 75° to the polar axis.

Calculate the coercivity of the particle in each case, taking $M_S = 2{\cdot}14$ Wb/m². (Fig. 8.11 may be referred to for Part (a).)

SUGGESTED READING

COOKE, A. H. 'Paramagnetic relaxation effects', *Rep. Prog. Phys.* **13,** 276 (1950).

STONER, E. C. 'Ferromagnetism. Magnetization curves', *Rep. Prog. Phys.* **13,** 83 (1950).

REFERENCES

[1] GORTER, C. J. *Paramagnetic Relaxation*, Elsevier, Amsterdam (1947).

[2] GORTER, C. J. and KRONIG, R. DE L. 'On the theory of absorption and dispersion in paramagnetic and dielectric media', *Physica, 's Grav.* **3,** 1009 (1936).

[3] SIXTUS, K. L. and TONKS, L. 'Propagation of large Barkhausen discontinuities', *Phys. Rev.* **37,** 930 (1931).

[4] LANDAU, L. and LIFSHITZ, E. 'Theory of dispersion of magnetic permeability in ferromagnetic bodies', *Phys. Z. Sowju.* **8,** 153 (1935).

[5] DIJKSTRA, L. J. and SNOEK, J. L. 'Propagation of large Barkhausen discontinuities in Fe-Ni alloys', *Philip Res. Rep.* **4,** 334 (1949).

[6] KERSTEN, M. 'Theory of ferromagnetic hysteresis and initial permeability', *Phys. Z.* **44,** 63 (1943).

[7] NÉEL, L. 'Principles of a new general theory of the coercive field', *Annls Univ. Grenoble*, **22,** 299 (1946).

[8] BECKER, R. 'Elastic strains and magnetic properties', *Phys. Z.* **33,** 905 (1932).

[9] CIOFFI, P. P. 'New high permeabilities in hydrogen-treated iron', *Phys. Rev.* **45,** 742 (1934).

[10] ARNOLD, H. D. and ELMEN, G. W. 'Permalloy, an alloy of remarkable magnetic properties', *J. Franklin Inst.* **195,** 621 (1923).

[11] STONER, E. C. and WOHLFARTH, E. P. 'A mechanism of magnetic hysteresis in heterogeneous alloys', *Phil. Trans. R. Soc.* A, **240,** 599, 1948.

[12] NÉEL, L., WEIL, L. and AUBRY, J., French Pat. Spec. Chambéry, 323 (1942); British Pat. Spec. 590, 392 (1947).

Chapter 9

FERRIMAGNETISM

9.1 Introductory

Ferromagnetism and antiferromagnetism have been discussed in Chapter 5 and ascribed to the forces of interaction between spinning electrons in neighbouring atoms. Ferromagnetism occurs in metals, according to the discussion in Section 5.6, when the atomic spacing exceeds a critical value. The interaction is then positive so that the spins are parallel causing spontaneous magnetic saturation and the formation of domains. On the other hand, in manganese and chromium with a closer spacing there is negative interaction so that, below the Néel temperature, the magnetic moments of atoms which are nearest neighbours are antiparallel. Again there is a domain structure but, in this case, with equal and opposite systems of atomic magnets so that the resultant magnetization is zero. These metals are antiferromagnetic. MnO shown in Fig. 5.9 and many other chemical compounds are also found to be antiferromagnetic. Such substances have properties similar to the paramagnetics, and their behaviour is described in Section 5.8. However, in the iron oxide Fe_3O_4 a third phenomenon called ferrimagnetism occurs.

9.2 The Iron Oxide Fe_3O_4

This oxide occurs naturally as the iron ore magnetite and in its impure state has an initial relative permeability of about 10 and a coercivity of about 16 A/cm. Lumps of the material, or lodestones, thus may have the properties of weak permanent magnets and, like compass needles, be capable of aligning themselves in the earth's field or of attracting pieces of iron. When purified the oxide has an initial relative permeability of 70 though its magnetic saturation value is only about one quarter of that of iron. On the other hand the electrical resistivity is of the order of 10^{-2} ohm cm which is more than 100 times greater than that of the metals. The possibility of producing a material with properties like a ferromagnetic and, at the same time, high electrical resistivity stimulated interest in iron oxide. The relatively low resistivity of the ferromagnetic metals precludes their use in high-frequency applications, such as in radio, because of the overwhelming effects of the induced eddy currents even when very thin sheets are used. SNOEK [1] and his co-workers were thus led to make exhaustive investigations on Fe_3O_4 and other oxides, and their work eventually resulted in the introduction of a number of useful oxide materials which are ferrimagnetic and are known in general as ferrites.

Fe_3O_4 crystallizes on a cubic system, cohesion being by ionic bonds, or electrostatic attraction between the oxygen and the metal ions. This type of bonding has been discussed with particular reference to sodium chloride in Section 3.1.2. The structure may be regarded primarily as a face-centred cubic system of oxygen ions, a system shown for the metal nickel in Fig. 3.6 (*A*). The iron ions then fit interstitially in particular positions in the oxygen lattice.

Reference to Appendix 1 will show that a free atom of oxygen has four electrons in the 2*p* sub-shell and is therefore two electrons short of the six required to fill this sub-shell, or the eight required to fill completely the *L* main shell of the atom. In the oxide each oxygen atom acquires two electrons which are transferred from the iron atoms. The oxygen lattice is therefore made up of negative oxygen ions each with two units of negative charge. We may denote the ion as O^{2-}. Since the oxygen ions now have completely full sub-shells they will have no magnetic moment and thus make no direct contribution to the magnetization of the crystal.

In one formula unit, Fe_3O_4, a total of eight electrons have to be transferred to the four oxygen atoms from the three iron atoms. One iron atom gives up two electrons and thus becomes a positive ion with a resultant positive charge of two units. We may refer to this atom as divalent and denote the ion as Fe^{2+}. The two remaining iron atoms each give up three electrons becoming positive ions with three units of charge. We may call them trivalent and denote by Fe^{3+}.

Since the oxygen ions have no magnetic moment the magnetic properties of the oxide are associated with the iron ions only. The magnetic moments of these ions turn out to be simply whole numbers of Bohr magnetons for two reasons. In the first place in the solid crystals there is almost complete orbital quenching so that the orbital contributions to the magnetism of the ions are negligible. Evidence for this has been given for paramagnetic materials in Section 4.2.4 and for the ferromagnetic metals in Sections 5.3 and 5.4. Secondly in the case of ferromagnetism a strong Weiss field exists which may be supposed to break down the Russell–Saunders coupling between spins. The units of magnetic moment in the ions are therefore spinning electrons in positions which are either parallel or antiparallel to the resultant magnetic field in the material. This has been discussed in Section 5.4.

Considering this in terms of quantum numbers we have when the orbitals are quenched $L = 0$ and $J = S$. When there is Russell–Saunders coupling between spins, as in a paramagnetic, the number of Bohr magnetons per ion is given by $p = 2\{S(S + 1)\}^{\frac{1}{2}}$. Theoretical and observed values for a number of ions are given in Table 4.2. When $L = 0$ and the Russell–Saunders coupling is also broken down the number of Bohr magnetons per ion is simply equal to the whole number $2S$.

The moments which we have to consider are therefore only those due to the uncompensated spins in the iron ions. The electronic structure of the iron atom is given in Appendix 1. The divalent iron ion gives up the two $4s$ electrons. The ion then has six electrons in the incompleted $3d$ sub-shell. Since this sub-shell requires ten electrons to fill it, the application of Hund's rules, discussed in Section 2.7, indicates that for the ion in its state of lowest energy the $3d$ sub-shell will have five electrons with positive spin and one with negative spin. The magnetic moment of the Fe^{2+} ion is then due to four uncompensated spins ($S = 2$) and will be $4\mu_B$. The trivalent iron ions give up the two $4s$ electrons and the $3d$ electron of negative spin so that the Fe^{3+} ions each have a moment of $5\mu_B$ ($S = 5/2$).

Because of the comparatively large spacing between the metal ions in the oxide crystal positive interaction between their spin moments, referred to in Section 9.1, might be expected. In that event the total magnetic moment of one formula unit of Fe_3O_4 should be $4 + 5 + 5 = 14\mu_B$. Experimentally, however, the value is found to be about $4{\cdot}1\mu_B$. An explanation of this discrepancy has been put forward by Néel. It arises from the particular siting of the iron ions in the oxygen lattice and the phenomenon called superexchange already briefly mentioned in Section 5.9.

9.3 Superexchange in Ionic Crystals

An explanation of the superexchange effect is best given by reference to the antiferromagnetic oxide MnO. In this oxide the Mn^{2+} ions are the carriers of the magnetic moments while the O^{2-} ions have no moment. The observed antiferromagnetism implies a negative interaction between the manganese ions. On the other hand the distance between these ions in the crystal is too great for a direct negative interaction to occur and, as just stated, a weak positive interaction would seem to be more likely. An explanation of this apparent anomaly is due mainly to KRAMERS [2] and to ANDERSON [3]. Superexchange, whereby negative interaction between widely spaced metal ions may occur, is in fact attributed to the presence of the intervening oxygen ions and to orbital overlaps between the dissimilar ions. The theories are somewhat tentative and will be discussed qualitatively as follows.

The crystal structure of MnO is shown in Fig. 5.9 and it will be seen that the antiparallel manganese ions are colinear with their neighbouring oxygen ions. The O^{2-} ions each have six $2p$ electrons in three antiparallel pairs. The electron density due to each pair has a dumb-bell shape, the charge clouds for the three pairs taking the forms shown in Fig. 3.2. The dumb-bell axis of each electron pair coincides with a line joining Mn^{2+} ions on either side of an oxygen ion. This will be appreciated, for example, by looking at the oxygen ion at the centre of the cube shown in Fig. 5.9, this ion having six near-neighbour manganese ions.

The outer electrons of the Mn^{2+} ions are in $3d$ sub-shells which are half filled with five electrons in each. According to Hund's rules these electrons have parallel spins and the ion has a moment of $5\mu_B$.

The phenomenon of superexchange is considered to be due to an overlap between the manganese $3d$ orbits and the oxygen $2p$ orbits with a continuous interchange of electrons between them. It appears that, for the overall energy of the system to be a minimum, the moments of the manganese ions on either side of the oxygen ion must be antiparallel. The manganese magnetic moments are thus, in effect, coupled through the intervening oxygen ion. The idea is illustrated in Fig. 9·1.

In Fig. 9.1(*a*) and 9.1(*c*) the outer electrons in a pair of Mn^{2+} ions, and in an intervening O^{2-} ion in the unexcited state, are shown by the arrows. One suggested mode of coupling is indicated in Fig. 9.1(*b*). The two electrons of a pair in the oxygen ion are simultaneously transferred, one to the left and the other to the right. If their directions of spin are unchanged then, again by Hund's rules, the moments of the two manganese ions must be antiparallel as shown. A transference to and fro is assumed to occur continuously.

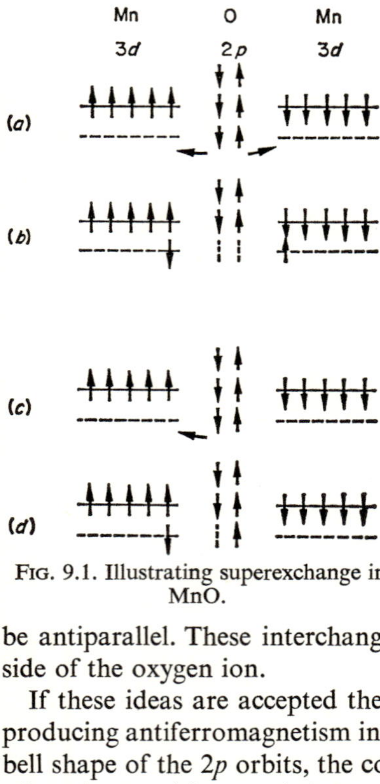

FIG. 9.1. Illustrating superexchange in MnO.

Another possibility is represented in Fig. 9.1(*d*). One electron only has been transferred to the manganese ion on the left. The oxygen ion now has a moment of $1\mu_B$ and if there is negative interaction between the oxygen ion and the right-hand manganese ion then again the moments of the manganese ions will be antiparallel. These interchanges also would occur continuously on either side of the oxygen ion.

If these ideas are accepted then the oxygen ions play an essential part in producing antiferromagnetism in the oxide. Moreover, because of the dumb-bell shape of the $2p$ orbits, the coupling mechanism should be most effective when the metal ions and the oxygen ions lie in one straight line, that is, the angle between the bonds is 180°, and this is the case with MnO.

According to NÉEL [4] negative interaction also occurs between the metal ions in Fe_3O_4 and in other ferrites to be discussed in this chapter. The crystal structure is now more complicated than in MnO. The spacings between near-neighbour oxygen and metal ions are not all the same and the bond angles also vary. The subject has been discussed in some detail by GORTER [5]. It is, however, concluded that, taking into account the spacings and bond angles, the negative interaction in the ferrites is due basically to the same superexchange mechanism as that just discussed for MnO.

9.4 The Crystal Structure of Fe_3O_4

A unit cell of the Fe_3O_4 crystal is represented by the cube the front face of which is shown by the broken lines in Fig. 9.2(*a*). This cell contains eight formula units each of which may be written in the form $Fe^{3+}(Fe^{2+}Fe^{3+})O_4^{2-}$, where the suffixes indicate the electronic state of the ions. The purpose of the brackets is to enclose the metal ions which occupy what are known as *B* sites in the crystal structure. The remaining iron ion occupies an *A* site.

The cell therefore contains 32 oxygen ions, 8 divalent iron ions on *B* sites, and 16 trivalent ions half of which are on *A* sites and half on *B* sites. As already mentioned the oxygen ions form a face-centred cubic structure and are represented in Fig. 9.2(*a*) by the large spheres. The *A* and *B* interstitial spaces in the oxygen lattice which accommodate the metal ions are different in size and this is illustrated in Fig. 9.2 at (*b*) and (*c*). The ions in the larger *B* spaces have six near-neighbour oxygen ions while the *A* ions have four near-neighbour oxygen ions as shown.

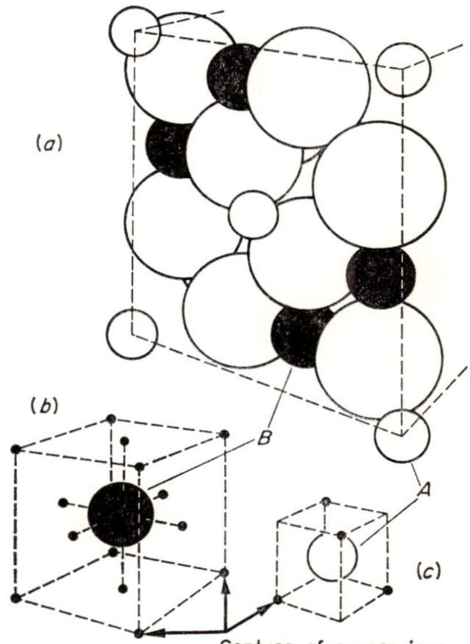

FIG. 9.2. Crystal structure of Fe_3O_4 showing *A* and *B* sites.

The disposition of the iron ions in the unit cell may be further described with the aid of Fig. 9.3. The unit cell is here shown subdivided into eight equal cubes in two groups of four. The *B* ions are contained in the four cubes as depicted in the upper part of the figure. As already stated half the *B* ions are divalent, shown by the black spheres, the other half being trivalent and shown by the larger white spheres in this figure. There are thus four ions in each cube in the upper part of Fig. 9.3. The remaining four cubes in the lower part of the figure each have one *A* ion added at the centre as shown. The smaller white spheres in the lower part of Fig. 9.3, and also in Fig. 9.2, represent those trivalent iron ions which are on *A* sites.

It is supposed that superexchange forces exist mutually between the *B* ions and this may be referred to as a *BB* interaction. Similarly between the *A* ions there is an *AA* interaction, and finally between ions on *A* and *B* sites there is an *AB* superexchange interaction. All three interactions are negative, or antiferromagnetic in character, but according to Néel the *AB* interaction predominates and overwhelms the *AA* and *BB* interactions. The resultant effect is that domains are formed in the oxide in which the magnetic moments of the *B* ions are antiparallel with those of the *A* ions. The saturation moment of the domain is thus due to a difference between the unequal *B* and *A* lattices. In one formula unit $Fe^{3+}(Fe^{2+}Fe^{3+})O_4^{2-}$ there is one divalent ion Fe^{2+} with a moment of $4\mu_B$ and one trivalent ion Fe^{3+} with a moment of $5\mu_B$ both on *B* sites, and on an *A* site one trivalent Fe^{3+} ion with a moment of $5\mu_B$. The moment per formula unit is therefore $4 + 5 - 5 = 4\mu_B$ and this would represent the saturation moment of the domain per formula unit at 0°K. Thus

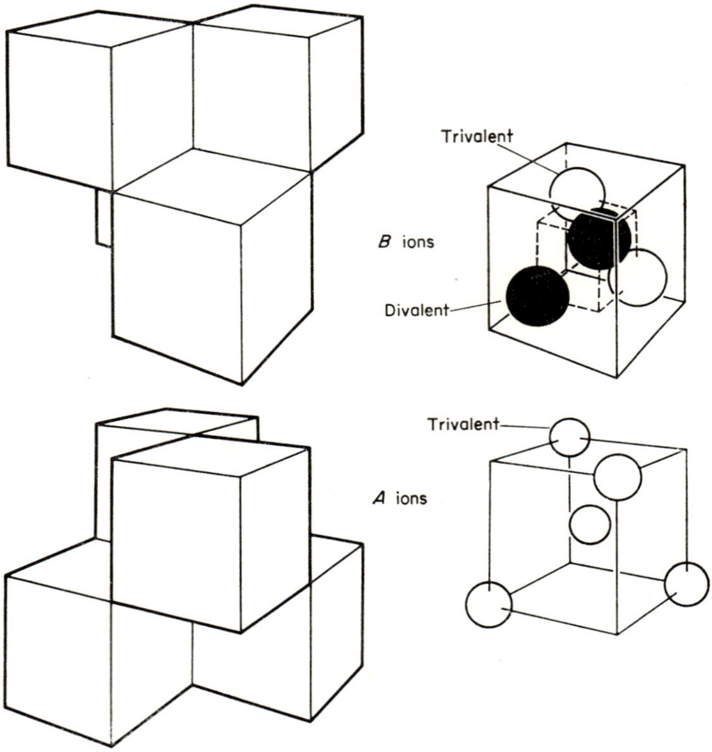

FIG. 9.3. Showing unit cell of Fe_3O_4 divided into eight cubes with *A* and *B* sites.

referring to Figs. 9.2 and 9.3 the resultant magnetic moment of the ferrimagnetic domains may be regarded as due to the divalent ions only, represented by the black spheres, since the trivalent ions on *A* and *B* sites, respectively, are equal and antiparallel and so cancel each other out magnetically. The theoretical value of $4\mu_B$ is in approximate agreement with experimental observation which shows also that the easy direction in the domain corresponds with a [111] crystal direction.

The crystal structure of Fe_3O_4 is similar to that of the mineral spinel, though with a minor but important difference. The chemical formula of spinel may be written $Mg^{2+}(Al^{3+}Al^{3+})O_4^{2-}$ where magnesium and aluminium ions now take the place of iron ions in the structure. Once again the bracket encloses ions on *B* sites and in the case of spinel the trivalent ions are all on *B* sites, as the formula indicates, and the divalent ions are on *A* sites. This is known as a normal spinel structure. Fe_3O_4, and other ferrites of cubic structure to be discussed, are said to have an inverse spinel structure, where as already described the divalent ions are on *B* sites and the trivalent ions are equally divided between *A* and *B* sites.

Fe_3O_4 has ferrimagnetic properties because of its inverse structure which leads to the formation of domains. Snoek and his co-workers found that oxides of inverse structure could be artificially produced in which divalent ions of another element, for example Mn, Ni, Co, Mg or Cu, could be substituted for the divalent Fe^{2+} ions in Fe_3O_4. An extensive range of ferrites could thus be made having the general formula $Fe^{3+}(M^{2+}Fe^{3+})O_4^{2-}$ where M^{2+} represents a divalent ion of any suitable element. The magnetic and electrical properties of these single ferrites are found to vary considerably. Moreover two or more single ferrites may be combined together in different proportions. They form solid solutions in one another and give a wide range of synthetic materials with different properties. A few of these mixed ferrites have useful properties and are now widely employed, more particularly in high-frequency applications.

9.5 Néel's Theory of Ferrimagnetism

NÉEL [4] has shown how to determine theoretically the domain saturation in relation to temperature of a ferrite having *A* and *B* systems of ions with negative *AA*, *BB* and *AB* interactions. The ideas involved are an extension of those used in ferromagnetism and the elements of the theory are as follows.

Figure 5.1 illustrates the occurrence of a spontaneous saturation value M_S in a ferromagnetic domain at temperature *T*. The magnetization curve of the material, treated first as a paramagnetic, is expressed for any temperature *T* by equation (4.9). This involves the Brillouin function $B(a)$ given by equation (4.8). Positive interaction between spins in the ferromagnetic case produces, in effect, the Weiss molecular field $H = N_W M_T$ and the point of intersection of the Weiss field line with the paramagnetic curve gives the point *P* in Fig. 5.1 and determines the saturation value of the domain at temperature *T*.

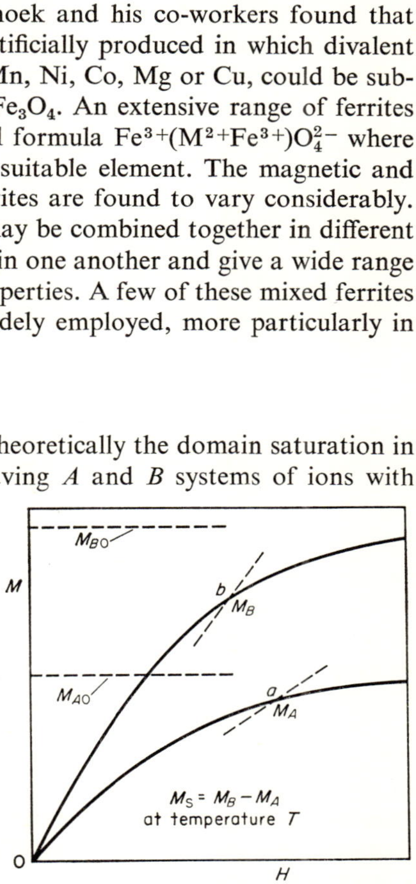

FIG. 9.4. Illustrating spontaneous saturation in *A* and *B* lattices of a ferrite.

In the ferrimagnetic crystal with two unequal lattices *A* and *B* we may again suppose that each lattice separately behaves like a paramagnetic. Each lattice is, however, acted upon by a field akin to the Weiss field, but the field in each case is rather more complicated since it is due to the negative super-exchange interactions. The superposition of the exchange fields upon the separate Brillouin magnetization curves of the A and B lattices results in spontaneous saturation values M_A and M_B, respectively, in the two lattices, but these are antiparallel and interdependent.

The points just mentioned are illustrated in Fig. 9.4. The paramagnetic magnetization curve for the A lattice is given by $M/M_{A0} = B(a)$, where $a = gJ\mu_B H/kT$. If we suppose, as in the ferromagnetic case, that the ionic moments are due to spin only with no Russell–Saunders coupling, then $g = 2$ and $J = 1/2$ and the magnetization curve for the A lattice becomes

$$M/M_{A0} = \tanh(\mu_B H/kT) \tag{9.1}$$

and for the B lattice

$$M/M_{B0} = \tanh(\mu_B H/kT) \tag{9.2}$$

where M_{A0} and M_{B0} are the saturation values at 0°K or at infinite field (see Sections 4.2.2, 5.2 and 5.4). Equations (9.1) and (9.2) are represented by the curves Oa and Ob in Fig. 9.4.

Let the superexchange field constant for the AB interaction be $-N$, for AA interaction let it be $-\alpha N$ and for the BB interaction $-\beta N$, where α and β are both less than unity. In the A lattice we suppose saturation occurs at the point a in Fig. 9.4 and the saturation value is M_A. Similarly in the B lattice the saturation value is M_B at point b, and M_A and M_B are antiparallel. The resultant saturation value of the ferrite at temperature T is then

$$M_S = M_B - M_A$$

At point a the spontaneously produced field acting in the direction of the magnetization in the A lattice is $H = NM_B - \alpha NM_A$ and $M = M_A$. Similarly at b for the B lattice $H = NM_A - \beta NM_B$ and $M = M_B$. These values of H and M must satisfy equations (9.1) and (9.2), respectively. Therefore we have

$$M_A/M_{A0} = \tanh\{(\mu_B/kT)(NM_B - \alpha NM_A)\} \tag{9.3}$$

and

$$M_B/M_{B0} = \tanh\{(\mu_B/kT)(NM_A - \beta NM_B)\} \tag{9.4}$$

At any temperature T, and knowing the quantities M_{A0}, M_{B0}, N, α and β, equations (9.3) and (9.4) can be solved for the two unknowns M_A and M_B. Whence a relation between M_S and T can be found.

With increasing temperature the saturation magnetization in both A and B lattices will fall and when either falls to zero the AB interaction disappears. Hence the saturation magnetization in the other lattice also falls to zero at the same temperature. There is thus a common Curie point and, in fact, the ferrite may be expected to have an M_S–T relationship like that of a ferromagnetic, although as mentioned below, some interesting dissimilarities are possible with certain ferrites.

However, the observed relation between M_S and T for the oxide Fe_3O_4 is shown in Fig. 9.5. In this case M_S at 0°K is 0·64 Wb/m² and the Curie temperature is 585°C (858°K). The broken line shows for comparison the theoretical curve for a ferromagnetic assumed to have the same saturation value at absolute zero and the same Curie temperature. This theoretical curve may be determined from equation (5.4), ($J = 1/2$), and is shown also in Fig. 5.2.

For any ferrite with interacting A and B lattices the shape of the saturation-temperature curve will depend upon the relative magnitudes of M_{A0} and

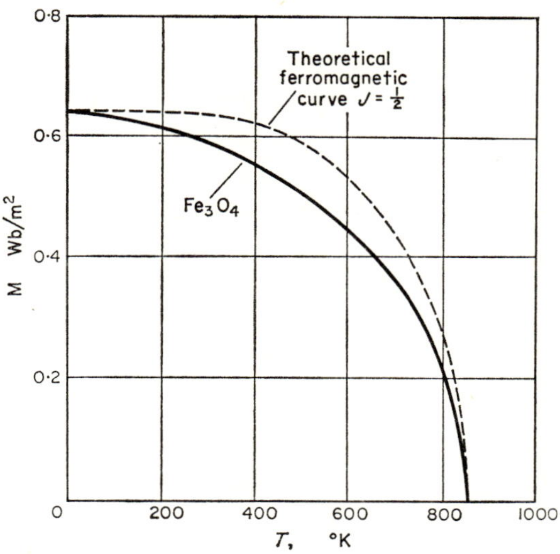

FIG. 9.5. Observed saturation intensity in Fe_3O_4 in relation to absolute temperature.

M_{B0}, and on the values of the coefficients α and β. In the A lattice at any temperature the total superexchange field is, as already stated,

$$H = NM_B - \alpha NM_A$$

or

$$H = (M_B/M_A - \alpha)NM_A$$

The effective Weiss field factor is no longer a constant, but at 0°K it would be $(M_{B0}/M_{A0} - \alpha)N$ for the A lattice, and similarly in the B lattice it would be $(M_{A0}/M_{B0} - \beta)N$. Whence if $M_{B0} > M_{A0}$ and if, for example, the BB interaction is strong, so that β is large, and the AA interaction is weak, making α small, the effective Weiss field factor at 0°K in the B lattice may be relatively small and relatively large in the A lattice. Now in a ferromagnetic the Curie temperature is proportional to the Weiss field constant (see equations (5.1) and (5.2)), while the A and B lattices of the ferrite have, as already pointed out, a common Curie temperature. The curve relating M_B and T for the B lattice at low temperatures may therefore be expected to fall more rapidly with increasing temperature than the M_A–T curve for the A lattice. This is illustrated in Fig. 9.6 at (*a*) and (*b*). It is thus theoretically possible for the saturation value of the ferrite, $M_S = M_B - M_A$, to become zero and to reverse sign before the Curie temperature is reached, as shown in Fig. 9.6(*c*).

On the other hand with $M_{B0} > M_{A0}$, but with α greater than β, it is possible for the effective Weiss field factor at 0°K to be greater for the A lattice than for the B lattice. The M_B–T curve could then fall more slowly with increasing low temperatures than the M_A–T curve, as shown in Fig. 9.6 at (*d*) and (*e*). In the example given in the figure the M_S–T curve for the ferrite, shown

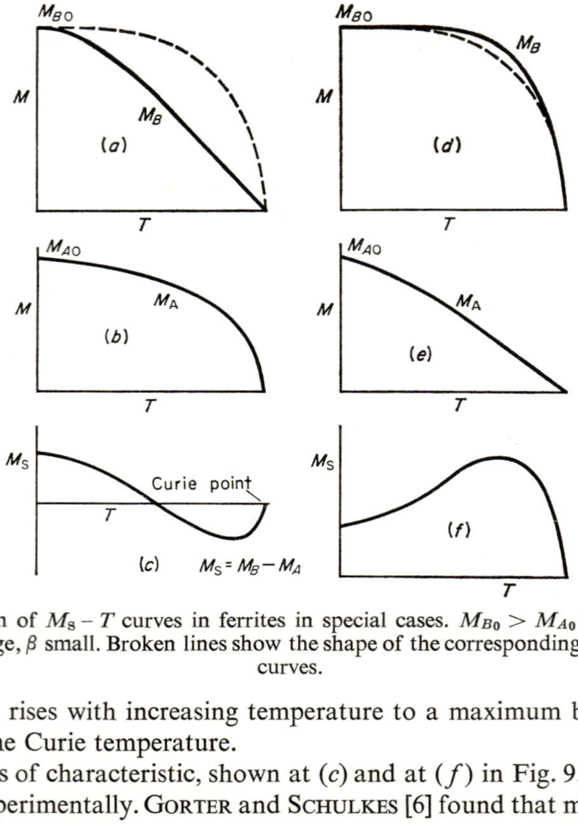

FIG. 9.6. Form of $M_S - T$ curves in ferrites in special cases. $M_{B0} > M_{A0}$: (*a*) β large, α small, (*b*) α large, β small. Broken lines show the shape of the corresponding ferromagnetic curves.

at (*f*), then rises with increasing temperature to a maximum before falling to zero at the Curie temperature.

Both types of characteristic, shown at (*c*) and at (*f*) in Fig. 9.6, have been observed experimentally. GORTER and SCHULKES [6] found that mixed ferrites, having the composition $Li_{0\cdot5}Fe_{2\cdot5-a}Cr_aO_4$, had saturation-temperature curves of the form shown in Fig. 9.6(*c*) when the quantity a was between 1·0 and 1·6. The observed M_S–T characteristic for the composition $Li_{0\cdot5}Fe_{1\cdot25}Cr_{1\cdot25}O_4$, that is when $a = 1\cdot25$, is shown by the curve a in Fig. 9.7. This result shows a Curie temperature at about 470°K, but M_S also falls to zero at 311°K, or 38°C. The direct measurement of M_S in relation to T does not of itself prove that M_S reverses sign at 38°C but that reversal occurs was demonstrated in the following way. A rod of the material was saturated at a low temperature. On removal of the magnetic field the rod retained a remanent magnetization. The remanent magnetization is due, of course, to the fact that the domain vectors are no longer randomly disposed but that a preponderance of the domain vectors will have a component of magnetization in one direction along the axis of the rod. The remanence was then measured with increasing temperature, found to fall to zero at 38°C, and then reversed sign in the rod as shown by curve *b* in Fig. 9.7. This showed that the domain saturation $M_S = M_B - M_A$ had itself reversed in sign, and was a convincing confirmation of the correctness of Néel's theory of ferrimagnetism.

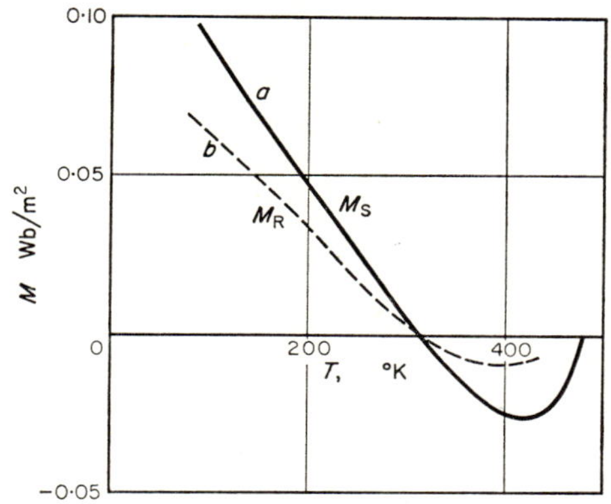

FIG. 9.7. Saturation intensity and remanence in relation to temperature for the mixed ferrite $Li_{0·5}Cr_{1·25}Fe_{1·25}O_4$.

9.6 Single Ferrites

As already mentioned in Section 9.4 artificial single ferrites may be made having the general formula $Fe^{3+}(M^{2+}Fe^{3+})O_4^{2-}$ where M^{2+} is a divalent ion of a suitable element. The general method of manufacture has been outlined by GORTER [5] and by SMIT and WIJN [7]. The raw materials in the form of oxides or carbonates are converted to fine particles by ball-milling, then mixed, dried and prefired to produce homogeneity. The material is then milled again and the powder die-pressed to the desired final shape. This is followed by sintering at temperatures in the range 1100°C to 1450°C. Shrinkage during sintering has to be allowed for and some porosity of the final product cannot be avoided. Close control of the oxygen content and avoidance of residual stresses are important to give the best electrical and magnetic properties.

When the divalent ions represented by M^{2+} in the chemical formula above are from different elements the properties of the resulting single ferrites also differ. Since the trivalent iron ions are equally distributed on *A* and *B* sites they cancel each other out magnetically, and the magnetic moment per formula unit is then theoretically the same as the magnetic moment of the divalent ion. This moment is obtained, as previously described, simply from the number of uncompensated spins in the ion, with no orbital contribution and no Russell–Saunders coupling.

Table 9.1 gives some of the properties of a number of single ferrites. The elements from manganese to copper are successive elements in the periodic table, and being present in the ferrites as divalent ions, are responsible for the resultant saturation magnetic moment of the domains. The electronic structure of the free atoms of these elements is given in Appendix 1. The atoms become divalent by losing the two electrons in the 4*s* sub-shell (one from 4*s* and one from the 3*d* sub-shell in the case of copper). Reference to Appendix 1, and the application of Hund's rules, give the positive and negative spins in the 3*d* sub-shells which are shown in the table, and the difference gives the theoretical number of Bohr magnetons per formula unit of the ferrite. The observed values of magnetic moment per formula unit are also given in the table and are seen to be in fairly good agreement with the theoretical values.

Magnesium ferrite is also given in the table. The outermost sub-shell of the Mg^{2+} ion is 2*p* and contains three electrons of positive and three of negative spin. The theoretical moment of the Mg^{2+} ion is therefore zero, although the observed value is $1{\cdot}1\mu_B$.

Zinc ferrite turns out to have a normal spinel structure rather than the inverse structure of the magnetic ferrites. The formula may be written $Zn^{2+}(Fe^{3+}Fe^{3+})O_4^{2-}$. The Zn^{2+} ions are on *A* sites and, as Table 9.1 shows, have zero magnetic moment. This leaves trivalent iron ions on *B* sites with a negative *BB* interaction between equal ions. Zinc ferrite might therefore be

TABLE 9.1

SOME PROPERTIES OF SINGLE FERRITES

Ferrite	M^{2+} Ion 3*d* Electron Spins			Moment per Formula Unit μ_B		$K_1 \times 10^{-4}$ J/m³	Saturation Magnetostriction $\times 10^6$			M_S Wb/m² 20°C	Curie Temperature °C	Initial Permeability μ_{BI}
	+ve	−ve	Difference	Theoretical	Observed		λ_S	λ_{111}	λ_{100}			
$MnFe_2O$	5	0	5	5	4·6	−0·28	−5			0·50	300	250
$FeFe_2O$	5	1	4	4	4·1	−1·1	+41	+81	−19	0·60	585	70
$CoFe_2O$	5	2	3	3	3·7	+38	−110	+120	−590	0·53	520	
$NiFe_2O$	5	3	2	2	2·3	−0·62	−26			0·34	585	10
$CuFe_2O$	5	4	1	1	1·3	−0·60	−10			0·17	455	70
$MgFe_2O$*	3	3	0	0	1·1	−0·25	−6			0·15	440	10
$ZnFe_2O$†	5	5	0	0	0	—	—	—	—	—	—	—

* 2*p* electron spins for Mg

† normal spinel structure for zinc ferrite

expected to be antiferromagnetic and thus to have a Néel point, though measurements show it to be paramagnetic only.

9.7 Mixed Ferrites

As already mentioned in Section 9.4 two or more single ferrites may be combined together in different proportions, forming solid solutions in one another. It has been found empirically by measurements on a wide variety of mixed ferrites that some have useful technological properties. Among the useful ferrites are nickel–zinc, manganese–zinc and magnesium–manganese combinations. Zinc ferrites have a normal spinel structure and hence, as pointed out in the preceding paragraph, are not ferrimagnetic. It is therefore remarkable to find that zinc ferrite can be a beneficial addition in a mixed ferrite. It has, indeed, been shown by GUILLAUD [8, 9] and by GORTER [10] that the addition of zinc ferrite will increase the saturation value of the mixed ferrite in a number of cases.

The chemical formula of a mixed ferrite containing zinc may be written $(Zn_a^{2+}Fe_{1-a}^{3+})\,(M_{1-a}^{2+}Fe_{1+a}^{3+})O_4^{2-}$. The symbol a represents the fraction of zinc ferrite in the combined ferrite. The divalent zinc ions, which have no magnetic moment, are on A sites in accordance with the normal spinel nature of zinc ferrite. The divalent M ions are on B sites, and the brackets in the formula separate the A and B ions. The trivalent iron ions each have a moment of $5\mu_B$. If the M ions have a moment of $m\mu_B$, then, per formula unit, the total moment in Bohr magnetons on B sites is $m(1-a)+5(1+a)$, and on A sites the total antiparallel moment is $5(1-a)$. If the resultant moment per formula unit is n_B, then by taking the difference of A and B moments,

$$n_B = 10a + m(1-a) \tag{9.5}$$

The relation between n_B and the fraction a of zinc ferrite present depends on m, the moment of the M ions in the mixed ferrite. The values to be assigned

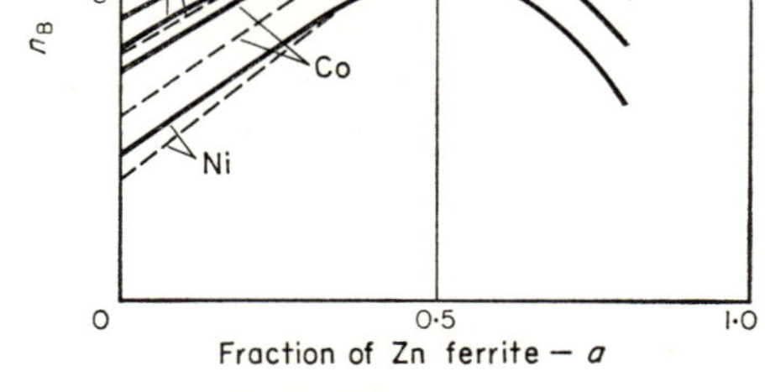

FIG. 9.8. Saturation magnetic moments of mixed ferrites containing zinc. Full lines, observed: broken lines, theory.

to m are 5 for Mn^{2+}, 4 for Fe^{2+}, 3 for Co^{2+} and 2 for Ni^{2+} (see Table 9.1). For these values of m equation (9.5) is represented by the broken lines in Fig. 9.8. Experimental values of the saturation value, n_B, for manganese, iron, cobalt and nickel ferrites, respectively, containing zinc, are shown by the full lines in the figure. It will be seen that the saturation value in each case rises, in fair agreement with equation (9.5), as the zinc ferrite content is increased. A maximum is reached when a is about 0·5 and the moment then falls with further increases in the amount of zinc ferrite present. As a approaches unity the ferrite reverts more nearly to the non-ferrimagnetic zinc ferrite of normal spinel structure, and hence the experimental curves may be expected to converge toward $n_B = 0$ when $a = 1$.

9.8 Resonance Effects in Ferrites

If measurements of the magnetic properties of a ferrite are made over a frequency range from zero up to very high frequencies, for example up to 10,000 Mc/s, interesting variations of permeability and loss in relation to frequency are encountered which are reminiscent of the resonances which may arise in a mechanical system subjected to forced vibrations of increasing frequency. The phenomena occurring are observable in the ferrites because the electrical resistivity of these materials is so high that eddy currents under alternating flux conditions are negligible. Any such magnetic effects in the ferromagnetic metals, which have a relatively low electrical resistivity, would in general be completely masked by eddy currents.

Examples of the experimental results achieved are shown in Figs. 9.9 and 9.10. At the high frequencies involved in these experiments the flux densities are very low and hence only values of the initial relative permeability were observed.

It is convenient to consider the relative permeability as the complex quantity $\mu'_R - j\mu''_R$ and the significance of the components μ'_R and μ''_R may be seen by referring back to Fig. 8.1 and to the discussion of the susceptibility of a paramagnetic material in Section 8.2. If Fig. 8.1 now refers to a ferrite magnetized at a particular frequency, then the flux density B is given by $B = (\mu' - j\mu'')H$ where μ' and μ'' are absolute values, and referring to Fig. 8.1(*b*) we may also put $B = \mu_0 H + M \cos \delta_M - jM \sin \delta_M$, or $B = (\mu_0 + \kappa' - j\kappa'')H$. Whence $\mu' = \mu_0 + \kappa'$ and $\mu'' = \kappa''$, or $\mu'_R - 1 = \kappa'/\mu_0$ and $\mu''_R = \kappa''/\mu_0$. The power loss per unit volume is, from equation (8.7), given by $P = \omega HM \sin \delta_M$ where δ_M is the phase angle between M and H and $\tan \delta_M = \kappa''/\kappa' = \mu''_R/(\mu'_R - 1)$. Frequently the loss angle is taken to be the phase angle between B and H. If this angle is δ_B then in Fig. 8.1(*b*) $\delta_B = 90 - \phi$ and the power loss may be written $P = \omega HB \sin \delta_B$ where now $\tan \delta_B = \mu''_R/\mu'_R$.

Values of μ'_R and μ''_R, due to WIJN, GEVERS and VAN DER BURGT [11], are given in Fig. 9.9 in relation to frequency for nickel-zinc ferrites of three different compositions. The curves indicate that there is an upper useful

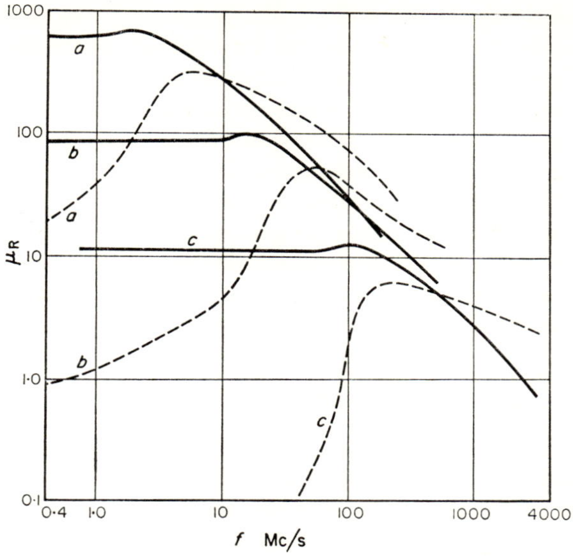

FIG. 9.9. Initial relative permeability of nickel-zinc ferrites of composition $Ni_{1-a}Zn_aFe_2O_4$: (*a*) $a = 0{\cdot}36$, (*b*) $a = 0{\cdot}64$, (*c*) $a = 0$. Full lines μ'_R : broken lines μ''_R.

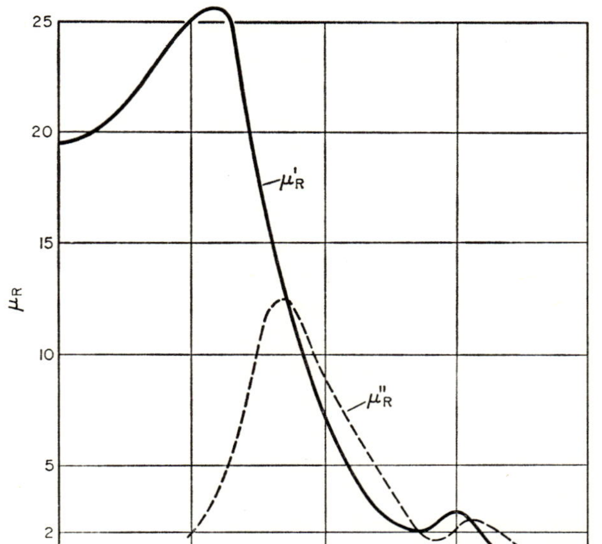

FIG. 9.10. Real and imaginary parts of initial complex permeability of a magnesium-iron ferrite in relation to frequency (RADO, WRIGHT and EMERSON).

frequency limit which shows itself by a sharp rise in μ''_R followed by a drop in μ'_R. The form of the curves indicates an internal resonance ascribed by SNOEK [12] to electron spin resonance. This would occur when the frequency of the applied alternating field coincided with the angular frequency of precession of electron spins, this precession occurring in the presence of an internal or an externally-applied steady field.

However, Fig. 9.10 gives results due to RADO, WRIGHT and EMERSON [13] for a magnesium-iron ferrite which shows two regions of resonance. Rado and his co-workers have produced evidence to show that the resonance at the lower frequency, about 50 Mc/s in this case, is due to an oscillation of the domain boundaries, while the higher frequency resonance, at about 1200 Mc/s, is due to the spin resonance proposed by Snoek.

9.8.1 *Domain Boundary Resonance*

In Section 8.5 the initial permeability of a ferromagnetic crystal was considered in relation to the idealized model shown in Fig. 8.8(*a*). If the field applied in that case is alternating at a high frequency then the domain boundaries would oscillate through small distances about their mean positions. Some force is required to move the boundaries and a certain stiffness is associated with them. The restoring force acting on the boundaries was supposed in Section 8.5 to be due to the magnetostriction and to the residual stresses in the crystal. Moreover movement of a boundary wall through the crystal will also cause a rearrangement of the axes of the spinning electrons in the volume swept out, and so the boundary must possess inertia. We can in fact attribute both stiffness and mass to the boundary wall, which may then respond to an alternating applied force like a mechanical system.

In order to determine the apparent mass of a domain boundary it is first necessary to have some physical picture of the mechanism whereby an applied magnetic field can exert a force or a pressure upon a boundary in order to make it move. This picture will emerge from the following analysis.

The nature of a domain boundary wall has been discussed in Section 7.3. In a 180° boundary wall the spin axes spiral around through 180° from one side of the wall to the other, and this is illustrated diagrammatically in Fig. 7.3(*c*). In Fig. 9.11(*a*) we will consider a 180° wall of unit surface area and thickness t_w with the domain magnetizations on either side of the wall parallel and antiparallel to the OZ axis as shown by the arrows in the figure. We will suppose, with the wall at rest, that there are equal increments of the angular rotation of the spin axes in successive layers of atoms across the thickness of the wall. As mentioned in Section 7.3 this is an approximation but for present purposes a convenient one. We consider a layer of thickness dy at a distance y from the XZ plane as shown in Fig. 9.11(*a*). If the angular rotation of the spin axes in this layer is ϕ then $\phi/\pi = y/t_w$.

At time $t = 0$ let a steady field H be suddenly applied in the OZ direction. The effect upon the spins in the elementary layer is to cause a precession of

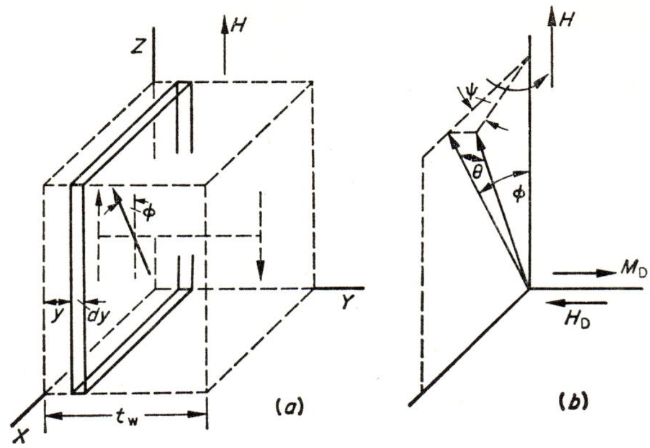

FIG. 9.11. Illustrating the mechanism of domain wall movement.

the spin axes about the direction of H. At time t suppose the spin axes have precessed through a small angle ψ as shown in Fig. 9.11(*b*), the angle between the spin axes and the XZ plane then being θ. The angular velocity of precession will be $d\psi/dt = \gamma H$ where $\gamma = g\mu_0 e/(2m) = \mu_0 e/m$ for spin ($g = 2$) (see Section 2.5). Therefore at time t, $\psi = \gamma Ht$ and $\theta = \psi \sin \phi = (\gamma H \sin \phi)t$. Since the spin axes are now inclined at a small angle θ to the XZ plane there is a component of intensity of magnetization in the elementary layer, and perpendicular to the plane of the wall, equal to $M_S\theta$ and $M_S\theta = M_S(\gamma H \sin \phi)t = M_S t\gamma H \sin (\pi y/t_w)$. The total component of magnetic moment of the wall of unit surface area and thickness t_w, in a direction normal to the wall is therefore

$$M_S t\gamma H \int_0^{t_W} \sin (\pi y/t_w) dy = \gamma 2 M_S t H t_w/\pi$$

The mean intensity of magnetization normal to the wall is then

$$M_D = 2M_S t\gamma H/\pi$$

The demagnetizing factor of the wall in the same direction is $N_D = 1/\mu_0$. The intensity M_D will therefore produce a demagnetizing field, parallel to YO in Fig. 9.11, of amount

$$H_D = N_D M_D = 2M_S t\gamma H/(\pi\mu_0)$$

The field H_D now produces a precession of the spin axes about the direction of H_D and we may regard this as the mechanism which causes the whole wall to move forward in the OY direction.

The demagnetizing field H_D produces a precessional angular velocity on the spin axes $-d\phi/dt = \gamma H_D$. Bearing in mind that the spinning electrons outside the wall on either side remain unmoved, rotation of the spin axes about the OY direction corresponds to a linear movement of the whole wall.

Taking account of the signs the linear velocity becomes

$$dy/dt = -(t_w/\pi)d\phi/dt = (t_w/\pi)\gamma H_D$$

$$\therefore\ dy/dt = \{(2t_w\gamma^2 M_S H)/(\pi^2\mu_0)\}t$$

Differentiating and rearranging gives

$$\{(\pi^2\mu_0)/(t_w\gamma^2)\}d^2y/dt^2 = 2M_S H \tag{9.6}$$

The total effective pressure, or force per unit area, on the wall is $2M_S H$, since if the wall moves a unit distance, then remembering that M_S reverses in the volume swept out, $2M_S H$ is the decrease in magnetic potential energy of the system per unit area in the field H. Therefore in equation (9.6) the apparent mass of the wall per unit area is

$$m_w = (\pi^2\mu_0)/(t_w\gamma^2) \tag{9.7}$$

Since $\mu_0 = 4\pi \times 10^{-7}$ H/m, $\gamma = \mu_0 e/m$ and $e/m = 1{\cdot}76 \times 10^{11}$ C/kg equation (9.7) becomes

$$m_w = t_w^{-1} \times 2{\cdot}54 \times 10^{-16}\ \text{kg/m}^2 \tag{9.8}$$

This is the apparent mass per unit area of a 180° boundary (MKS units, t_w in metres), and it may be noted that this mass is inversely proportional to the wall thickness.

To find what the resonant frequency of a boundary of mass m_w might be if the boundary were set into oscillation, it is necessary to know the restoring force on the boundary for a small displacement. We will consider the idealized model of a ferromagnetic crystal shown in Fig. 8.8(*a*). Referring to the discussion in Section 8.5 the restoring force on a boundary for displacement x is given by $(3\lambda_{100}/2)\sigma_m \sin(\pi x/l)$, and when x is small this force becomes $\{3\pi\lambda_{100}\sigma_m/(2l)\}x$. Equation (8.16) gives the initial relative permeability of the model in terms of the saturation magnetostriction λ_{100} and the internal stress σ_m and using this equation the restoring force on the boundary becomes $\{M_S^2/(\mu_0\mu_{RI}l)\}x = k_1 x$.

The motion of a boundary of mass m_w acted upon by a restoring force $k_1 x$ is given by the equation $m_w d^2x/dt^2 + k_1 x = 0$. The boundary would oscillate with a simple harmonic motion and with an angular frequency $\omega_R = (k_1/m_w)^{\frac{1}{2}}$. Substituting for k_1 and, for a 90° boundary, putting m_w as half the value given in equation (9.7), we have for the resonant frequency

$$f_R = 1{\cdot}26 \times 10^4 \times M_S(t_w/l\mu_{RI})^{\frac{1}{2}}\ \text{Mc/s} \tag{9.9}$$

In this equation t_w is the thickness of the corresponding 180° wall, l is the domain width and μ_{RI} the initial relative permeability at a very low frequency. Tentatively putting in some numerical values of the right order of magnitude, namely $M_S = 0{\cdot}3$ Wb/m², $t_w = 10^{-7}$ m, $l = 10^{-4}$ m, equation (9.9) becomes

$$f_R = 120/(\mu_{RI})^{\frac{1}{2}}\ \text{Mc/s} \tag{9.10}$$

Taking values of μ_{RI} from the low-frequency values of μ'_R in Fig. 9.9 and f_R from the peaks of the broken lines representing μ''_R, and similarly for the lower frequency resonance in Fig. 9.10, it will be found that equation (9.10) is in moderate agreement with the experimental results, and it might therefore be concluded that the resonances in Fig. 9.9 and the lower frequency resonance in Fig. 9.10, are domain boundary effects.

The results in Figs. 9.9 and 9.10 were obtained when the specimens were acted upon by an alternating applied field. The observed values of μ''_R show also that some loss mechanism was operating. Assuming viscous damping on the domain boundaries, that is a damping force proportional to the boundary velocity, the differential equation for oscillating boundaries may now be written $m_w\, d^2x/dt^2 + k_2\, dx/dt + k_1 x = M_S H \sin \omega t$ where k_2 is a constant and H is the amplitude with ω the angular frequency of the applied field. By taking the steady state solution of this equation, and noting the

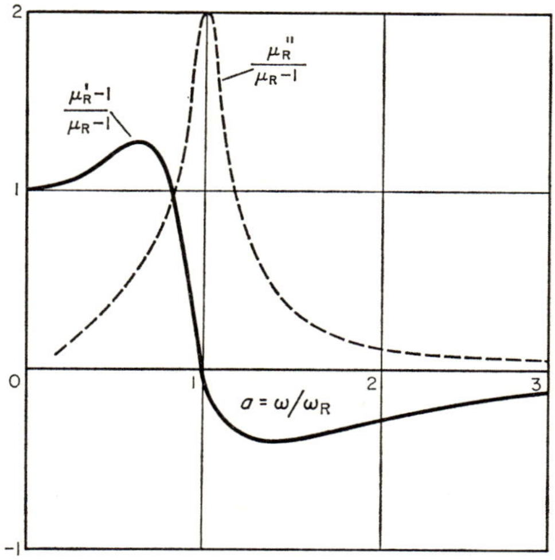

FIG. 9.12. Theoretical relation between real and imaginary components of permeability. Domain boundary considered as a mass oscillating in a damping medium.

components of x which are in phase and in quadrature, respectively, with the applied field it is easy to show that

$$(\mu'_R - 1)/(\mu_R - 1) = (1 - a^2)/\{(1 - a^2)^2 + b^2a^2\} \qquad (9.11)$$

and

$$\mu''_R/(\mu_R - 1) = ba/\{(1 - a^2)^2 + b^2a^2\} \qquad (9.12)$$

where μ_R is the zero frequency initial relative permeability as before, $a = \omega/\omega_R$ and $b = k_2/(m_w\omega_R)$. Equations (9.11) and (9.12) are plotted in Fig. 9.12 for the case when $b = 0{\cdot}5$, and the shape of these theoretical curves may be compared with the experimental results in Figs. 9.9 and 9.10.

9.8.2 *Spin Resonance*

As already mentioned the resonance effect observed in Fig. 9.10 at the higher frequency is believed to be associated with the electron spins rather than with domain boundaries. The effect is most easily noticed in the ferrites because of their high electrical resistivity and the absence of appreciable eddy currents, although the phenomenon was first found by GRIFFITHS [14] on a thin film of nickel.

The resonant frequency depends, among other things, upon the shape of the specimen, and the theory has been developed by KITTEL [15] for the case when an alternating field is applied at right-angles to a strong steady field.

We consider a single-domain specimen in the form of a general ellipsoid with the principal axes coinciding with the coordinate axes as shown in Fig. 9.13(*a*). A strong steady field H_Z is applied along the OZ axis and the domain will have a saturation value M_S in a direction which we suppose is inclined at a small angle to OZ. Taking M_S to be due to spins only the spin axes will remain parallel with each other, and the effect of a field in the OZ direction will be, not to pull the axes into alignment with the field, but to cause them

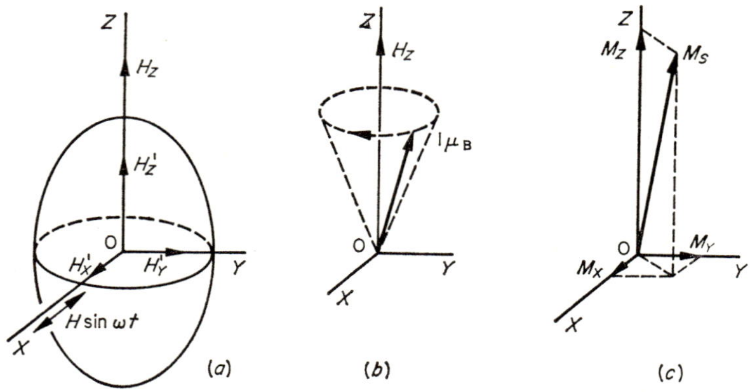

FIG. 9.13. Illustrating spin resonance in an ellipsoidal specimen.

to precess about the OZ axis. This is indicated for an individual spin of moment $1\mu_B$ in Fig. 9.13(*b*). The precession of the parallel system of spins in the domain about OZ will induce components of alternating magnetization in the OX and OY directions. We will moreover apply an alternating field $H \sin \omega t$ in the OX direction, and we are in fact now concerned with the magnetic properties of the specimen in this direction.

At any instant of time t let the components of M_S along the axes be M_X, M_Y and M_Z as shown in Fig. 9.13(*c*). Let the demagnetizing factors for the general ellipsoid be N_X, N_Y and N_Z. The components of magnetic field strength acting upon the material will now be

$$H_X' = H \sin \omega t - N_X M_X$$
$$H_Y' = -N_Y M_Y$$
$$H_Z' = H_Z - N_Z M_Z$$

Since the spins are acted upon by three mutually perpendicular fields, the spin axes will have components of precession about the three coordinate axes. Let the angular momenta, per unit volume, associated with these precessions be P_X, P_Y and P_Z. The rate of change of angular momentum about the OX axis is dP_X/dt, and from the properties of the gyroscope this must be equal to the total torque acting upon M_S about the OX axis. Taking proper care of the signs this torque is $H'_Z M_Y - H'_Y M_Z$. Therefore

$$dP_X/dt = H'_Z M_Y - H'_Y M_Z \tag{9.13}$$

Now $M_X/P_X = \gamma = \mu_0 e/m$ (see Section 5.3). Whence substituting for P_X, H'_Z and H'_Y in equation (9.13) we have

$$dM_X/dt = \gamma\{H_Z + (N_Y - N_Z)M_Z\}M_Y \tag{9.14}$$

Similarly for the OY axis we have

$$dP_Y/dt = -H'_Z M_X + H'_X M_Z$$

which becomes

$$dM_Y/dt = \gamma[-\{H_Z + (N_X - N_Z)M_Z\}M_X + M_Z H \sin \omega t] \tag{9.15}$$

Since H_Z is large compared with the other components of field in this case, we will treat M_Z to be a constant, so that differentiating equation (9.14) gives

$$d^2M_X/dt^2 = \gamma\{H_Z + (N_Y - N_Z)M_Z\}dM_Y/dt \tag{9.16}$$

Substituting in equation (9.16) for dM_Y/dt from equation (9.15) and rearranging gives

$$\begin{aligned} d^2M_X/dt^2 + \gamma^2\{H_Z + (N_Y - N_Z)M_Z\}\{H_Z + (N_X - N_Z)M_Z\}M_X \\ = \gamma^2\{H_Z + (N_Y - N_Z)M_Z\}M_Z H \sin \omega t \end{aligned} \tag{9.17}$$

Equation (9.17) shows how the intensity of magnetization M_X in the OX direction varies in response to the externally-applied field $H \sin \omega t$ in that direction. The natural angular frequency is given by

$$\omega_R = \gamma[\{H_Z + (N_Y - N_Z)M_Z\}\{H_Z + (N_X - N_Z)M_Z\}]^{\frac{1}{2}} \tag{9.18}$$

and the steady state condition is given by

$$M_X = \gamma^2\{H_Z + (N_Y - N_Z)M_Z\}M_Z H\{1/(D^2 + \omega_R^2)\} \sin \omega t$$

This becomes

$$M_X = [M_Z/\{H_Z + (N_X - N_Z)M_Z\}]\,[1/\{1 - (\omega/\omega_R)^2\}]H \sin \omega t \tag{9.19}$$

In this analysis losses have been neglected and, as equation (9.19) shows, M_X is in phase with, or in phase opposition to the applied field depending on whether ω is less or greater than ω_R. If κ_X is the absolute susceptibility in the OX direction then $\kappa_X = \kappa'_X$ (see Fig. 8.1), with $\kappa''_X = 0$ and, from equation (9.19),

$$\kappa'_X = [M_Z/\{H_Z + (N_X - N_Z)M_Z\}]\,[1/\{1 - (\omega/\omega_R)^2\}] \tag{9.20}$$

κ'_X is thus a function of the shape of the specimen and of the angular frequency ω of the applied field.

If the specimen is in the form of a sphere, when $N_X = N_Y = N_Z$, then from equation (9.18)

$$\omega_R = \gamma H_Z = \mu_0(e/m)H_Z$$

If, for example, $H_Z = 400$ A/cm $= 4 \times 10^4$ A/m, with $\mu_0 = 4\pi \times 10^{-7}$ H/m and $e/m = 1{\cdot}76 \times 10^{11}$ C/kg, the resonant frequency is

$$f_R = \omega_R/(2\pi) = 1400 \text{ Mc/s}$$

On the other hand for a flat plate in the XZ plane, we may put $N_X = N_Z = 0$ and $N_Y = 1/\mu_0$. Then

$$\omega_R = \gamma\{(H_Z + M_Z/\mu_0)H_Z\}^{\frac{1}{2}} = (e/m)\{(\mu_0 H_Z + M_Z)\mu_0 H_Z\}^{\frac{1}{2}}$$

or $$\omega_R = (e/m)(\mu_0 B_Z H_Z)^{\frac{1}{2}} \tag{9.21}$$

If the plate were of iron with $H_Z = 4 \times 10^4$ A/m, $B_Z = 2{\cdot}2$ Wb/m², then $f_R = \omega_R/(2\pi) = 9300$ Mc/s.

The absolute susceptibility of either a spherical or a flat specimen is, from equation (9.20),

$$\kappa_X' = (M_Z/H_Z)[1/\{1 - (\omega/\omega_R)^2\}$$

and at zero frequency $\kappa_X' = M_Z/H_Z = \kappa_{X0}$

$$\therefore \ \kappa_X'/\kappa_{X0} = 1/\{1 - (\omega/\omega_R)^2\} \tag{9.22}$$

If μ_{RX}' is the real component of relative permeability in the OX direction for angular frequency ω and μ_R is the corresponding value at zero frequency, the equation (9.22) may be written

$$(\mu_{RX}' - 1)/(\mu_R - 1) = 1/\{1 - (\omega/\omega_R)^2\} \tag{9.23}$$

This expression is plotted in Fig. 9.14 and it will be seen that the real component of permeability may be negative when the frequency of the applied

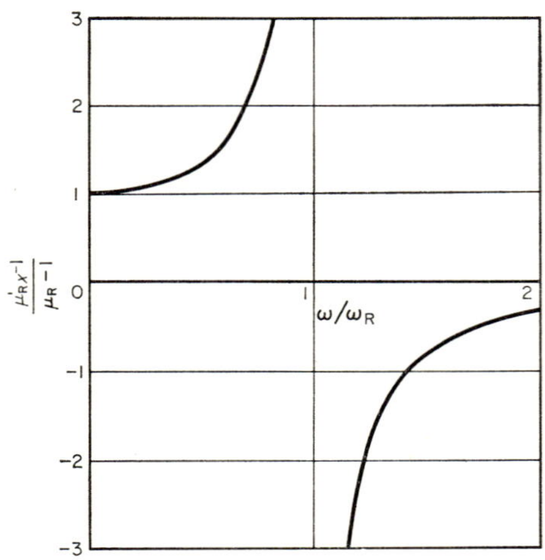

FIG. 9.14. Theoretical relation between real component of permeability and frequency with spin resonance.

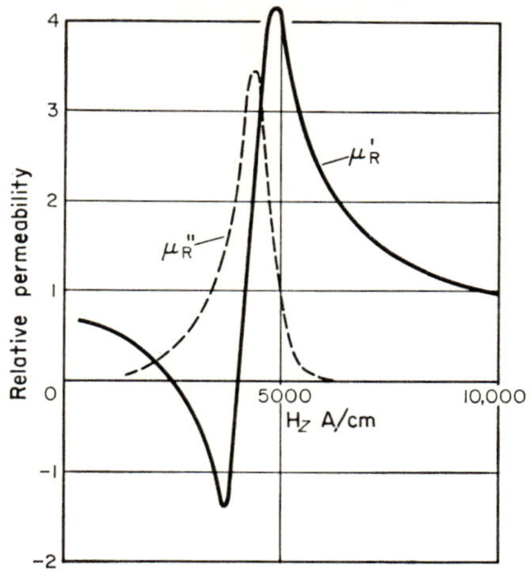

FIG. 9.15. Real and imaginary components of relative permeability of nickel film in relation to a steady field H_Z, with a fixed frequency of 24,400 Mc/s (STANDLEY).

field is greater than the resonant frequency. Negative values of $\mu'_{RX} - 1$ simply mean that M_X is in phase opposition to the applied field.

Figure 9.15 gives some experimental results due to STANDLEY [16] for a thin nickel film. The frequency of the applied field was maintained at 24,400 Mc/s and the field H_Z was varied. A resonant condition is shown by the results when H_Z is about 4400 A/cm, where μ''_{RX} reaches a peak. The negative values of μ'_{RX} may also be noted.

We may use the numerical values in this case to check equation (9.21) which may be written

$$(\mu_0 H_Z + M_S)H_Z = (2\pi f_R m/e)^2 \, (1/\mu_0)$$

Putting $M_S = 0{\cdot}64$ Wb/m² for nickel, with $f_R = 24{\cdot}4 \times 10^9$ c/s and solving for H_Z gives $H_Z = 4{\cdot}8 \times 10^5$ A/m which is in good agreement with the observed value at resonance in the figure.

In the analysis of spin resonance just given the presence of a steady field H_Z has been assumed. Spin resonance is, however, observed in practice, as in the case of Rado's results in Fig. 9.10, without an externally-applied field. However, any internal effect which causes the magnetization in a domain to point in a particular direction fixes an axis about which precession may occur. An equivalent steady field is, in fact, provided by the magnetic anisotropy of the crystals. For the [100] direction of a crystal of cubic structure this equivalent field is given by $H_A = 2K_1/M_S$ (see Chapter 6 and Table 6.1). For iron, for example, $H_A = 2 \times 4{\cdot}4 \times 10^4/2{\cdot}14 = 4{\cdot}1 \times 10^4$ A/m = 410 A/cm, and these fields are of the right order of magnitude to account for the observed resonance effects.

PROBLEMS

9.1. Assuming the oxygen ions in a ferrite crystal to be hard spheres of radius r in contact, calculate the radius of interstitial spherical ions in contact on A sites and also on B sites.

9.2. Estimate the apparent mass per unit area of an oscillating 90° boundary and the resonant frequency of the boundary, given the domain width $= 0{\cdot}01$ cm, the domain wall thickness $= 4 \times 10^{-6}$ cm, $M_S = 0{\cdot}35$ Wb/m^2 and $\mu_{RI} = 20$.

(See Section 9.8.1.)

9.3. A small ferrite prolate spheroid has a demagnetizing factor along the polar axis of $\mu_0 N_a = 0{\cdot}2$. A steady magnetic field of 500 A/cm is applied along this axis. Calculate the resonant frequency for an alternating field applied at right-angles to the polar axis.

SUGGESTED READING

GORTER, E. W. 'Some properties of ferrites in connection with their chemistry', *Proc. Inst. Radio Engrs*, **43,** 1945 (1955).

SMIT, J. and WIJN, H. P. *Ferrites*, Philips Technical Library, Eindhoven (1959).

REFERENCES

[1] SNOEK, J. L. '*New Developments in Ferromagnetic Materials*, Elsevier, Amsterdam (1947).

[2] KRAMERS, H. A. 'Interaction between magnetic atoms in a paramagnetic crystal', *Physica*, *'s Grav.* **1,** 182 (1934).

[3] ANDERSON, P. W. 'Molecular field theory of antiferromagnetism', *Phys. Rev.* **79,** 350, 705 (1950).

[4] NÉEL, L. 'Magnetic properties of the ferrites, ferrimagnetism and antiferromagnetism', *Annls Phys.* **3,** 137 (1948).

[5] GORTER, E. W. 'Some properties of ferrites in connection with their chemistry', *Proc. Inst. Radio Engrs*, **43,** 1945 (1955).

[6] GORTER, E. W. and SCHULKES, J. A. 'Reversal of spontaneous magnetization as a function of temperature in LiFeCr spinels', *Phys. Rev.* **90,** 487 (1953).

[7] SMIT, J. and WIJN, H. P. *Ferrites*, p. 216, Philips Technical Library, Eindhoven (1959).

[8] GUILLAUD, C. and CREVEAUX, H. 'Ferromagnetic properties of Co–Zn and of Mn–Zn Ferrites, *C. r. Acad. Sci., Paris*, **230,** 1458 (1950).

[9] GUILLAUD, C. and SAGE, M. 'Magnetic properties of Mg–Zn ferrites', *C. r. Acad. Sci., Paris*, **232,** 944 (1951).

[10] GORTER, E. W. 'Saturation magnetization and crystal chemistry of ferrimagnetic oxides', *Philips Res. Rep.* **9,** 295, 321, 403 (1954).

[11] WIJN, H. P., GEVERS, M. and VAN DER BURGT, C. M. 'Note on the high-frequency dispersion in nickel-zinc ferrites', *Rev. mod. Phys.* **25,** 91 (1953).

[12] SNOEK, J. L. 'Dispersion and absorption in magnetic ferrites at frequencies above one megacycle', *Physica, 's Grav.* **14,** 207 (1948).

[13] RADO, G. T., WRIGHT, R. W. and EMERSON, W. H. 'Ferromagnetism at very high frequencies: two mechanisms of dispersion in a ferrite', *Phys. Rev.* **80,** 273 (1950).

[14] GRIFFITHS, J. H. E. 'Anomalous high-frequency resistance of ferromagnetic metals', *Nature, Lond.*, **158,** 670 (1946).

[15] KITTEL, C. 'Theory of ferromagnetic resonance absorption', *Phys. Rev.* **73,** 155 (1948).

[16] STANDLEY, K. J., 'Ferromagnetic resonance at microwave frequencies', *Sci. Prog., Lond.* **38,** 231 (1950).

Chapter 10
TECHNOLOGICAL ASPECTS OF MAGNETISM

10.1 **Introductory**

The electrical industry owes its existence to the occurrence in nature of ferromagnetism and, in particular, to the cheapness and abundance of iron, which is the most important ferromagnetic element. Magnetic circuits are an essential part of every electrical generator, transformer and motor, and of much other electrical equipment, and the economic provision of electrical power depends especially upon the high magnetic saturation value and other magnetic properties of iron. In electrical power plant iron is invariably alloyed with small percentages of silicon. For applications where high magnetic permeability in weak fields is required, the nickel-iron alloys are used. A great variety of steels and alloys containing iron, nickel and cobalt in combination with other elements has been developed for use in permanent magnets, while during the last few decades the ferrimagnetic oxide materials have been introduced. The ferrites are virtually electrical insulators and consequently have wide use in radio, and as high-frequency devices up into the microwave region. Some ferrites also have strong permanent-magnet properties.

It is beyond the scope of this book to discuss the available useful magnetic materials in detail, but some are listed in Appendices 2 and 3 and a few specialist books on the subject are given as suggested reading at the end of this chapter. However, in the following sections some of the more interesting aspects of magnetic materials and of magnetic devices will be considered.

10.2 **Iron and Silicon-Iron Sheet Materials**

The magnetic circuits of the equipment used for the generation and distribution of electrical power are built almost exclusively of iron, or of iron alloyed with silicon in amounts up to about 4·5% by weight. This is not only because of the comparatively low cost of iron but also because of the high operating magnetic flux densities attainable in iron and silicon-iron. (See, for example, the saturation values of iron, cobalt and nickel in Table 5.1.)

In modern turbo-alternators the solid steel rotor or field system carries a steady magnetic flux and, for a two-pole rotor, is driven at a speed in revolutions per second equal to the frequency. This frequency has been standardized for power supply in Great Britain at 50 c/s, although in the United States,

Japan and elsewhere both 50 c/s and 60 c/s are common. The flux in the magnetic circuits of the stator of the machine is then alternating at the supply frequency. Similar considerations apply in electric motors while power transformer cores carry only an alternating flux.

Since an alternating flux induces eddy currents, or Foucault currents, in the material, it is necessary to build the magnetic circuits of thin laminations with a light surface coating of insulation. This is because of the skin effect whereby, due to the eddy currents, the magnetic flux tends to concentrate near the surface of the material. Not only does this surface concentration of flux, when it occurs, lead to heavy eddy-current energy losses but the magnetic utilization of the material below the surface becomes poor. These effects increase as the frequency increases. However, at a frequency of 50 c/s, and keeping the sheet thickness down to about 0·33 mm (0·013 inch), the value commonly adopted in these applications, skin effects are minimized and the flux density under normal operating conditions is, to a first approximation, nearly uniform across the thickness of the sheet. It is therefore standard practice in electrical power equipment to use iron or silicon-iron sheet materials of about the thickness just mentioned, although in particular circumstances laminations thinner or thicker than this may be employed.

10.2.1 *Purified Iron*

It has already been mentioned in Section 8.4 that to obtain high permeability and low hysteresis losses in a ferromagnetic material the crystals should be as free as possible from imperfections. In practice this means that the material should have a low content of impurities, and should be carefully heat-treated to remove internal cold-working strains, so that there is the minimum amount of distortion of the crystal lattice. Very extensive investigations by YENSEN [1] and others have shown that foreign elements like carbon, sulphur, oxygen and nitrogen held interstitially in the crystal lattice of iron, even in small amounts, are particularly harmful. It was found too, in the early days, that in some of the iron or mild steel laminations then used in transformer cores a slow but frequently large increase in hysteresis loss occurred with time. This 'magnetic ageing' is attributed to a metallurgical instability in which carbon, oxygen, and nitrogen, held initially in supersaturated solid solution, have the effect of increasing the lattice distortion by being slowly precipitated in the form of iron carbide, oxide and nitride. However, in modern materials with a careful control of the impurity content magnetic ageing is negligible.

Iron and silicon-iron sheet materials used in electrical equipment have a total impurity content, exclusive of silicon, of less than about 0·4% and in the grain-oriented material discussed in Section 10.2.3 the purity is better than this. A good guide to magnetic quality is given by the observed values of the maximum relative permeability μ_{RM} and by the hysteresis loss per cycle w_H for an alternating flux of a given amplitude of flux density $\pm B_{max}$. Typical values for the iron laminations used, for example, in electric motors, are $\mu_{RM} = 5000$ and $w_H = 70$ J/m³/cycle for $B_{max} = 0{\cdot}5$ Wb/m². The corresponding values in hot-rolled silicon-iron sheets with about 4% silicon are $\mu_{RM} = 9000$ and $w_H = 35$ J/m³/cycle.

CLEAVES and HIEGEL [2], however, produced pure iron by a chemical process with a total impurity content of less than 0·013% and obtained a value of $\mu_{RM} = 88{,}400$. For this iron they also quoted the following room-temperature values: density = 7·874 g/cm³, electrical resistivity = 9·71 microhm cm and magnetic saturation value $M_S = 2{\cdot}158$ Wb/m².

Still higher values of permeability for iron have been obtained by CIOFFI [3]. By maintaining small ring specimens at a temperature of about 1475°C in a hydrogen atmosphere for a long period—up to about 18 hours—the interstitial impurities present were reduced to very small amounts, the total

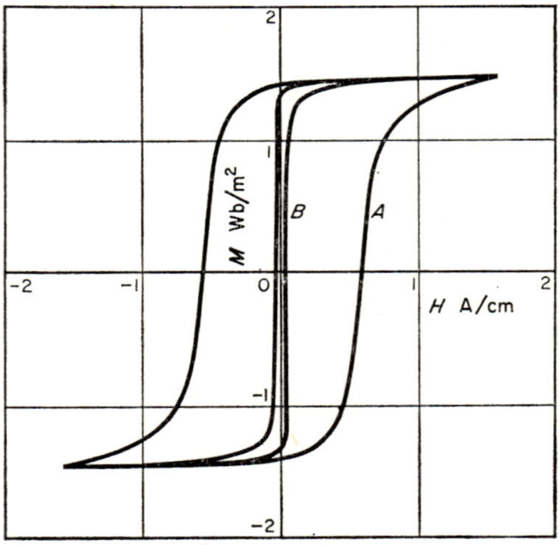

FIG. 10.1. Effect of high-temperature treatment in hydrogen on the hysteresis loss of iron: *A* before and *B* after treatment (CIOFFI).

impurity content being about 0·04%, that is to about one tenth of the value in the commercial iron laminations. This treatment was followed by a prolonged slow cool to eliminate thermal strains. The magnetic properties observed in this case were: initial relative permeability $\mu_{RI} = 14{,}000$, $\mu_{RM} = 280{,}000$ and $w_H = 19$ J/m³/cycle for $B_{max} = 1{\cdot}4$ Wb/m². Figure 10.1 shows hysteresis loops obtained where *A* was before treatment and *B* after the high-temperature anneal in hydrogen. The special treatment reduced the hysteresis loss to about one twentieth of its original value.

CIOFFI, WILLIAMS and BOZORTH [4] applied a similar treatment to a [100] 'picture-frame' single crystal specimen of iron and reported for the [100] direction a value $\mu_{RM} = 1{\cdot}43 \times 10^6$. GOERTZ [5], on the same type of single crystal specimen but of iron containing 6·5% silicon, obtained a value of $\mu_{RM} = 3{\cdot}8 \times 10^6$ for the [100] direction.

10.2.2 *Effects of Alloying Iron with Silicon*

The important benefits obtainable by alloying iron with silicon in sheet material for use in electrical equipment were first noted by BARRETT, BROWN and HADFIELD [6] in 1900. At the frequencies used in electrical power supply these alloys still have no competitor.

The density, magnetic saturation value M_S and the first anisotropy constant K_1 all decrease with increasing percentages of silicon as shown in Fig. 10.2(*a*) and (*b*). Fig. 10.2(*c*) shows how, according to Schulze, the magnetostriction of polycrystalline silicon-iron is related to the silicon content, and

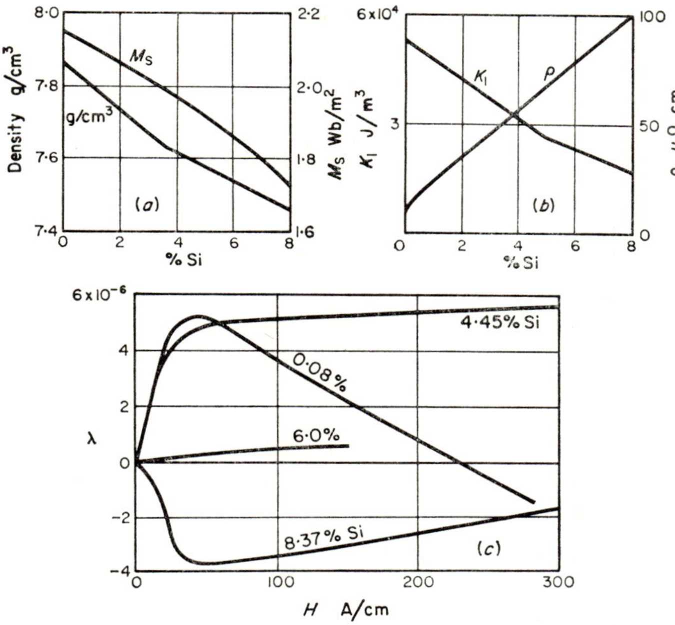

FIG. 10.2. Effects of silicon in iron on density, saturation value M_S, anisotropy constant K_1 resistivity ρ and magnetostriction λ. Silicon percentages by weight.

some values of the saturation magnetostriction of silicon-iron single crystals, due to CARR [7], have already been given in Fig. 6.19(*d*). The saturation value λ_{100} becomes zero for about 6·3% silicon, and the magnetostriction of the polycrystalline alloys is negative at higher percentages of silicon. From a practical point of view silicon has the beneficial effects of raising the permeability of the alloys in the region below the knee of the magnetization curve and of reducing the hysteresis loss for a given value of B_{max}. Typical magnetization curves and hysteresis loops are shown in Fig. 10.3(*a*) and (*b*).

Typical total iron losses at a frequency of 50 c/s for the ordinary non-oriented polycrystalline sheet materials of a nominal thickness of 0·33 mm are shown in Fig. 10.3(*c*). These losses include, as already mentioned, a

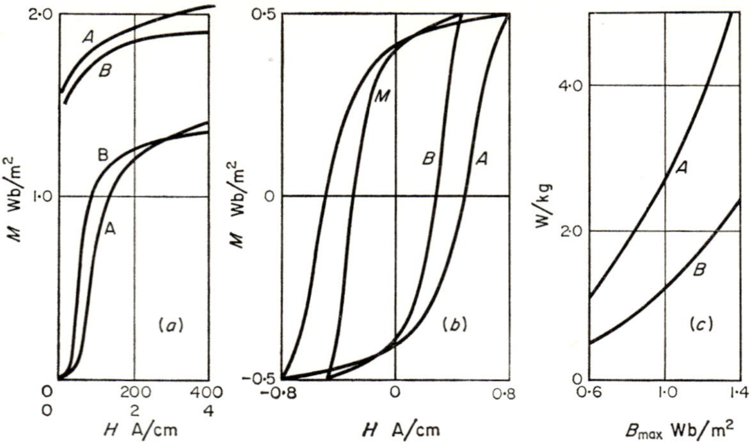

FIG. 10.3. Magnetic properties of polycrystalline (non-oriented) 0·33 mm sheets: *A*–iron, *B*—4% silicon-iron. (*a*) magnetization curves, (*b*) hysteresis loops, (*c*) total iron loss at 50 c/s.

component due to the eddy currents which in turn depends upon the electrical resistivity of the material. The variation of resistivity with silicon content is shown in Fig. 10.2(*b*). For the commercially produced iron used in electrical equipment, having about 0·2% silicon, the resistivity is about 14 $\mu\Omega$ cm. With 4% silicon added the resistivity is increased fourfold to about 55 $\mu\Omega$ cm. The eddy-current loss is theoretically inversely proportional to the resistivity in thin sheets so that the addition of 4% silicon to iron should reduce the eddy-current loss component correspondingly. The question of eddy-current losses, real or apparent, will, however, be referred to further in Section 10.2.4.

10.2.3 *Grain-Oriented Silicon-Iron*

Single crystals of iron or of silicon-iron have very strong directional magnetic properties with the highest permeability and lowest hysteresis loss in a cube-edge or [100] direction of the crystal. The results given in Fig. 10.3 are for polycrystalline sheets in which the disposition of the crystals is not particularly related to the plane of the sheet or the direction of rolling. Since the constituent crystals are almost randomly disarranged the magnetic properties vary only to a minor extent with direction in the sheet.

It has, however, been found possible by a suitable manufacturing process to produce silicon-iron sheet material with a very high degree of 'preferred orientation' of the crystals. In this case, in the finished material, the crystals tend to be arranged with (110) planes parallel to the surface of the sheet and [001] crystal axes parallel with the direction in which the sheet was rolled. The polycrystalline material then has directional magnetic properties similar to those of a single crystal in a (110) plane. In practice this means that, in the direction of rolling, the sheet material has a high permeability and low hysteresis loss and, in fact, low total iron losses under alternating conditions.

The material is thus particularly suitable for transformer cores, having longitudinal stream lines of magnetic flux, and it is also found useful in turbogenerator stator cores.

The commercially produced material has about 3·2% silicon and is generally known as 'cold-reduced grain-oriented silicon steel'. A method of manufacture was first described in detail by Goss [8] in 1935. The material is made in continuous strip mills. Starting with coils of hot-rolled strip about 2 mm in thickness the material is cold-reduced in successive passes to its

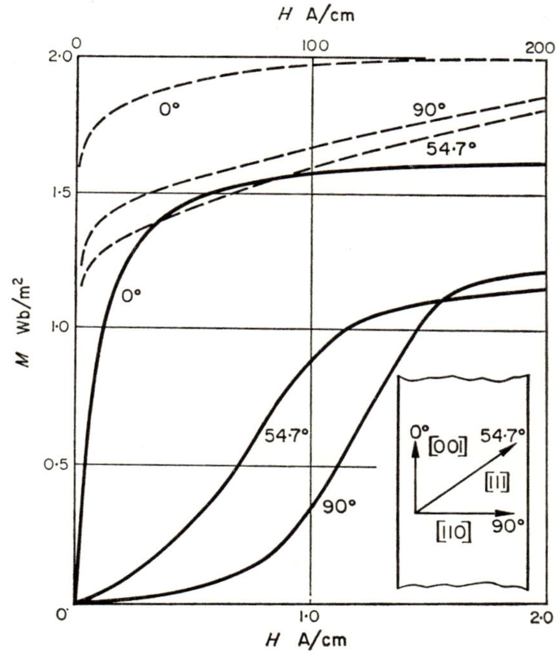

FIG. 10.4. Magnetization curves of cold-reduced, grain-oriented, 3·2% silicon-iron sheet, showing directional properties. Upper scale for broken curves.

final thickness of 0·33 mm. The material in this severely cold-worked condition is then heat-treated in a hydrogen atmosphere at about 1100°C. During this heat treatment the material recrystallizes with a final crystal, or grain size of 1 or 2 mm and with an approximately parallel alignment of crystals in the sheet of the kind just mentioned.

Magnetization curves obtained for strips cut at 0°, 90° and 54·7° to the rolling direction are shown in Fig. 10.4. The more perfect the alignment of the crystals with the preferred position, the more closely would these magnetization curves be expected to correspond with those for the [001], [110] and [111] directions of a single crystal, and the curves in Fig. 10.4 may be compared with those in Fig. 6.6 for an iron crystal.

Figure 10.5(*a*) shows hysteresis loops for three directions in the sheet, the tips of all these loops being at $B_{max} = \pm 1{\cdot}6$ Wb/m². The comparatively very low hysteresis loss for the 0° direction will be noted as well as the curious shapes of the 54·7° and 90° loops. Figure 10.5(*b*) shows the hysteresis loss, obtained from the areas of the loops, and these results again emphasize the low loss in the rolling direction [9].

Figure 6.15 shows the form of the torque curves obtained on single crystal, ellipsoidal specimens, cut in a (110) plane from crystals having positive anisotropy constants K_1 and K_2 and the curves are described by equation (6.23). If a disc specimen is cut from the grain-oriented material and the torque

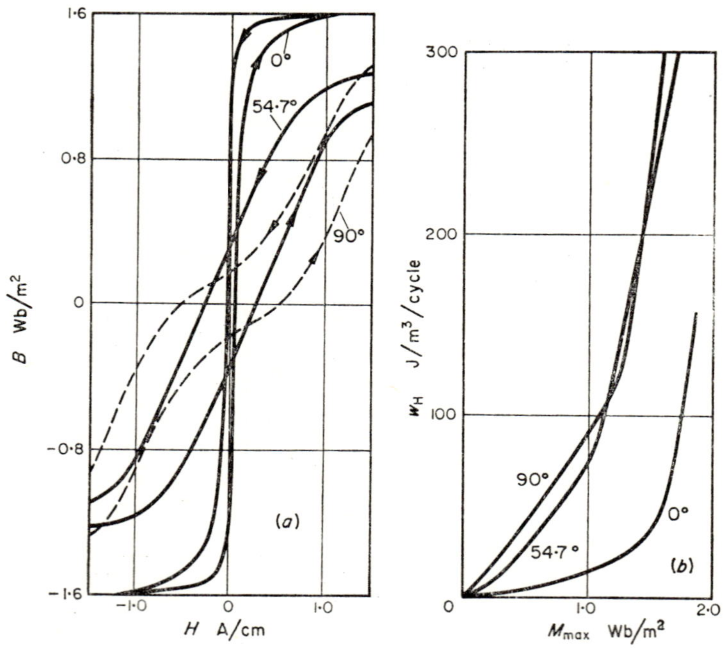

FIG. 10.5. Hysteresis loops and losses for the same material as in Fig. 10.4.

curve measured in a saturating magnetic field the torque curve is found to have very nearly the same form as the curves in Fig. 6.15. If K_1 and K_2 are known for the material of the disc a single crystal torque curve may be plotted from equation (6.23), and this is also the torque curve to be expected if the alignment of the crystals in the disc were 100% perfect, since the polycrystalline disc would then be indistinguishable from a single crystal. In practice the crystals in the disc will have some spread about the preferred position so that the observed torques per unit volume are somewhat less than the theoretical values for a single crystal. The ratio between the torques in the observed and theoretical curves is a good guide to the efficiency of the cold-rolling process of manufacture, and measurement of torque curves on polycrystalline discs provides a quick and convenient practical test of the degree of preferred orientation achieved.

10.2.4 *Eddy Currents and Anomalous Iron Losses*

In transformer cores and other electrical equipment the laminations work under alternating conditions. The ideal operating condition is when the total magnetic flux in the lamination is sinusoidal. Eddy currents are induced in the material, which cause an eddy-current energy loss, and which also have a demagnetizing effect tending to reduce the flux density from its surface value to a minimum value at the centre of the lamination. Both effects are of some practical importance and are interrelated.

However, exact calculation of the internal conditions in a lamination, either of the eddy currents or of the internal flux density, cannot be made because of the extreme nonlinearity between B and H, this nonlinearity being observable in the material as a continuous series of hysteresis loops. In fact if the total flux in a lamination is sinusoidal, the flux density at any internal point will be of distorted wave form, and the same applies to the internal eddy currents and to the effective field strength at any point. Or if the applied field is sinusoidal, only the field strength at the surface of the lamination is undistorted.

An approximation to the conditions in a lamination can, however, be derived theoretically, if a linear relation between B and H is assumed. The hysteresis loop then collapses into a straight line, and the permeability is assumed to be a constant, with a value which is some sort of average of the observed permeability taken over the working range of flux density. This assumption ignores the nonlinearity of B in relation to H, so that for a sinusoidal total flux, or a sinusoidal applied field, all other quantities are also sinusoidal. Some experimental work on analogues [10], built to simulate thin ferromagnetic laminations, and in which the nonlinearities are reproduced, seems to show that the actual eddy-current energy losses in real laminations are unlikely to be greatly different from the losses calculated assuming constant permeability. This comparison, however, assumes that the material is a homogeneous medium, and disregards any possible complications which may arise from the discontinuities associated with the ferromagnetic domain structure. The effect of the domains will be referred to later.

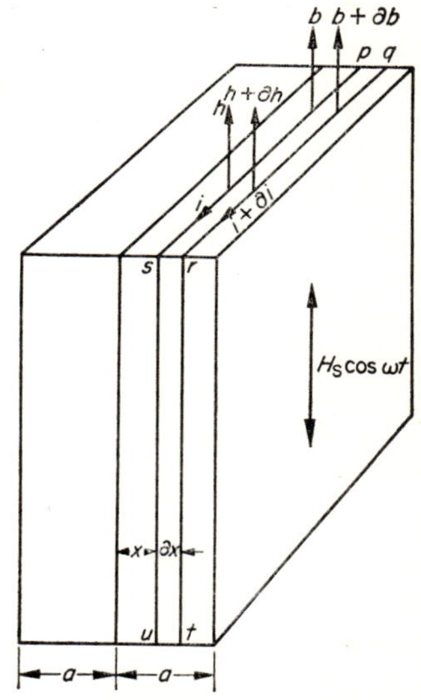

FIG. 10.6. Unit area of a lamination of thickness $2a$.

Figure 10.6 represents a unit area of a lamination of thickness $2a$, in a magnetic field $H_S \cos \omega t$, applied vertically in the figure. We consider a layer of thickness ∂x at a distance x from the central plane of the lamination. Let the instantaneous values of eddy-current density, field strength and flux density at x from the centre be i, h and b respectively, these quantities being functions of both x and time t. We assume the material has a constant absolute permeability μ, and electrical resistivity ρ. Then $b = \mu h$ and the flux density at the surface of the lamination is $b_S = \mu H_S \cos \omega t = B_S \cos \omega t$.

The flux in the element of width ∂x is $b.\partial x$ and, for the directions of the arrows shown, the electromotive force round the circuit $pqrs$ in the figure is

$$\rho(i + \partial i) - \rho i = \frac{\partial(b.\partial x)}{\partial t} = \partial x.\frac{\partial b}{\partial t}$$

$$\therefore \rho.\frac{\partial i}{\partial x} = \frac{\partial b}{\partial t} \tag{10.1}$$

The current in the elemental area $srtu$ is $i.\partial x$ and the magnetomotive force round this circuit is

$$i.\partial x = (h + \partial h) - h$$

$$\therefore i = \frac{\partial h}{\partial x} = \frac{1}{\mu}.\frac{\partial b}{\partial x} \tag{10.2}$$

$\therefore$ from equations (10.1) and (10.2)

$$\frac{\partial^2 b}{\partial x^2} = \frac{\mu}{\rho}.\frac{\partial b}{\partial t} \tag{10.3}$$

The instantaneous eddy-current loss in watts per unit volume at distance x from the centre is $i^2\rho$ and the average value of this is obtained by integrating over one period of time $T = 2\pi/\omega$, and will be a function of x. Integrating this function in turn over the limits of $x = \pm a$ gives the mean eddy-current loss per unit volume of the whole lamination. The mathematical operations required for this case of a flat lamination have been given in detail by RUSSELL [24], and in particular a solution of equation (10.3) is required. Some of the results obtained are as follows.

The instantaneous flux density b at any point inside the lamination is given by

$$b = B_S \left(\frac{\cosh 2px + \cos 2px}{\cosh 2pa + \cos 2pa}\right)^{\frac{1}{2}}.\cos(\omega t - \gamma) \tag{10.4}$$

where

$$\tan \gamma = \frac{\sinh p(a-x).\sin p(a+x) + \sinh p(a+x).\sin p(a-x)}{\cosh p(a-x).\cos p(a+x) + \cosh p(a+x).\cos p(a-x)}$$

and $p = (\pi\mu f/\rho)^{\frac{1}{2}}$

If B_x is the amplitude of the sine wave of flux density at any point inside the lamination at x from the centre then, from equation (10.4),

$$\frac{B_x}{B_S} = \left(\frac{\cosh 2px + \cos 2px}{\cosh 2pa + \cos 2pa}\right)^{\frac{1}{2}} \tag{10.5}$$

However, the sine waves corresponding to B_x will vary in relative time phase angle since the phase angle γ in equation (10.4) is a function of x. When this is taken into account the amplitude B_{max} of the flux density as observed by a search coil round the whole lamination is given by

$$\frac{B_{max}}{B_S} = \left(\frac{\cosh 2pa - \cos 2pa}{\cosh 2pa + \cos 2pa}\right)^{\frac{1}{2}} \Big/ (\sqrt{2}\, pa) \qquad (10.6)$$

Ewing defined an 'equivalent depth of uniform magnetization' d by the expression $d.B_S = a.B_{max}$. Thus d is the thickness of each of two layers, one

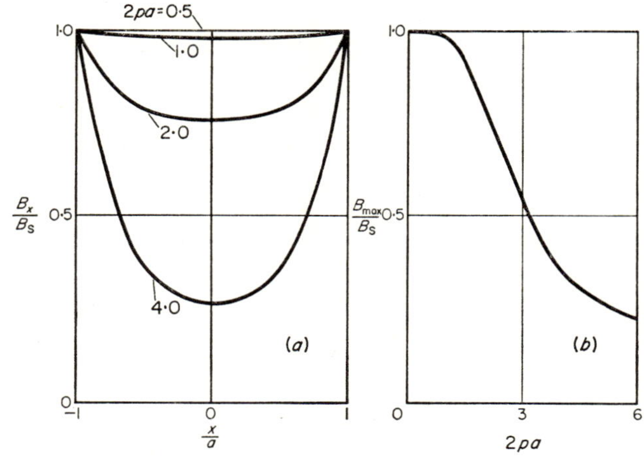

FIG. 10.7. Flux penetration into sheets of thickness $2a$ in relation to $2pa$ where $p = (\pi\mu f/\rho)^{\frac{1}{2}}$.

on either side of the sheet, which if uniformly magnetized to the surface flux density B_S would give the same effective total flux as that in the whole sheet. Whence from equation (10.6)

$$d = \left(\frac{\cosh 2pa - \cos 2pa}{\cosh 2pa + \cos 2pa}\right)^{\frac{1}{2}} \Big/ (\sqrt{2}\, p) \qquad (10.7)$$

For a thick plate, that is when $2pa$ is large, equation (10.7) reduces to

$$d = \{\rho/(\mu\omega)\}^{\frac{1}{2}} \qquad (10.8)$$

The curves in Fig. 10.7(*a*) have been derived using equation (10.5) and show the amplitude of flux density B_x at any point in the sheet for several values of $2pa$. Figure 10.7(*b*) from equation (10.6) shows how the effective flux density B_{max}, with a given surface value B_S, decreases as $2pa$ is increased.

Figure 10.8 shows how the equivalent depth of uniform magnetization d varies with frequency for the particular case, for example, of a cold-reduced, grain-oriented silicon-iron, with $\rho = 48 \times 10^{-6}\ \Omega$.cm and $2a = 0{\cdot}33$ mm, with several assumed values for the relative permeability. It will be seen that skin effects are negligible up to $f = 100$ c/s in this case so long as $\mu_R < 10{,}000$.

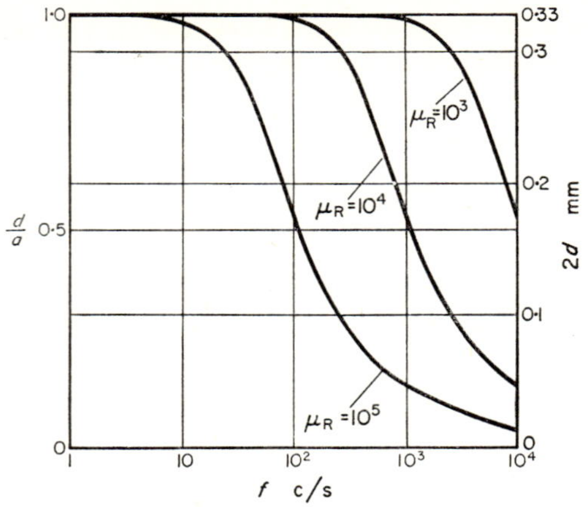

FIG. 10.8. Equivalent depth of uniform magnetization d in relation to frequency for a 3·2% silicon-iron sheet, assuming $2a = 0{\cdot}33$ mm, $\rho = 48 \times 10^{-8}$ Ω.m, and for several assumed values of permeability.

The mean eddy-current power produced in the lamination per unit volume becomes

$$P = \{\rho p B_S^2/(2a\mu^2)\}\left(\frac{\sinh 2pa - \sin 2pa}{\cosh 2pa + \cos 2pa}\right) \tag{10.9}$$

When $pa > \pi$ the quantity in the right-hand bracket of equation (10.9) approaches unity and the power becomes

$$P = \rho p B_S^2/(2a\mu^2) \tag{10.10}$$

For a thin sheet when pa is small the quantity in the right-hand bracket of equation (10.9) approaches $(2pa)^3/3$ and equation (10.9) becomes

$$P = (\pi . 2a . f . B_S)^2/(6\rho) \tag{10.11}$$

In practice the flux density of importance is not the surface value B_S but the effective value for the whole sheet B_{max}. Whence, using equations (10.6) and (10.9), the eddy-current power per unit volume becomes

$$P = (\rho p^3 a B^2_{max}/\mu^2)\left(\frac{\sinh 2pa - \sin 2pa}{\cosh 2pa - \cos 2pa}\right) \tag{10.12}$$

When pa is small, the quantity in the right-hand bracket of equation (10.12) now approaches $2pa/3$ and thus, for a thin sheet,

$$P = (\pi . 2a . f . B_{max})^2/(6\rho) \tag{10.13}$$

Equation (10.13) is the formula normally used to calculate the eddy-current losses in thin sheets at low frequencies, for example for $2a = 0{\cdot}33$ mm and $f = 50$ c/s, and according to this equation the energy loss per cycle due to eddy currents, for a given value B_{max}, is a linear function of the frequency f. Equation (10.13) may be rewritten

$$P/(f . B_{max}^2) = (\pi . 2a)^2 f/(6\rho) \tag{10.14}$$

The more complete expression which is not limited to small values of pa is obtained from equation (10.12) and reduces to

$$P/(f.B^2_{\max}) = (\pi.pa/\mu)\left(\frac{\sinh 2pa - \sin 2pa}{\cosh 2pa - \cos 2pa}\right) \qquad (10.15)$$

The straight line given by equation (10.14) and the curve given by equation (10.15) are compared in Fig. 10.9 for the same conditions as in Fig. 10.8 with $\mu_R = 10^4$. It will be seen that skin effects do not appreciably alter the eddy-current losses in this case, for a given value of $B_{\max}$, until the frequency exceeds about 500 c/s, so that for all practical purposes equations (10.13) and (10.14) are adequate below this frequency.

Typical iron losses in watts per kilogram in three standard grades of lamination used in electrical engineering are shown in Fig. 10.10 for a sinusoidal

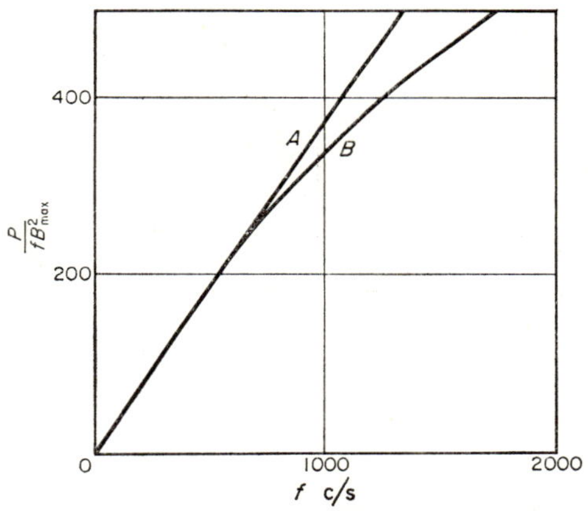

FIG. 10.9. Eddy-current energy loss per cycle $[P/(f B^2_{\max})]$. A-from equation (10.14) for thin sheets, B-from equation (10.15) allowing for skin effects. $2a = 0{\cdot}33$ mm, $\rho = 48 \times 10^{-8}$ Ω.m and $\mu_R = 10^4$.

flux at $f = 50$ c/s. The sheet thickness in each case is approximately 0·33 mm and the materials at (*a*), (*b*) and (*c*) are, respectively, commercial iron as used in electric motors, hot-rolled 4% silicon-iron, and cold-reduced, grain-oriented 3·2% silicon-iron magnetized in the low-loss rolling direction. The upper curve in each case is the total iron loss. The lowest curve is the hysteresis component of the total loss, based upon the assumption that the hysteresis loss per cycle is the same at 50 c/s as the value deduced from the static hysteresis loops. The middle curve in each case is obtained by adding to the hysteresis component the eddy-current loss calculated using equation (10.13), this calculated component frequently being referred to as the classical eddy-current loss. As seen in the figure the sum of the eddy-current and hysteresis components is less than the observed total iron loss by an amount, shaded black in the figure, called the anomalous loss. This component which, in the older types of material shown at (*a*) and (*b*), is only about 8% or 10% of the total at 50 c/s, amounts to about one half of the total loss for the grain-oriented material, and its origin is a matter of speculation [11]. The apparent eddy-current loss, that is the sum of the classical and anomalous components, divided by the classical eddy loss is called the anomaly factor and represented by η. The approximate values of η in the examples given in Fig. 10.10 at (*a*), (*b*) and (*c*) are, respectively, about 1·5, 2·0, and 3·5. In the grain-oriented material in particular, η varies with B_{max}, with f and especially with direction in the sheet.

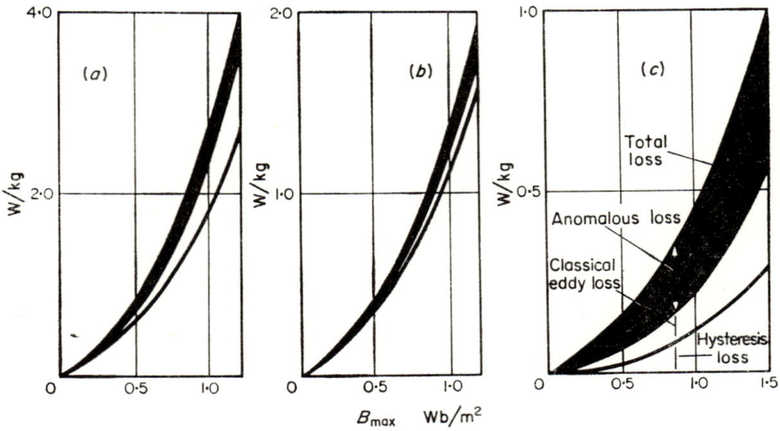

FIG. 10.10. Comparison of components of iron loss in three materials, 0·33 mm thick at 50 c/s: (*a*) iron, (*b*) hot-rolled 4% silicon-iron, (*c*) cold-reduced, grain-oriented, 3·2% silicon-iron in the rolling direction.

Figure 10.11 shows some results for the grain-oriented material. At (*a*) the components of loss for the rolling direction, expressed in joules per kilogram per cycle, are shown in relation to frequency for a value of B_{max} maintained at 1·6 Wb/m². At (*b*) the corresponding results are for a direction at 90° to the rolling direction with $B_{max} = 1{\cdot}3$ Wb/m². The variation of η in relation to B_{max} is shown in Fig. 10.11(*c*) for magnetization in several directions in the sheet at 50 c/s, and the high values of η will be noted.

The fact that any ferromagnetic material is made up of self-saturated domains, means that the microscopic magnetic flux pattern in the material is not smooth and continuous in the manner envisaged in calculating the classical eddy-current loss. Magnetization proceeds by a movement of domain boundaries and, if the domains are relatively large, the eddy currents induced in the neighbourhood of the moving boundaries will differ from the simple classical pattern. It has been proposed that the eddy-current anomaly is due to this, the whole of the apparent eddy-current loss then being due to these local eddy currents at moving domain boundaries. Anomaly factors based on this interpretation have frequently been calculated assuming simple idealized domain arrangements in a thin sheet. The calculated values depend in

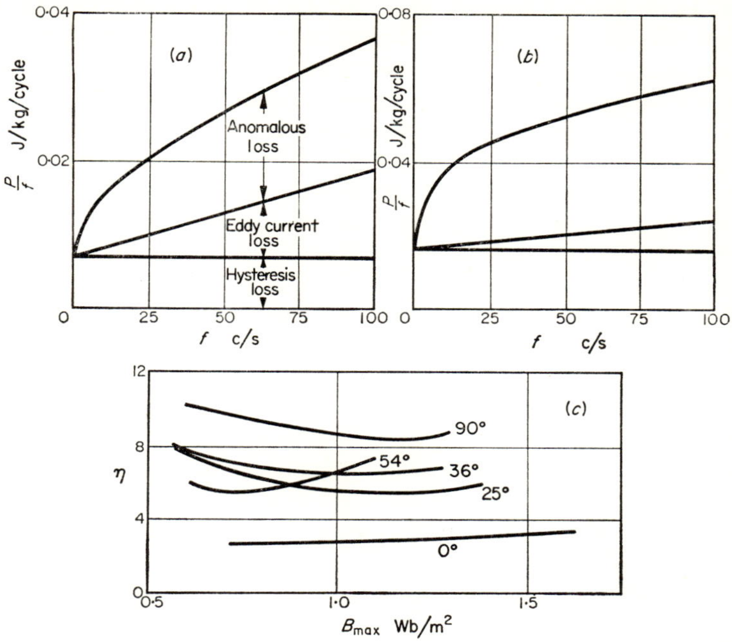

FIG. 10.11. Separation of components of loss for cold-reduced, grain-oriented material in relation to frequency: (*a*) along rolling direction, $B_{max} = 1{\cdot}6$ Wb/m², (*b*) at 90° to the rolling direction, $B_{max} = 1{\cdot}3$ Wb/m², (*c*) anomaly factors for different directions at 50 c/s.

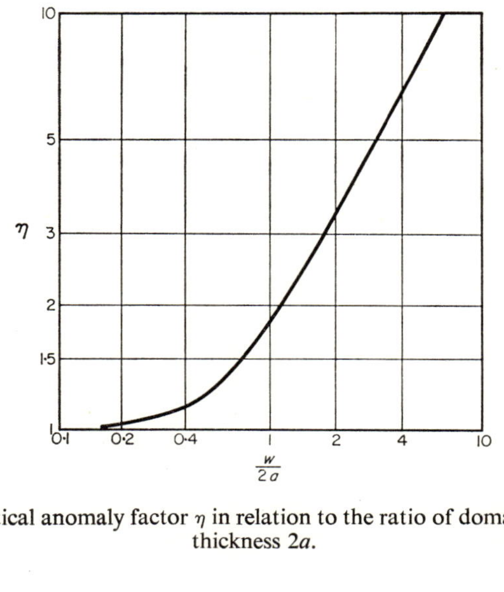

FIG. 10.12. Theoretical anomaly factor η in relation to the ratio of domain width w to sheet thickness $2a$.

general on the ratio of domain size, that is distance between domain boundaries, to the sheet thickness. The results are well summarized in Fig. 10.12, which is due to LEE [12], and which shows η in relation to the ratio $w/(2a)$ where w is the domain width. It seems difficult, however, to explain the observed results, especially for example those in Figs. 10.10 and 10.11, on this basis, for the domain widths would need to be two or more times the sheet thickness to obtain values of $\eta > 3$, say, and the domains in fact do not appear to be as large as this. It is also necessary to explain the increase in η with decreasing frequency, particularly for frequencies below about 10 c/s, and also the large variation of η with direction in the sheet shown in Fig. 10.11(*c*). The loss mechanism operating does not, indeed, appear to be well understood.

10.3 Nickel-Iron Alloys

An interesting and commercially important series of ferromagnetic materials is obtained by alloying nickel and iron together in various proportions. The Curie temperature and the magnetic saturation value M_S at room temperature are shown in Fig. 10.13 in relation to the percentage of nickel in the alloys. It will be seen that the Curie point goes below room temperature for alloys with about 27% nickel, and that M_S has a maximum when the nickel content is about 50%.

The longitudinal magnetostriction has been discussed in Chapter 6, and reference to Fig. 6.19(*c*) shows that the magnetostriction changes sign in alloys with about 82% nickel. Not only the magnetostriction but the anisotropy constants K_1 and K_2 also are small for alloys of about this composition as may be seen in Fig. 10.13. The figure also shows the variation of the electrical resistivity with nickel content.

High values of permeability may be expected when the magnetostriction is small. This has been discussed in some detail in Section 8.5. Low values of the constants K_1 and K_2 should also be helpful in improving the magnetic properties. A comparison of the theoretical values of μ_{RI}, dependent on the magnetostriction in Becker's theory, with the values for the nickel-iron alloys observed by Elmen, has already been made and the results are shown in Fig. 8.9(*b*). Highest values of permeability were found to occur for 78·5% nickel, a material referred to as Permalloy A. This observation and the facts that M_S has a maximum value for about 50% nickel, and the electrical resistivity a maximum for about 36% nickel, have led to the development of a number of commercially useful materials the properties of a few of which are given in Appendix 2.

The very high values of the initial and maximum permeabilities attainable, for example in Permalloy A and Mumetal, will be noted in the Appendix. BOOTHBY and BOZORTH [13], however, found that these values could be increased still more by very great care in manufacture. By melting in vacuum, pouring in helium or nitrogen, heat-treating in hydrogen at 1300°C and

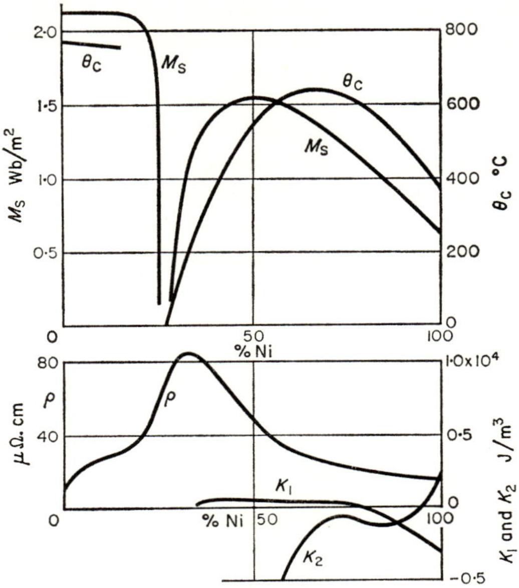

FIG. 10.13. Some properties of the nickel-iron alloys.

cooling at a critical rate, they obtained values of $\mu_{RI} = 10^5$ and $\mu_{RM} = 10^6$. Further particulars of this material, under the name Supermalloy, are given in Appendix 2.

The relative permeability in relation to flux density is given in Fig. 10.14 for several sheet materials all about 0·33 mm thickness. The superiority of the nickel-iron alloys, at lower flux densities and particularly in very weak fields,

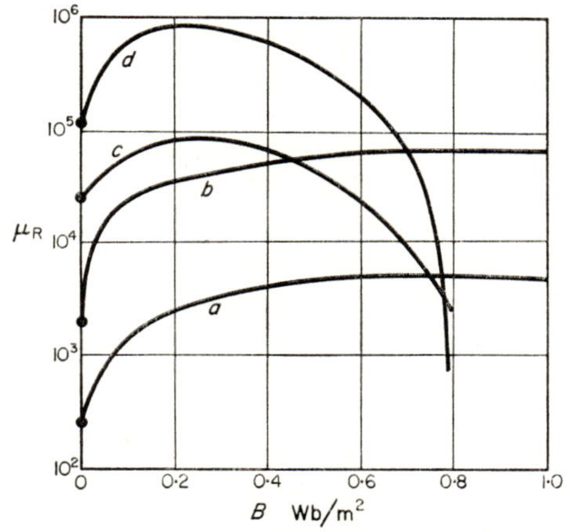

FIG. 10.14. Permeability of several materials compared: sheets 0·33 mm thick. (*a*) iron, (*b*) cold-reduced, grain-oriented, 3·2% silicon-iron in the rolling direction, (*c*) Mumetal, (*d*) Supermalloy.

over iron and silicon-iron, will be clear from the figure. However, because of the relatively low saturation values of the nickel-iron alloys, this superiority extends up to about $B = 0{\cdot}7$ Wb/m^2 only.

The addition of cobalt to iron has the effect of raising the saturation value M_S. The maximum increase obtainable over pure iron is about 12%, when the cobalt content is about 50% [14]. In practice a small amount of some other element such as vanadium has to be added to make the alloy malleable, and this addition reduces the saturation value otherwise attainable. An example, under the name Permendur, is given in Appendix 2 with a value $M_S = 2{\cdot}36$ Wb/m^2 compared with $M_S = 2{\cdot}12$ Wb/m^2 for ordinary commercial iron laminations.

Included in Appendix 2 are also a few commercial grades of ferrite, noteworthy in this immediate comparison for their exceedingly high electrical resistivity and relatively low saturation values.

10.4 Permanent Magnet Materials

In moving-coil measuring instruments, in loud speakers and in many other components in electrical engineering, permanent magnets are employed, the function of which in general is to maintain a desired magnetic flux density in

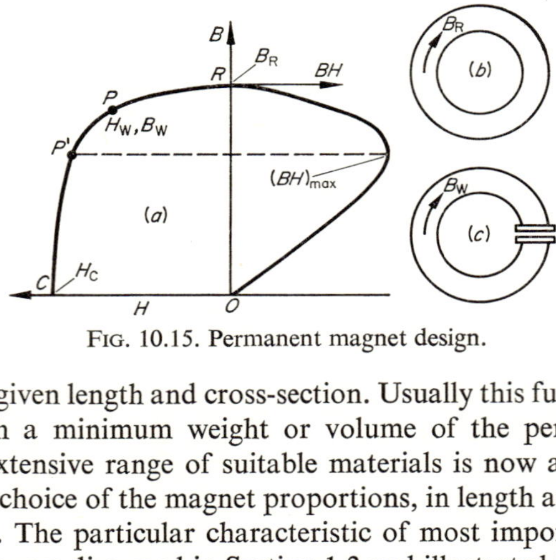

FIG. 10.15. Permanent magnet design.

an air gap of a given length and cross-section. Usually this function should be performed with a minimum weight or volume of the permanent magnet material. An extensive range of suitable materials is now available, and so permits a wide choice of the magnet proportions, in length and cross-section, to be achieved. The particular characteristic of most importance is the demagnetization curve discussed in Section 1.3 and illustrated in Fig. 1.3.

If a closed ring of a ferromagnetic material, as shown in Fig. 10.15(*b*), is magnetized circumferentially to saturation and the applied field is then removed the flux density B_R, or retentivity, remains in the material. If the applied field is now reversed in direction the relation between the flux density B and the field H will be represented by a demagnetization curve of the form shown by *RPC* in Fig. 10.15(*a*), B falling to zero when the demagnetizing field equals the coercivity H_C. The product BH may also be plotted in relation to B as shown in the figure and this product will have a maximum value $(BH)_{max}$ corresponding to some point P' on the demagnetization curve. High values of B_R, H_C and particularly of $(BH)_{max}$ are generally desirable in a permanent magnet material.

If, instead of applying a reverse field, the ring shown in Fig. 10.15(*b*) carrying the remanent flux density B_R, had an air gap introduced as shown at (*c*), this also would cause the flux density in the material to fall, for example, to a point P on the demagnetization curve. Let the length of the air gap be g and the pole face area A_G with a flux density B_G in the gap. Also let the length and cross-section of the material be l and A, respectively, and the flux density and demagnetizing field in the material be B_W and H_W. If flux leakage is neglected then

$$B_W A = B_G A_G \tag{10.16}$$

The free poles induced on the pole faces produce a magnetomotive force on the gap $= H_G g = B_G g/\mu_0$ and a demagnetizing magnetomotive force in the material $H_W l$. Since there is no externally-applied field

$$H_W l = B_G g/\mu_0 \tag{10.17}$$

From equations (10.16) and (10.17) the volume of the magnet material is

$$V = Al = B_G^2 A_G g/(B_W H_W \mu_0) \tag{10.18}$$

Hence for an air gap of given dimensions requiring a given flux density, the volume of the magnet is a minimum when the product $B_W H_W$ is a maximum, that is when $B_W H_W = (BH)_{max}$. In that case the working point on the demagnetization curve would be P' in Fig. 10.15(*a*), and theoretically this can be arranged, as may be seen from equations (10.16) and (10.17), by selecting suitable values for A and l. In practice flux leakage, affecting the uniformity of flux density both in the air gap and in the magnet material, would have to be taken into account. None the less the quantity $(BH)_{max}$ is an important figure of merit for any permanent magnet material.

Prior to 1932 the only available permanent magnet materials were steels, that is iron alloys depending on carbon for their particular properties. Relatively high coercivity, accompanied by mechanical hardness, is attributed in these materials to a martensitic structure induced by suitable heat treatment including quenching in which iron carbide in a fine state of incipient division introduces strain centres on an atomic scale. The coercivity of these martensitic steels may be explained on the basis of high internal stresses. The subject has been discussed in Section 8.5 and an expression for H_C is given in equation (8.18). The martensitic steels were improved by additions of tungsten, chromium and cobalt, and the properties of several of the materials are given in Appendix 3.

The permanent magnet materials most extensively used now, however, are not steels but alloys containing iron, nickel and aluminium, referred to as diffusion or dispersion-hardened alloys, and first described by MISHIMA

[15] in 1932. Mishima's alloy has a composition approximating to the formula Fe_2NiAl and the magnetic properties are sensitive to the rate of cooling during heat treatment and to the composition. The coercivities attainable in alloys of this type are, as already mentioned in Section 8.5, too great to be accounted for by the strain-magnetostriction theory. To apply the theory successfully would require the assumption of internal stresses greater than the breakdown strength of the material. However, in the high coercivity state, the alloy appears to consist of two metallurgical phases, comprising a matrix of FeNiAl which is only weakly ferromagnetic and small islands or particles of a second phase which is almost pure iron. Assuming these particles to be of single-domain size the small-particle theory of high coercivity might well apply, as already suggested in Section 8.6. Table 8.2 gives possible theoretical values of H_C, based on the theory, and greater values than these have not been observed in practice. The original Mishima alloy, known as Alni, was further improved by additions of copper and cobalt in an alloy called Alnico [16]. Particulars of the alloys are given in Appendix 3.

Following some preliminary work by OLIVER and SHEDDEN [17] in 1938 it was found by VAN URK [18] that, when the cobalt content of the Alnico type of alloy was increased, the material could be made very strongly anisotropic magnetically by cooling the alloy from a temperature of about 1250°C in a unidirectional magnetic field greater than about 2400 A/cm. For the selected direction in the alloy it was found that the retentivity B_R was approximately doubled, and the value of $(BH)_{max}$ was increased about three times, when compared with the values attainable if there was no field during heat treatment. Trade names for these alloys are Ticonal, Alcomax and

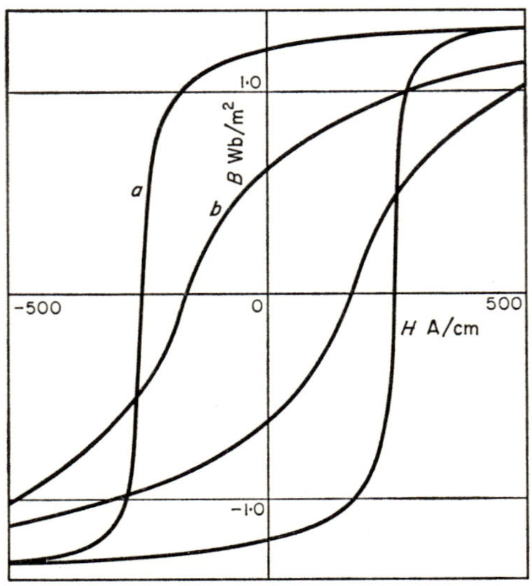

FIG. 10.16. Hysteresis loops of Alnico V showing anisotropy due to magnetic heat treatment (JELLINGHAUS): (*a*) in preferred direction, (*b*) at 90° to preferred direction.

Alnico V. Figure 10.16 gives two hysteresis loops for Alnico V due to JELLINGHAUS [19] which show the remarkable difference in the magnetic properties in the preferred direction and laterally. The properties of some of the alloys are given in Appendix 3.

According to WENT [20] and to NESBITT and WILLIAMS [21] the magnetic anisotropy of these alloys may be explained as follows. The alloy consists of two phases, NiAl which is not ferromagnetic, and ferromagnetic FeCo. Both phases are body-centred cubic with nearly the same lattice constants so that crystals may grow containing both phases together. During cooling in a strong magnetic field near to a [100] direction of the crystals there is a preferential recrystallization, in which the non-ferromagnetic NiAl phase forms along (001) and (010) planes but not along the (100) plane that is nearly perpendicular to the field. Between the NiAl layers small plates of FeCo occur elongated in the general direction of the field. The fine-particle theory would then easily account for the high coercivity in the preferred direction, and the much lower value perpendicular to this direction, even in the case of the polycrystalline material.

Better properties in the magnetically-annealed anisotropic materials can be achieved if the crystals in the material are not randomly disposed but have their cube-edge crystal axes aligned in a preferred direction. It is possible also to grow large elongated or columnar crystals by suitable technique during cooling of the material from the melt, although the production process is difficult. A comparison of the demagnetization curves obtained for alloys of the same composition is given in Fig. 10.17 for the following conditions: (a) random crystals, no magnetic field during heat treatment, (b) random

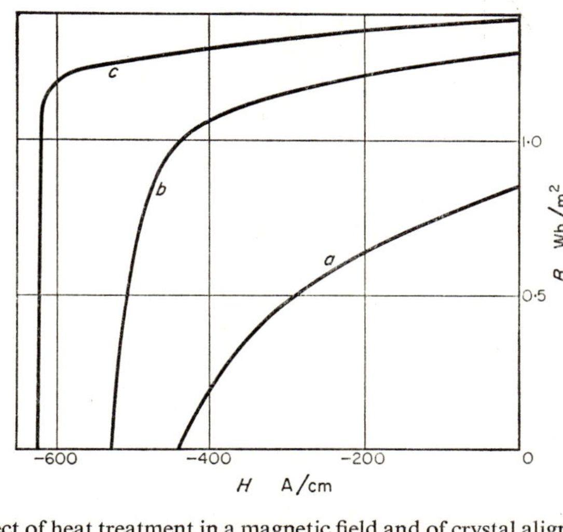

FIG. 10.17. Effect of heat treatment in a magnetic field and of crystal alignment on Alcomax (GOULD): (*a*) random crystals, no field during heat treatment, (*b*) random crystals, heat-treated in field, (*c*) aligned columnar crystals, heat-treated in field.

crystals, with a magnetic field during heat treatment, (c) aligned columnar crystals, heat treated in a magnetic field. The composition of the alloy in this example was 14% Ni, 7·5% Al, 25% Co, 3% Cu, 0·7%Nb with remainder Fe. The improvements obtainable from heat treatment in a magnetic field and from crystal alignment will be clear. The values of $(BH)_{max}$ in these three cases, respectively, were 14, 43 and 70 J/m^3 [25].

A great many other alloys of limited application have also been produced some of which are more easily machined than the very hard diffusion-hardened alloys. Examples are given in Appendix 3 in the miscellaneous group. Iron and iron-cobalt powder-magnet materials, already referred to in Section 8.6, are also included in the Appendix.

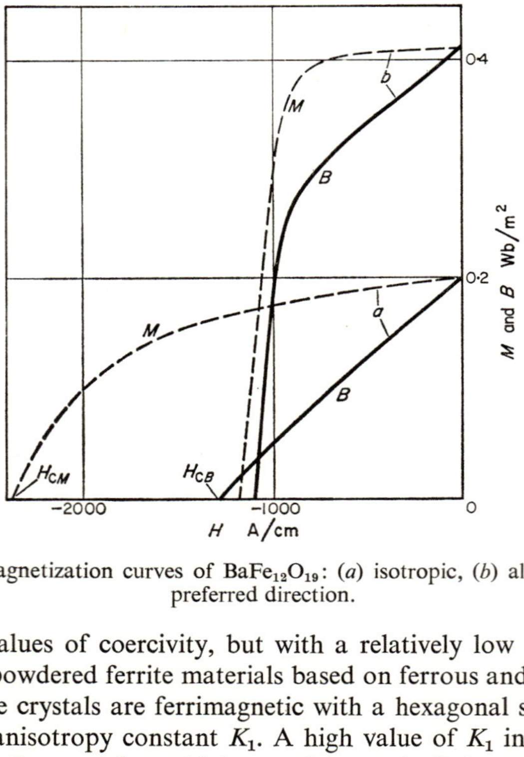

FIG. 10.18. Demagnetization curves of $BaFe_{12}O_{19}$: (*a*) isotropic, (*b*) aligned crystals, in preferred direction.

Very high values of coercivity, but with a relatively low retentivity, are obtainable in powdered ferrite materials based on ferrous and barium oxide. In this case the crystals are ferrimagnetic with a hexagonal structure and a high uniaxial anisotropy constant K_1. A high value of K_1 in single-domain particles gives the particles a high coercivity and, if the easy axes of the particles are also aligned in the magnet material, values of B_R approaching the saturation intensity may be achieved. Figure 10.18 illustrates this for $BaFe_{12}O_{19}$ in the isotropic, and in the crystal-oriented conditions. The demagnetization curves are plotted in relation to both B and M and show, for the isotropic material in particular, the substantial difference between H_{CB} and H_{CM}.

10.5 **Phenomena in Ferrites**

Ferrites are now used at high frequencies, for example for rod aerials at radio frequencies and for various devices in wave guides at microwave frequencies. The particular physical property which makes these applications possible is the exceptionally high electrical resistivity so that the eddy-current complications occurring in the ferromagnetic metals do not occur.

At very high frequencies, and especially in the microwave region, the corresponding wavelengths are small and comparable with or less than the dimensions of the ferrite component. It is necessary then to take account of the propagation of the waves inside the material and the effects of surface boundary conditions and internal reflections. Interesting effects occur when a strong steady magnetic field is applied in a direction either perpendicular to or along the direction of travel of the waves. Electromagnetic waves have mutually perpendicular electric and magnetic fields in a plane normal to the direction of travel. Consequently both the permittivity and the permeability of the medium are involved. The effective relative permittivity of the ferrites at microwave frequencies may be comparatively high, of the order of 20 for example.

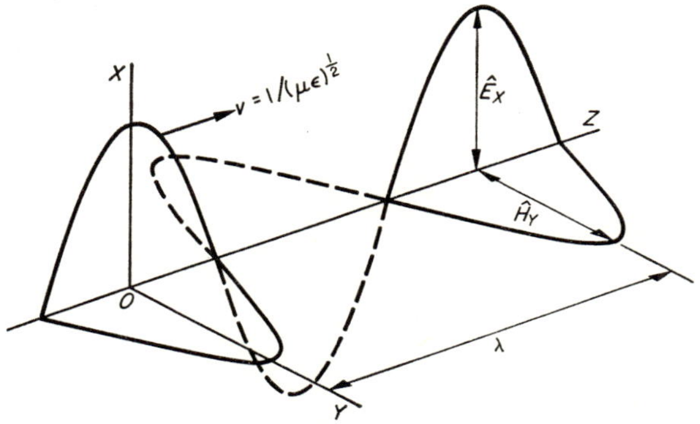

FIG. 10.19. Propagation of a plane electromagnetic wave in an infinite isotropic medium.

We will consider the propagation of a plane electromagnetic wave in a ferrite medium of infinite size, and the effect of a strong, steady magnetic field superimposed in the direction of travel of the wave. Some of the phenomena exhibited by ferrites at very high frequencies may be understood from this simple case. Lengthy mathematical analyses will not be given but only some of the results of these analyses.

In Fig. 10.19 a rectangular system of Cartesian coordinates is taken in an infinite, isotropic, loss-free medium of absolute permeability and permittivity μ and ϵ, respectively. We consider a plane wave in the XY plane in the figure travelling along the OZ axis with a velocity v. The nature of the plane wave is that at any instant of time the electric and magnetic fields are, respectively, the same in magnitude and direction at any point in any plane parallel to

XOY. From well-known theory based on Maxwell's equations we may write the electric field parallel to OX, and the magnetic field parallel to OY, as functions of t and z thus:

$$e_X = \hat{E}_X \cos(\omega t - \beta z) \quad \text{and} \quad h_Y = \hat{H}_Y \cos(\omega t - \beta z)$$

In Fig. 10.19 the disposition of the electric and magnetic fields along the OZ direction is represented for the instant of time $t = 0$. For a later time the sinusoidal distribution of the fields will have moved along OZ. The velocity of propagation is $v = 1/(\mu\epsilon)^{\frac{1}{2}} = c/(\mu_R\epsilon_R)^{\frac{1}{2}}$, where c is the velocity of light in free space and μ_R, ϵ_R, are the relative values of permeability and permittivity, respectively. The phase constant, or wavelength constant, is $\beta = \omega/v = (\omega/c)(\mu_R\epsilon_R)^{\frac{1}{2}}$ and the wavelength in the medium is $\lambda = 2\pi/\beta = v/f = (c/f)(\mu_R\epsilon_R)^{\frac{1}{2}}$.

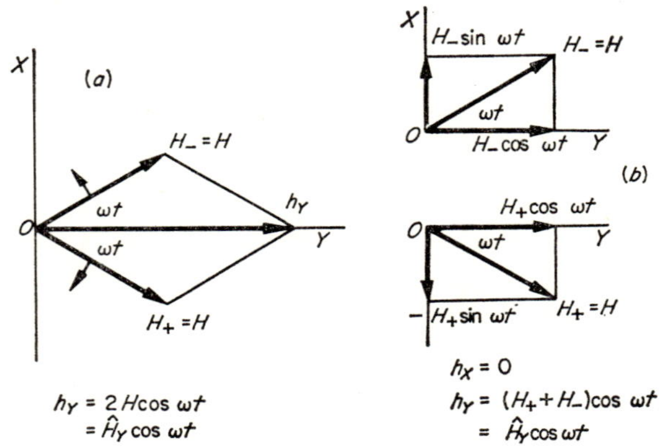

FIG. 10.20. Equivalence of a field $h_Y = \hat{H}_Y \cos \omega t$ (*a*) to the field due to two equal and oppositely-rotating vectors H_+ and H_-, or (*b*) to the field from two systems of components in space and time quadrature.

In this case the electric field and magnetic field vectors remain in the XZ and YZ planes, respectively, and the travelling wave is said to be plane polarized.

It is now important, for present purposes, to notice the following. If we consider, for example, the XY plane were $z = 0$ we have $h_Y = \hat{H}_Y \cos \omega t$. The same magnetic field in this plane, in instantaneous magnitude and direction, may be obtained in either of two alternative ways illustrated in Fig. 10.20.

Figure 10.20(*a*) represents two equal and oppositely-rotating vectors, H_+ and H_-. H_+ and H_- are constant in magnitude, that is independent of time or distance, and each rotates at angular velocity ω. If at time t each has rotated through an angle ωt from the initial datum position OY, the resultant field if $H_+ = H_- = H$ is $h_Y = 2H \cos \omega t$ and then $h_Y = \hat{H}_Y \cos \omega t$ if $H_+ = H_- = (1/2)\hat{H}_Y$.

In Fig. 10.20(*b*) two systems of alternating fields are represented, which are in space and time quadrature. For the system comprising H_- sin ωt along OX and H_- cos ωt along OY, since $\sin^2 \omega t + \cos^2 \omega t = 1$, the resultant is clearly the constant vector H_- rotating counterclockwise with constant angular velocity ω. Similarly the system $-H_+$ sin ωt and H_+ cos ωt is equivalent to the vector H_+ rotating clockwise with constant angular velocity ω. The resultant of the two systems in Fig. 10.20(*b*) is therefore again $h_Y = (H_+ + H_-) \cos \omega t = \hat{H}_Y \cos \omega t$. In symbolic form the two systems may be denoted as $H_X = \pm jH_Y$ where the subscripts indicate the space quadrature and $\pm j$ gives the time quadrature.

Hence in this case in any plane parallel to the XY plane any alternating field in the plane is equivalent to two equal and constant vectors rotating in opposite directions at the synchronous speed ω. Or alternatively it may have substituted for it two systems of space and time quadrature components.

In Fig. 10.19 let a strong steady field H_Z be now applied in the direction OZ. H_Z is supposed strong enough to saturate the ferrite so that the magnetization along OZ is M_S. Components of magnetic field perpendicular to OZ due to the travelling wave are assumed relatively small. Crystal anisotropy in the material is neglected and, since the medium is of infinite extent, shape anisotropy is also absent.

The magnetization M_S arises from electrons spinning on parallel axes and the effect of H_Z is to produce a precession of the spin axes about OZ with an angular velocity $\omega_0 = \gamma H_Z$ where, for spin, $\gamma = \mu_0 e/m = 2{\cdot}21 \times 10^5$ MKS units. Looking in the positive direction OZ the direction of the precession is clockwise and the angular velocity depends only on H_Z. If for example $H_Z = 1000$ A/cm $= 10^5$ A/m (1250 oersted), then $\omega_0 = 2{\cdot}21 \times 10^{10}$ radian/s, and the corresponding frequency $f_0 = \omega_0/2\pi = 3510$ Mc/s. If now in the XY plane of Fig. 10.19 the frequency of the microwave field h_Y was also f_0, and the field was replaced by two rotating vectors H_+ and H_-, oppositely rotating at angular velocity ω_0, it is apparent that the H_+ vector would be rotating in the same direction and at the same speed as the precession and there would be a condition of resonance. On the other hand the H_- vector rotating at the same speed but in the opposite direction would not be in resonance with the precession. Hence the effective permeabilities of the ferrite to the components H_+ and H_- would not be the same, and this would be so, in fact, at any frequency.

If therefore, in the general case, the propagation of the H_+ and H_- vectors separately in the OZ direction is considered, it follows that although their frequencies and speeds of rotation remain the same, the different effective permeabilities that they encounter will give them different linear velocities along OZ and so different wavelengths in the material. At any point z the vector resultant of H_+ and H_- will now no longer remain in the YZ plane but will rotate about OZ as the wave progresses. At frequencies above the resonant frequency this rotation is counterclockwise and below the resonant frequency the rotation is clockwise, that is in the same direction as the precession. The rotations relate to a viewer looking along OZ in the positive direction with H_Z acting in the same direction.

The mathematical theory of an electromagnetic wave in a gyromagnetic medium was developed by LANDAU and LIFSHITZ [22] in 1935. From the theory an expression for the permeability in tensor form can be written down (POLDER [23], 1949) and this in turn can be written into Maxwell's equations and the rotation of the field vectors during propagation determined. (See, for example, textbooks by Lax and Button and by Waldron in Suggested Reading at the end of this chapter.)

The relation between intensity of magnetization and magnetic field, according to Landau and Lifshitz, may be put in the form

$$\frac{d\mathbf{M}}{dt} = \gamma(\mathbf{M} \times \mathbf{H}) - \frac{\alpha_T \gamma}{|M|}[\mathbf{M} \times (\mathbf{M} \times \mathbf{H})]$$

where **M** and **H** are the vector intensity of magnetization and field strength, respectively, and $\gamma = \mu_0 e/m$. The second term takes account of losses where $\alpha_T = 1/(\omega T)$ and T may be called a relaxation time. The mechanism whereby the losses are produced is not specified.

It is convenient to consider the complex susceptibility and complex permeability as discussed already in Sections 8.2 and 9.8. In the present case we are interested in these quantities in terms of the effective response of the ferrite to the two oppositely-rotating magnetic fields which comprise the actual alternating field in the medium.

We may put the complex relative susceptibility for rotation in the positive direction as $\kappa_+ = \kappa'_+ - j\kappa''_+$ and the relative permeability as $\mu_+ = \mu'_+ - j\mu''_+$ where $\mu'_+ = 1 + \kappa'_+$ and $\mu''_+ = \kappa''_+$. Similarly for the negative direction of rotation $\mu'_- = 1 + \kappa'_-$ and $\mu''_- = \kappa''_-$.

The following results are obtained

$$\mu_0(\mu'_\pm - 1) = \frac{\gamma M_S(\omega_0 \mp \omega)T^2}{(\omega_0 \mp \omega)^2 T^2 + 1} \tag{10.19}$$

$$\mu_0 \mu''_\pm = \frac{\gamma M_S T}{(\omega_0 \mp \omega)^2 T^2 + 1} \tag{10.20}$$

where $\gamma = \mu_0 e/m$, $\omega_0 = \gamma H_Z$ and T is a relaxation time.

The curves in Fig. 10.21 have been calculated from equations (10.19) and (10.20) for a ferrite assuming $M_S = 0{\cdot}2$ Wb/m² and a relaxation time $T = 2 \times 10^{-10}$ second. In fig. 10.21(*a*) a fixed frequency of 10,000 Mc/s is assumed and the variation of the relative permeability components plotted in relation to H_Z. The resonance line width of the μ''_+ curve, that is the width of the curve at half maximum value, is given by $\Delta H = 2/(\gamma T)$ and in this particular case $\Delta H = 450$ A/cm. The resonance curve obviously becomes sharper as T increases. The applied frequency of 10,000 Mc/s is in resonance when $H_Z = 2840$ A/cm, and in Fig. 10.21(*b*) it is assumed that H_Z is kept

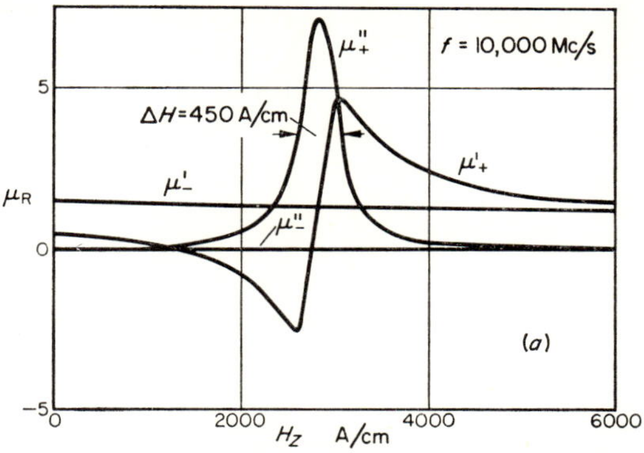

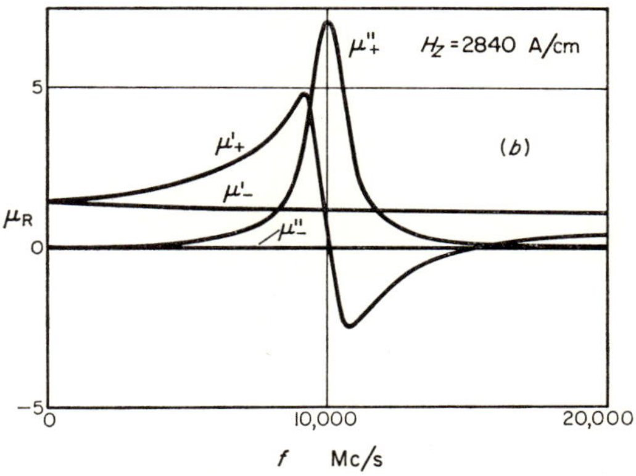

FIG. 10.21. Effective permeability components of oppositely-rotating waves in a ferrite: $M_S = 0{\cdot}2$ Wb/m², $T = 2 \times 10^{-10}$ s. (*a*) $f = 10{,}000$ Mc/s and H_Z varied, (*b*) $H_Z = 2840$ A/cm and f varied.

at this value and the frequency varied. It will be noticed that when the applied frequency exceeds the resonant frequency the pure magnetizing or dispersive component μ''_+ falls below unity and indeed becomes negative.

The rotating waves will propagate in the *OZ* direction with the respective propagation constants

$$\Gamma_+ = \alpha_+ + j\beta_+ \quad \text{and} \quad \Gamma_- = \alpha_- + j\beta_-$$

where α_+ and α_- are attenuation constants while β_+ and β_- are the wave-length or phase constants. The opposing angular velocities of the rotating waves are equal, but the linear velocities and hence the wavelengths in the ferrite are different. We have the velocities $v_+ = \omega/\beta_+$, $v_- = \omega/\beta_-$ and the wavelengths $\lambda_+ = 2\pi/\beta_+$, $\lambda_- = 2\pi/\beta_-$. The following expressions give $\alpha_\pm$ and $\beta_\pm$.

$$\alpha_\pm = (\omega/c)(\epsilon_R/2)^{\frac{1}{2}}[(\mu'^2_\pm + \mu''^2_\pm)^{\frac{1}{2}} - \mu'_\pm]^{\frac{1}{2}} \tag{10.21}$$

$$\beta_\pm = (\omega/c)(\epsilon_R/2)^{\frac{1}{2}}[(\mu'^2_\pm + \mu''^2_\pm)^{\frac{1}{2}} + \mu'_\pm]^{\frac{1}{2}} \tag{10.22}$$

where ϵ_R is the relative permittivity of the ferrite, $c = 3 \times 10^8$ m/s is the

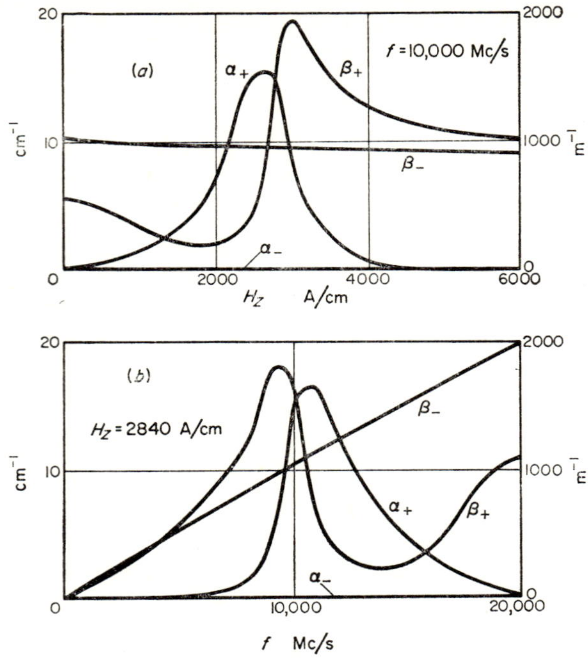

FIG. 10.22. Attenuation and phase constants for propagation of oppositely-rotating waves in a ferrite for the same conditions as in Fig. 10.21 and with $\epsilon_R = 16$.

velocity of light in free space and the permeability components are also relative values. The propagation constant components are then in $(\text{metre})^{-1}$ units.

Calculated values of α_+, α_-, β_+ and β_-, from equations (10.21) and (10.22), are given in Fig. 10.22 for the same conditions as in Fig. 10.21 and with $\epsilon_R = 16$.

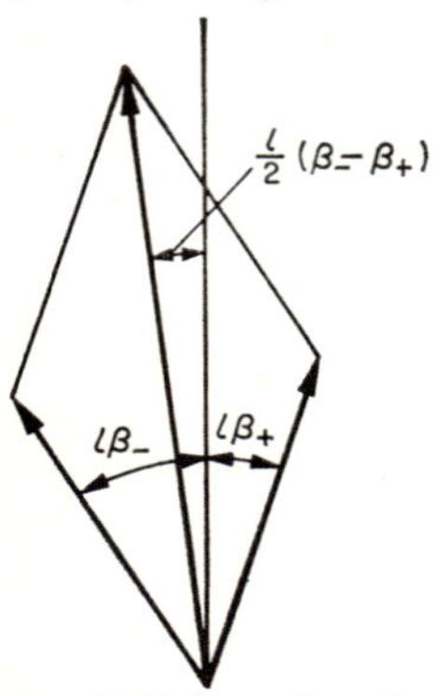

FIG. 10.23. Rotation of resultant vector with oppositely-rotating vectors in a ferrite.

Since the phase constants β_+ and β_- are not the same for a given field H_Z and frequency f, the oppositely-rotating waves have, as already mentioned, different wavelengths and move with different linear velocities along OZ. If attenuation is disregarded, the tips of the rotating vectors may be considered to move, respectively, along right- and left-hand spirals of different pitch, which are synchronously rotating in opposite directions. If we start at a point along OZ where the vectors are in alignment, then at a distance l from this datum point, the respective vectors have rotated in opposite directions through angles $l\beta_+$ and $l\beta_-$. This is illustrated in Fig. 10.23. The resultant vector will have rotated in the negative direction through the angle

$$\theta = l(\beta_- - \beta_+)/2$$

We can estimate the rotation per unit length, when losses are small, from Fig. 10.22. For example, in Fig. 10.22(*a*) when $f = 10{,}000$ Mc/s and $H_Z = 500$ A/cm, α_+ and α_- are small, and $\beta_- = 10{\cdot}0\ \text{cm}^{-1}$, $\beta_+ = 4{\cdot}7\ \text{cm}^{-1}$. Then $\theta/l = (10{\cdot}0 - 4{\cdot}7)/2 = 2{\cdot}65$ rad/cm $= 152°$/cm. Again in Fig. 10.22(*b*), taking $H_Z = 2480$ A/cm and $f = 20{,}000$ Mc/s, α_+ and α_- are small, $\beta_- = 19{\cdot}8\ \text{cm}^{-1}$, $\beta_+ = 11{\cdot}2\ \text{cm}^{-1}$ and $\theta/l = (19{\cdot}8 - 11{\cdot}2)/2 = 4{\cdot}3$ rad/cm $= 252°$/cm.

The rotational effect is non-reciprocal in that if the electromagnetic wave passing along *OZ* is reflected back along *ZO*, then looking along *ZO* with *H* still in the original direction *OZ*, the angle θ continues to increase in the same sense. The phenomenon is known as Faraday rotation, since in 1845 Faraday discovered that if a beam of light, plane-polarized by means of a Nicol's prism, passed through a block of glass in a magnetic field directed along the axis of the beam, the axis of polarization of the light rotated through a small angle, and that on reflection of the light along the original path the angle at the starting point was doubled. For light in heavy flint glass in a field of 2000 A/cm (2500 oersted) the rotation amounts to about 3·5°/cm. The Faraday rotation per unit length in a ferrite at microwave frequencies is therefore considerably greater than the corresponding value with light in a transparent medium.

Ferrite Faraday rotation devices have been developed for use in wave guides and there are other useful ferrite devices somewhat similarly employed. The reader is referred to the specialist microwave treatises for a detailed discussion.

PROBLEMS

10.1. An iron-alloy sheet is 0·33 mm thick, the relative permeability, assumed constant, is 5000 and the electrical resistivity is 50 $\mu\Omega$.cm. If the amplitude of the sinusoidal flux density, as observed by a search coil round the specimen, is $B_{\text{max}} = 1{\cdot}0\ \text{Wb/m}^2$, calculate the amplitude of the flux density at the surface of the sheet, and the amplitude of the flux density at the centre of the sheet, when the frequency is (a) 500 c/s and (b) 2000 c/s.

10.2. A sample of silicon-iron sheet 0·33 mm thick, and electrical resistivity 50 $\mu\Omega$.cm, is found to have a total iron loss at a frequency of 50 c/s and $B_{\text{max}} = 1{\cdot}5\ \text{Wb/m}^2$ of 1·20 W/kg. If the hysteresis component of iron loss is 0·37 W/kg and the density of the material is 7·6 g/cm^3, calculate the anomaly factor.

10.3. A permanent magnet in the form shown in Fig. 10.15(*c*) has an air gap of 0·8 cm and a working pole-face area of 16 cm^2. The desired flux density in the gap is 0·5 Wb/m^2. The magnet material is to be worked at $B_W = 0{\cdot}7$

Wb/m^2, $H_W = -250$ A/cm. Neglecting all fringing effects calculate the required length and cross-section of the magnet material.

SUGGESTED READING

BOZORTH, R. M. *Ferromagnetism*, Van Nostrand, New York (1951).

BRAILSFORD, F. *Magnetic Materials*, Methuen, London (1960).

HADFIELD, D. *Permanent Magnets and Magnetism*, Iliffe, London (1962).

LAX, B. and BUTTON, K. J. *Microwave Ferrites and Ferrimagnetics*, McGraw-Hill, New York (1962).

MCCAIG, M. 'Recent developments in permanent magnetism', *J. appl. Phys.* **35,** 958 (1964).

SMIT, J. and WIJN, H. P. *Ferrites*, Philips Technical Library, Eindhoven (1959).

WALDRON, R. A. *Ferrites*, Van Nostrand, London (1961).

REFERENCES

[1] YENSEN, T. D. 'Magnetic properties of ternary Fe–Si–C alloys', *Trans. Am. Inst. elect. Engrs*, **43,** 145 (1924).

[2] CLEAVES, H. E. and HIEGEL, J. M. 'Properties of high purity iron', *J. Res. natn. Bur. Stand.* **28,** 643 (1942).

[3] CIOFFI, P. P. 'New high permeabilities in hydrogen-treated iron', *Phys. Rev.* **45,** 742 (1934).

[4] CIOFFI, P. P., WILLIAMS, H. J. and BOZORTH, R. M. 'Single crystals with exceptionally high magnetic permeabilities', *Phys. Rev.* **51,** 1009 (1937).

[5] GOERTZ, M. 'Iron-silicon alloys heat treated in a magnetic field', *J. appl. Phys.* **22,** 964 (1951).

[6] BARRETT, W. F., BROWN, W. and HADFIELD, R. A. 'Electrical conductivity and magnetic permeability of various alloys of iron', *Scient. Trans. R. Dubl. Soc.* **7,** 67 (1900); *J. Instn elect. Engrs*, **31,** 674 (1902).

[7] CARR, W. J. *See* BOZORTH, R. M. *Ferromagnetism*, p. 649, Van Nostrand, New York (1951).

[8] GOSS, N. P. 'New development in electrolytic strip steels characterized by fine grain structure approaching properties of a single crystal', *Trans. Am. Soc. Metals*, **23,** 515 (1935).

[9] BRAILSFORD, F. and ABU-EID, Z. H. M. 'Effects of tensile stress on the magnetic properties of grain-oriented silicon-iron laminations', *Proc. Instn elect. Engrs*, **110,** 751 (1963).

[10] BRAILSFORD, F. and BURGESS, J. M. 'Internal waveform distortion in silicon-iron laminations for magnetization at 50 c/s', *Proc. Instn elect. Engrs*, **108C,** 458 (1961).

[11] BRAILSFORD, F. and FOGG, R. 'Anomalous iron losses in cold-reduced, grain-oriented transformer steel', *Proc. Instn elect. Engrs*, **111,** 1463 (1964).

[12] LEE, E. W. 'Eddy-current losses in thin ferromagnetic sheets', *Proc. Instn elect. Engrs*, **105C,** 337 (1958).

[13] BOOTHBY, O. L. and BOZORTH, R. M. 'New magnetic material of high permeability', *J. appl. Phys.* **18,** 173 (1947).

[14] ELMEN, G. W. 'Magnetic alloys of iron, nickel and cobalt', *J. Franklin Inst.* **207,** 583 (1929); *Electl. Engng N.Y.* **54,** 1292 (1935).

[15] MISHIMA, T. 'Nickel-aluminium steel for permanent magnets', *Ohm*, **19,** 353 (1932); *Stahl Eisen*, **53,** 79 (1933).

[16] HORSBURGH, G. D. L. and TETLEY, F. W. British Pat. Spec. 431,660; 439,543 (1935).

[17] OLIVER, D. A. and SHEDDEN, J. W. 'Cooling of permanent magnet alloys in a constant magnetic field', *Nature, Lond.* **142,** 209 (1938).

[18] URK, A. T. VAN, 'Use of modern steels for permanent magnets', *Philips tech. Rev.* **5,** 29 (1940).

[19] JELLINGHAUS, W. 'Magnetization process of an Fe–Ni–Al–Co–Cu magnet alloy having preferred orientation', *Z. Metallk.* **39,** 52 (1948).

[20] HOSELITZ, K. 'Recent progress in the field of permanent magnets' (contribution by J. J. WENT, see p. 457). *J. Phys. Radium, Paris*, **12,** 448 (1951).

[21] NESBITT, E. A. and WILLIAMS, H. J. 'Shape and crystal anisotropy of Alnico 5', *J. appl. Phys.*, **26,** 1217 (1955).

[22] LANDAU, L. and LIFSHITZ, E. 'On the theory of dispersion of magnetic permeability in ferromagnetic bodies', *Phys. Z. Sowju.* **8,** 153 (1935).

[23] POLDER, D. 'On the theory of ferromagnetic resonance', *Phil. Mag.* **40,** 99 (1949).

[24] RUSSELL, A. *Alternating Currents*, Cambridge University Press (1914).

[25] GOULD, J. E. 'Progress in permanent magnet materials', *Proc. Instn elect. Engrs*, **106,** A, 493 (1959).

Appendix 1

ATOMIC STRUCTURES

Shell		K	L		M			N				Normal State
n		1	2		3			4				
Sub-shell		1s	2s	2p	3s	3p	3d	4s	4p	4d	4f	
l		0	0	1	0	1	2	0	1	2	3	
Z	Element											
1	H	1										$^2S_{\frac{1}{2}}$
2	He	2										$^1S_0$
3	Li	2	1									$^2S_{\frac{1}{2}}$
4	Be	2	2									$^1S_0$
5	B	2	2	1								$^2P_{\frac{1}{2}}$
6	C	2	2	2								$^3P_0$
7	N	2	2	3								$^4S_{\frac{3}{2}}$
8	O	2	2	4								$^3P_2$
9	F	2	2	5								$^2P_{\frac{3}{2}}$
10	Ne	2	2	6								$^1S_0$
11	Na	2	8		1							$^2S_{\frac{1}{2}}$
12	Mg	2	8		2							$^1S_0$
13	Al	2	8		2	1						$^2P_{\frac{1}{2}}$
14	Si	2	8		2	2						$^3P_0$
15	P	2	8		2	3						$^4S_{\frac{3}{2}}$
16	S	2	8		2	4						$^3P_2$
17	Cl	2	8		2	5						$^2P_{\frac{3}{2}}$
18	A	2	8		2	6						$^1S_0$
19	K	2	8		2	6		1				$^2S_{\frac{1}{2}}$
20	Ca	2	8		2	6		2				$^1S_0$
21	Sc	2	8		2	6	1	2				$^2D_{\frac{3}{2}}$
22	Ti	2	8		2	6	2	2				$^3F_2$
23	V	2	8		2	6	3	2				$^4F_{\frac{3}{2}}$
24	Cr	2	8		2	6	5	1				$^7S_3$
25	Mn	2	8		2	6	5	2				$^6S_{\frac{5}{2}}$
26	Fe	2	8		2	6	6	2				$^5D_4$
27	Co	2	8		2	6	7	2				$^4F_{\frac{9}{2}}$
28	Ni	2	8		2	6	8	2				$^3F_4$
29	Cu	2	8		2	6	10	1				$^2S_{\frac{1}{2}}$
30	Zn	2	8			18		2				$^1S_0$
31	Ga	2	8			18		2	1			$^2P_{\frac{1}{2}}$

Shell		K	L	M	N				O			P	Normal State
n		1	2	3	4				5			6	
Sub-shell					4*s*	4*p*	4*d*	4*f*	5*s*	5*p*	5*d*	6*s*	
l					0	1	2	3	0	1	2	0	
Z	Element												
32	Ge	2	8	18	2	2							$^3P_0$
33	As				2	3							$^4S_{\frac{3}{2}}$
34	Se				2	4							$^3P_2$
35	Br				2	5							$^2P_{\frac{3}{2}}$
36	Kr				2	6							$^1S_0$
37	Rb				2	6			1				$^2S_{\frac{1}{2}}$
38	Sr				2	6			2				$^1S_0$
39	Yt				2	6	1		2				$^6D_{\frac{3}{2}}$
40	Zr				2	6	2		2				$^3F_2$
41	Nb				2	6	4		1				$^6D_{\frac{1}{2}}$
42	Mo				2	6	5		1				$^7S_3$
43	Ma				2	6	6		1				
44	Ru				2	6	7		1				$^5F_5$
45	Rh				2	6	8		1				$^4F_{\frac{9}{2}}$
46	Pd				2	6	10						$^1S_0$
47	Ag				2	6	10		1				$^2S_{\frac{1}{2}}$
48	Cd				2	6	10		2				$^1S_0$
49	In				2	6	10		2	1			$^2P_{\frac{1}{2}}$
50	Sn				2	6	10		2	2			$^3P_0$
51	Sb				2	6	10		2	3			$^4S_{\frac{3}{2}}$
52	Te				2	6	10		2	4			$^3P_2$
53	I				2	6	10		2	5			$^2P_{\frac{3}{2}}$
54	Xe				2	6	10		2	6			$^1S_0$
55	Cs				2	6	10		2	6		1	$^2S_{\frac{1}{2}}$
56	Ba				2	6	10		2	6		2	$^1S_0$
57	La				2	6	10		2	6	1	2	$^2D_{\frac{3}{2}}$
58	Ce				2	6	10	1	2	6	1	2	$^3H_4$
59	Pr				2	6	10	2	2	6	1	2	
60	Nd				2	6	10	3	2	6	1	2	
61	Il				2	6	10	4	2	6	1	2	
62	Sm				2	6	10	6	2	6		2	$^7F_0$

Shell		K	L	M	N				O			P			Q	Normal State
n		1	2	3	4				5			6			7	
Sub-shell					4s	4p	4d	4f	5s	5p	5d	6s	6p	6d	7s	
l					0	1	2	3	0	1	2	0	1	2	0	
Z	Element															
63	Eu	2	8	18	2	6	10	7	2	6		2				$^8S_{\frac{7}{2}}$
64	Gd				2	6	10	7	2	6	1	2				$^9D_2$
65	Tb				2	6	10	8	2	6	1	2				
66	Dy				2	6	10	9	2	6	1	2				
67	Ho				2	6	10	10	2	6	1	2				
68	Er				2	6	10	11	2	6	1	2				
69	Tu				2	6	10	13	2	6		2				
70	Yb				2	6	10	14	2	6		2				
71	Lu				32				2	6	1	2				$^2D_{\frac{3}{2}}$
72	Hf								2	6	2	2				$^3F_2$
73	Ta								2	6	3	2				
74	W								2	6	4	2				$^5D_0$
75	Re								2	6	5	2				$^6S_{\frac{5}{2}}$
76	Os								2	6	6	2				
77	Ir								2	6	9					
78	Pt								2	6	9	1				$^3D_3$
79	Au								2	6	10	1				$^2S_{\frac{1}{2}}$
80	Hg								2	6	10	2				$^1S_0$
81	Tl								2	6	10	2	1			$^2P_{\frac{1}{2}}$
82	Pb								2	6	10	2	2			$^3P_0$
83	Bi								2	6	10	2	3			$^4S_{\frac{3}{2}}$
84	Po								2	6	10	2	4			
85	At								2	6	10	2	5			
86	Rn								2	6	10	2	6			$^1S_0$
87	Fr								2	6	10	2	6		1	
88	Ra								2	6	10	2	6		2	$^1S_0$
89	Ac								2	6	10	2	6	1	2	
90	Th								2	6	10	2	6	2	2	
91	Pa								2	6	10	2	6	3	2	
92	U								2	6	10	2	6	4	2	

Appendix 2

PROPERTIES OF SOFT MAGNETIC MATERIALS

Material and approximate Chemical Analysis: % by weight	ρ μΩ.cm	μ_{RI}	μ_{RM}	M_S Wb/m²	w_H at B_{max} =0·5 Wb/m² J/m³/cycle
Purified specimens:					
99·98% Fe: Cleaves and Heigel	9·71		$88{\cdot}4\times10^3$	2·158	
99·96% Fe: Cioffi		14×10^3	280×10^3		19 (B_{max} = 1·4 Wb/m²)
99·98% Fe: Cioffi, Williams and Bozorth: single crystal, [100]			$1{\cdot}43\times10^6$		
93·5% Fe, 6·5% Si: Goertz: single crystal, [100]			$3{\cdot}8\times10^6$		
Iron: motor grade: 99·6% Fe	14	250	5×10^3	2·12	70
Silicon-iron, hot-rolled: 96% Fe, 4% Si	55	500	9×10^3	1·95	35
Silicon-iron, cold-reduced, grain-oriented, in rolling direction, 96·8% Fe, 3·2% Si	48	2×10^3	70×10^3	2·0	4·3
Permalloy A: 78·5% Ni, 21·5% Fe	16	12×10^3	90×10^3	1·08	5
Mo-Permalloy: 78·5% Ni, 17·7% Fe, 3·8% Mo	55	20×10^3	75×10^3	0·85	4·5
Mumetal: 77% Ni, 14% Fe, 5% Cu, 4% Mo	62	25×10^3	90×10^3	0·80	4
Supermalloy: 79% Ni, 15% Fe, 5% Mo, 0·5% Mn	65	10^5	10^6	0·79	0·05
Hipernik: 50% Ni, 50% Fe	46	4×10^3	70×10^3	1·60	
Radiometal: 50% Ni, 50% Fe	45	2×10^3	25×10^3	1·60	22
Rhometal: 36% Ni, 64% Fe	85	$1{\cdot}8\times10^3$	7×10^3	0·9	45
Permendur: 49% Fe, 49% Co, 2% V	28	800	5×10^3	2·36	
	ρ Ω.cm				
Ferrites:					
Manganese-zinc, A1	80	750		0·34	
Manganese-zinc, A4	50	1200		0·36	
Nickel-zinc, B1	10	650		0·29	
Nickel-zinc, B5	10	18		0·19	
Magnesium-manganese	10			0·21	

Appendix 3

PROPERTIES OF PERMANENT MAGNET MATERIALS

Material and approximate Chemical Analysis % by weight: remainder Fe	$H_{C(B)}$ A/cm	B_R Wb/m²	$(BH)_{max}$ J/m³
Martensitic steels			
Carbon steel: 1% C	40	0·95	1·4
Tungsten steel: 0·7% C, 6% W, 0·3% Cr	52	1·05	2·4
35% cobalt steel: 35% Co, 3·5% Cr, 3% W, 1% C	200	0·90	7·6
Diffusion-hardened alloys			
Isotropic			
Alni: 13% Al, 24% Ni, 3·5% Cu	380	0·62	10
Alnico: 10% Al, 17% Ni, 12% Co, 6% Cu	400	0·80	13·6
Anisotropic			
Random crystals			
Alnico V: 13·5% Ni, 8% Al, 24% Co, 3% Cu	520	1·27	44
Crystals partly aligned			
Alcomax III, s.c.: 13·5% Ni, 8% Al, 24% Co, 3% Cu, 1% Nb	560	1·3	46·5
Crystals aligned			
Columnax: 13·5 Ni, 8% Al, 24% Co, 3% Cu, 1% Nb	600	1·35	62
Powder materials			
Iron	480	0·50	8
Iron-cobalt: 30% Co	400	0·80	14
Caslox (isotropic): $BaFe_{12}O_{19}$	1400	0·21	8
Magnadur II (anisotropic) $BaFe_{12}O_{19}$	1350	0·38	24
Miscellaneous			
Cunico: 50% Cu, 21% Ni, 29% Co	530	0·34	6·4
Vicalloy: 52% Co, 11% V	240	0·90	8
Comalloy: 17% Mo, 12% Co	180	1·05	8·8

Appendix 4

SOME PHYSICAL CONSTANTS

Velocity of light in free space	$c = 2{\cdot}99790 \times 10^{8}$ m.s^{-1}
Permeability of free space	$\mu_0 = 4\pi \times 10^{-7}$ H.m^{-1}
Permittivity of free space	$\epsilon_0 = 8{\cdot}8542 \times 10^{-19}$ F.m^{-1}
Mass of the proton	$m_P = 1{\cdot}6725 \times 10^{-27}$ kg
Mass of the electron	$m = 9{\cdot}1066 \times 10^{-31}$ kg
Charge on the electron	$e = 1{\cdot}60206 \times 10^{-19}$ C
Planck's constant	$h = 6{\cdot}6252 \times 10^{-34}$ J.s
	$h/(2\pi) = 1{\cdot}0544 \times 10^{-34}$ J.s
	$e/m = 1{\cdot}7591 \times 10^{11}$ C.kg^{-1}
Rydberg constant	$R_\infty = 10{,}973{,}731$ m^{-1}
Rydberg constant, hydrogen	$R_H = 10{,}967{,}770$ m^{-1}
Rydberg constant, helium	$R_{He} = 10{,}972{,}238$ m^{-1}
Bohr magneton	$\mu_B = 1{\cdot}1653 \times 10^{-29}$ Wb.m
Boltzmann's constant	$k = 1{\cdot}38044 \times 10^{-23}$ J.°K^{-1}
Avogadro's number	$N = 6{\cdot}0249 \times 10^{23}$ per gramme mole
Gas constant	$R = Nk = 8{\cdot}3170$ J/gramme mole/°K
Gyromagnetic constant	$\mu_0 e/(2m) = 1{\cdot}1053 \times 10^{5}$ A^{-1}. m.s^{-1}

ANSWERS TO PROBLEMS

Chapter 1

1.1. (a) $H = 0$, $B = M = 0{\cdot}0377$ Wb/m^2.
(b) $H = 2500$ A/m, $B = 0{\cdot}0314$ Wb/m^2, $M = 0{\cdot}0283$ Wb/m^2.
1.2. $\mu_0 N_X = 0{\cdot}168$, $\mu_0 N_Y = 0{\cdot}324$, $\mu_0 N_Z = 0{\cdot}508$: sum = unity.
1.4. $M = 0{\cdot}228$ Wb/m^2, $l_1 = 0{\cdot}955$, $m_1 = n_1 = 0{\cdot}212$.
1.5. (a) 2·07 N.m: (b) −2·72 J: (c) 1·36 J.
1.6. (a) −5·66 J: (b) 2·82 J: (c) 1·47 J.
1.7. (a) 7160 A/cm: (b) 18,300 A/cm.
1.8. 8240 A/cm, 63·7° to normal.
1.9. 6850 A/cm, 46° to the *a*-axis.

Chapter 2

2.1. (a) $0{\cdot}529 \times 10^{-10}$ m, $2{\cdot}117 \times 10^{-10}$ m, $52{\cdot}9 \times 10^{-10}$ m.
(b) Total: $-217{\cdot}9 \times 10^{-20}$ J, $-54{\cdot}5 \times 10^{-20}$ J, $-2{\cdot}18 \times 10^{-20}$ J.
K.E.: $217{\cdot}9 \times 10^{-20}$ J, $54{\cdot}5 \times 10^{-20}$ J, $2{\cdot}18 \times 10^{-20}$ J.
P.E.: $-435{\cdot}8 \times 10^{-20}$ J, $-109{\cdot}0 \times 10^{-20}$ J, $-4{\cdot}36 \times 10^{-20}$ J.
(c) $0{\cdot}0219 \times 10^8$ m.s^{-1}, $0{\cdot}0109 \times 10^8$ m.s^{-1}, $0{\cdot}00219 \times 10^8$ m.s^{-1}.
1·054 mA, 0·132 mA, 1·054 A.
(d) 99,600 A/cm, 3110 A/cm, 0·996 A/cm.
(e) $1\mu_B$, $2\mu_B$, $10\mu_B$.
2.2. $6{\cdot}576 \times 10^{15}$ s^{-1}, $0{\cdot}822 \times 10^{15}$ s^{-1}, $2{\cdot}466 \times 10^{15}$ s^{-1}.
2.3. $59{\cdot}18 \times 10^{15}$ s^{-1}, $7{\cdot}40 \times 10^{15}$ s^{-1}, $22{\cdot}19 \times 10^{15}$ s^{-1}.
2.4. $5{\cdot}26 \times 10^{10}$ s^{-1}, $4{\cdot}96 \times 10^{10}$ s^{-1}, $5{\cdot}11 \times 10^{10}$ s^{-1}.
$5{\cdot}261 \times 10^{7}$ s^{-1}, $5{\cdot}229 \times 10^{7}$ s^{-1}, $5{\cdot}245 \times 10^{7}$ s^{-1}.
2.5. 1, 3, 4, 5/4, $5{\cdot}53 \times 10^{10}$ rad/s.
2.6. $^6S_{\frac{5}{2}}$, 2.

Chapter 3

3.1. (a) 0·52, (b) 0·68, (c) 0·74, (d) 0·74.
3.2. $\langle 100 \rangle$, $\langle 111 \rangle$, $\langle 110 \rangle$, $\langle 2\bar{1}\bar{1}0 \rangle$.

Chapter 4

4.1. $9{\cdot}1 \times 10^{-7}$.
4.2. $3{\cdot}5 \times 10^{-4}$ Wb/m^2, $2{\cdot}9 \times 10^{6}$ A/cm.

Chapter 5

5.1. 1·65 Wb/m^2.

5.2. 8·84 × 10^6 rad/s.

5.3. 12 × 10^6 A/cm, 16 × 10^6 A/cm, 7·1 × 10^6 A/cm.

Chapter 6

6.1. $M_S = 2$ Wb/m^2, $K_1 = 4 \times 10^4$ J/m^4.

6.3. $K_1 = 36 \times 10^4$ J/m^3, $K_2 = 18 \times 10^4$ J/m^3.

6.4. $K_1 = 3{\cdot}2 \times 10^4$ J/m^3, $K_2 = 2 \times 10^4$ J/m^3.

6.5. (a) $\lambda_S = (\lambda_{111}/4)(1 - \cos 2\theta + 2\sqrt{2} \sin 2\theta)$.
Zero, $-19{\cdot}14 \times 10^{-6}$, -20×10^{-6}, -10×10^{-6}.
(b) $\lambda_S = (\lambda_{100}/4)(1 + 3 \cos 2\theta)$.
20×10^{-6}, 5×10^{-6}, zero, -10×10^{-6}.

Chapter 7

7.1. 0·493 J, −0·987 J, 0·493 J.

7.2. 5700 + 410 A/cm, 2·54 J, 0·046 J.

7.3. 0·6 erg/cm^2, 1·2 erg/cm^2.

7.4. 3·1 × 10^{-6} cm.

Chapter 8

8.1. (a) 3 × 10^{-8} erg, 3 × 10^{-6} erg.
(b) 1·3 × 10^{-5} erg, 1·3 × 10^{-2} erg.

8.2. 3000, 1·2 A/cm.

8.3. 1020 A/cm, 2190 A/cm.

Chapter 9

9.1. 0·225*r*, 0·414*r*.

9.2. 0·32 × 10^{-8} kg/m^2, 42 Mc/s.

9.3. 3520 Mc/s.

Chapter 10

10.1. 1·08 Wb/m^2, 0·99 Wb/m^2.
1·85 Wb/m^2, 0·92 Wb/m^2.

10.2. 3·1.

10.3. 12·7 cm, 11·4 cm^2.

INDEX

NAME INDEX